Advances in Intelligent Systems and Computing

Volume 1122

The series "Advances in Intelligent Systems and Computing" contains publications on theory, applications, and design methods of Intelligent Systems and Intelligent Computing. Virtually all disciplines such as engineering, natural sciences, computer and information science, ICT, economics, business, e-commerce, environment, healthcare, life science are covered. The list of topics spans all the areas of modern intelligent systems and computing such as: computational intelligence, soft computing including neural networks, fuzzy systems, evolutionary computing and the fusion of these paradigms, social intelligence, ambient intelligence, computational neuroscience, artificial life, virtual worlds and society, cognitive science and systems, Perception and Vision, DNA and immune based systems, self-organizing and adaptive systems, e-Learning and teaching, human-centered and human-centric computing, recommender systems, intelligent control, robotics and mechatronics including human-machine teaming, knowledge-based paradigms, learning paradigms, machine ethics, intelligent data analysis, knowledge management, intelligent agents, intelligent decision making and support, intelligent network security, trust management, interactive entertainment, Web intelligence and multimedia.

The publications within "Advances in Intelligent Systems and Computing" are primarily proceedings of important conferences, symposia and congresses. They cover significant recent developments in the field, both of a foundational and applicable character. An important characteristic feature of the series is the short publication time and world-wide distribution. This permits a rapid and broad dissemination of research results.

**** Indexing: The books of this series are submitted to ISI Proceedings, EI-Compendex, DBLP, SCOPUS, Google Scholar and Springerlink ****

More information about this series at http://www.springer.com/series/11156

Neeta Nain · Santosh Kumar Vipparthi
Editors

4th International Conference on Internet of Things and Connected Technologies (ICIoTCT), 2019

Internet of Things and Connected Technologies

Springer

Editors
Neeta Nain
Department of Computer Science
and Engineering
Malaviya National Institute of Technology
Jaipur, Rajasthan, India

Santosh Kumar Vipparthi
Department of Computer Science
and Engineering
Malaviya National Institute of Technology
Jaipur, Rajasthan, India

ISSN 2194-5357 ISSN 2194-5365 (electronic)
Advances in Intelligent Systems and Computing
ISBN 978-3-030-39874-3 ISBN 978-3-030-39875-0 (eBook)
https://doi.org/10.1007/978-3-030-39875-0

This Springer imprint is published by the registered company Springer Nature Switzerland AG
The registered company address is: Gewerbestrasse 11, 6330 Cham, Switzerland

ICIoTCT 2019 Preface

ICIoTCT is a platform to discuss advances in the Internet of Things (IoT) and Connected Technologies (various protocols, standards, etc.). The recent adoption of a variety of enabling wireless communication technologies like RFID tags BLE, ZigBee, etc. and embedded sensor and actuator nodes, and various protocols like CoAP, MQTT, DNS, etc. has made IoT to step out of its infancy.

Now smart sensors can collaborate directly with the machine without human involvement to automate decision making or to control a task. Smart technologies including green electronics, green radios, fuzzy neural approaches and intelligent signal processing techniques play important roles for the developments of the wearable health care systems. This conference aims at providing a forum to discuss the recent advances in enabling technologies and applications for IoT.

In addition to the main track, several special sessions were collocated with the conference, covering a wide range of related topics:

- Internet of Things: Modeling and Framework
- Internet of Things: Advances and Applications
- Internet of Things: Hardware and Architecture Dependencies
- Advances in Computational Intelligence, Computer Vision and Cloud Computing
- Recent Trends and Application in Science, Engineering and Health Care
- Security for IoT-Enabled Web, Big Data and Cloud Services
- Related Trends and Services in IoT and Conventional Technologies

About 105 original submissions were received from around the world. A total of 41 papers are selected for presentation and publication in the conference proceedings. In order to maintain high quality, a peer review process was carried out by each special session and conference. Each paper received at least two, and several received up to five reviews. The papers are evaluated on their relevance to special sessions or main conference topics, scientific correctness and clarity of presentation.

The success of the conference owes to the dedication of many who have contributed in many different ways to select a fine scientific mix of papers and exciting social events for the conference program. First, we acknowledge the commitment and hard work of the special session and chairs who have kept the scientific program in focus and made the discussions interesting and invaluable. We recognize the commitment and contributions of the program committee members and the extra reviewers for evaluating the papers on a very tight time schedule. We could not have done it without them.

Next, our heartfelt gratitude goes to our academic sponsor institutions for their cooperation, support and assistance: Malaviya National Institute of Technology Jaipur, India, Center for Global Management, California State University, San Bernardino, and research group at IAASSE, USA. Most importantly, we thank the authors for submitting and trusting their work to the conference.

We thank the Conference Honorary Chair, Prof. Udaykumar R Yaragatti (Director, MNIT Jaipur). We are most grateful to our Keynote Speakers Dr. Kumar Padmanabh, Senior Scientist in a research laboratory of British Telecom in Abu Dhabi, and Prof. Nishchal Kumar Verma, IIT Kanpur, for their enlightening talks and support. Last, we thank all the members of the local organizing committee. We hope the scientific program of ICIoTCT 2019 lives up to your expectations. In addition to the scientific program, we really hope that you found time to discover and appreciate the beautiful sights and events in and around Jaipur.

Neeta Nain
Santosh Kumar Vipparthi
General Chairs

Organization

Program Chairs

Neeta Nain — Malaviya National Institute of Technology Jaipur, India
Santosh Kumar Vipparthi — MNIT Jaipur, India

ICIoTCT2019 Acknowledgements

Malaviya National Institute of Technology Jaipur, India

Center for Global Management, California State University, San Bernardino

International Association of Academicians (IAASSE), CA, USA

Contents

A Structural Feature Based Automatic Vehicle Classification System at Toll Plaza

Vivek Singh, Amish Srivastava, Snehal Kumar, and Rajib Ghosh(✉)

National Institute of Technology Patna, Patna 800005, India
rajib.ghosh@nitp.ac.in

Abstract. Most of the existing toll collection systems in vehicle toll plazas in India are manual in nature. Automation of toll collection systems at toll plazas will make the system a lot faster and fraud-free. The primary task for building such a system is to classify the vehicles arriving at toll plazas because accordingly the amount of toll varies. Most of the existing works in this regard have focused on tracking and detecting of on-road vehicles, but very few of them tried to classify the vehicles. This article presents a novel machine learning based approach to detect vehicles arriving in toll plazas along with their types or classes. In this approach, various structural features are extracted from each vehicle before feeding those features to different classifiers. An exhaustive experiment has been performed on a large self-generated dataset using five different classifiers - Gaussian naive Bayes, Multinomial naive Bayes, Logistic regression, Random forest and Support Vector Classifier (SVC). An encouraging accuracy of 96.15% is obtained from the present system.

Keywords: Traffic monitoring · Vehicle type identification · Toll plaza · Structural features · Machine learning

1 Introduction

Most of the current toll collection systems in different vehicle toll plazas in India are human-controlled. Presence of humans in toll collection leads to chances of corruption and other fraudulent activities in toll plaza. As per the available data[1], Indian government earned more than Rs. 6700 crore through toll fee in 2015–16 financial year. It includes only human-controlled toll collection systems without any surveillance over them and without any proper full proof and reliable bookkeeping. Presently, bookkeeping is done at toll plazas by the staffs themselves working there. So, if they misplace even 10% of daily revenue then country will be losing around Rs. 670 crores in a year.

One possible solution of this problem is to install a surveillance system which will maintain the record of total earned revenue properly on a central database

[1] http://164.100.47.190/loksabhaquestions/annex/11/AS110.pdf.

N. Nain and S. K. Vipparthi (Eds.): ICIoTCT 2019, AISC 1122, pp. 1–10, 2020.
https://doi.org/10.1007/978-3-030-39875-0_1

system and be independent from the control of employees of toll plaza. A reasonable number of works [1,3,4,5–7,9] are available towards on-road tracking and detecting of vehicles, but a very little number of studies [17,18] exist on vehicle classification which is essential to develop an automated toll collection system at toll plazas because toll amount varies with types or classes of vehicles such as light weight, medium weight and heavy weight. The present work proposes a novel machine learning based approach to detect vehicles in toll plazas along with their types or classes. The overall framework of the proposed system is shown in Fig. 1. Initially, after acquiring the natural scene color video images of moving vehicles passing by the toll plaza using a high quality camera, each of its frame gets converted into gray-scale form. Next, background subtraction is performed between two consecutive frames to remove the stationary and unwanted objects and keep the vehicular portion only. Various structural features are then extracted from the vehicular portion. Finally, these features are studied using five different classifiers - Gaussian naive Bayes, Multinomial naive Bayes, Logistic regression, Random forest and Support Vector Classifier (SVC), to classify the vehicles in their respective classes. The novelty of the present work lies in the fact that it classifies each tracked vehicle using machine learning approach based on several structural features.

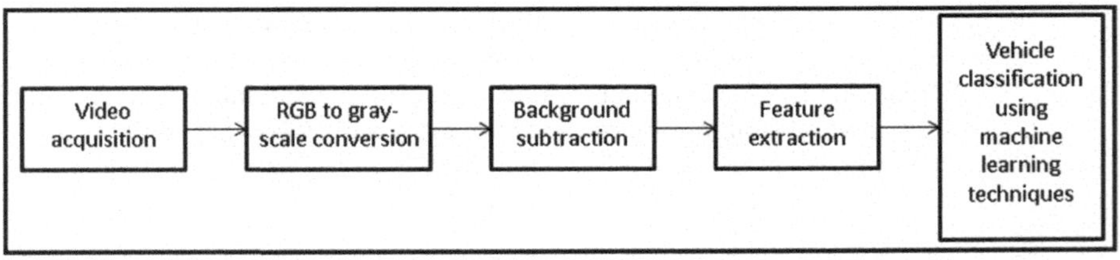

Fig. 1. Overall framework of the proposed system.

The remaining portions of this paper are arranged as follows. Section 2 details the related works. Method of data collection and dataset creation is discussed in Sect. 3. The proposed method is presented in Sect. 4. Classification technique and experimental results are discussed in Sect. 5. Section 6 presents the conclusion of this paper and future works that can be done in this research area.

2 Literature Survey

As already mentioned, although a reasonable number of studies are available towards on-road tracking and detecting of vehicles, but to the best of our knowledge, very few studies exist on vehicle classification which is essential to develop an automated toll collection system at toll plazas. Few of the related works are discussed below.

Tai et al. [1] proposed a tracking system for traffic monitoring and accident detection at road intersections. Active contour model approach was used to get the locations of motorcycles and automobiles in real time. Next, image measurement was incorporated with Kalman filtering techniques to track individual vehicle motion. Michalopoulos et al. [2] presented recent worldwide developments concerning the use of vehicle detection for wireless large-scale data collection and implementation of advanced traffic control and management schemes such as vehicle guidance/navigation. The most common approach to vehicle detection is using active sensors such as radar based, laser based and acoustic based. Radar based active sensor was used in [3], whereas laser based and acoustic based sensors were used in [4] and [5] respectively for on-road vehicle detection. Various Hypothesis Generation (HG) based methods have been proposed in the literature for on-road vehicle detection. One of those methods is *knowledge-based* method. In this method, a priori information or knowledge is used to hypothesize vehicle locations in an image. These information can be taken from symmetry, color, shadow, geometrical features or texture. Images of vehicles observed from rear or frontal views are in general symmetrical in the horizontal and vertical directions. This observation has been used as a cue for vehicle detection in several studies [6,7]. Several studies have explored the usage of color information to detect lanes/roads [8], or segment vehicles from background [9,10]. Usage of shadow information underneath a vehicle to detect on-road vehicles was found in few studies [11,12]. In another study, Bertozzi et al. [13] proposed a corner-based method to detect vehicle locations as vehicles in general have a rectangular shape with four corners. One template, corresponding to each of the four corners, were considered to locate all of the corners in the image. A search method was then followed to find out the matching corners. On the other hand, Matthews et al. [14] used edge detection method to find strong vertical edges. In order to locate the left and right positions of a vehicle, the local maximum peaks of the vertical profile of the edge image was computed followed by smoothing using a triangular filter. Sivaraman et al. [15] proposed a general active-learning framework for robust on-road vehicle recognition and tracking. This framework takes an active-learning approach to build vehicle-recognition and tracking systems. Chan et al. [16] proposed an automatic system to detect preceding vehicles on the highway under various lighting and different weather conditions. To adapt to different characteristics of vehicle appearance at daytime and nighttime, four cues including underneath, vertical edge, symmetry and taillight are fused for the preceding vehicle detection. Tropartz et al. [17] presented a vehicle classification system using infrared overhead laser sensors at toll plazas in New York City. Abdelbaki et al. [18] presented a laser intensity image based automatic vehicle classification system. Ghosh et al. [19] presented one investigation on vehicle number plate recognition using region-of-interest based filtering method.

3 Dataset Details

The dataset required for the present work was created by us. To collect video samples of moving vehicles passing by the toll plaza, a camera was fixed on middle portion of the wall of a toll plaza counter in Himachal Pradesh. A total of 200 video samples were collected for each type or class of vehicle (light weight, medium weight and heavy weight). Training and testing data are separated in 3:1 ratio. Each video sample is then splitted into its constituent frames. Frames not containing vehicle image are then manually removed. This manual filtering is done only during training and testing of the system. Once the system is deployed, the camera remains on all the time, keeps on recording the video and stores a specific number of frames in the memory (other frames are discarded). When a vehicle arrives in front of the toll plaza the tax collector presses the button to click the image of vehicle and then this image and previously stored frames are used for background subtraction. So, when tax collector presses the capture button last frame with the subtracted background is sent to the system. The detailed statistics of the dataset is shown in Table 1.

Table 1. Detailed statistics of the dataset used for experiment

Total samples	Training samples	Testing samples
600	450	150

4 Proposed Method

The input to the present system are videos of moving vehicles arriving at toll plazas and the outcome is the type (viz. heavy weight, medium weight and light weight) of those vehicles. The steps followed to achieve this objective are outlined below.

1. Image Acquisition
2. RGB to grayscale conversion
3. Removal of stationary objects
4. Feature extraction
5. Feeding features to prediction model.

The detailed framework of the proposed method is shown in Fig. 2.

4.1 RGB to Grayscale Conversion

The captured input frames of the video is in RGB format. Initially, these RGB frames are converted into grayscale form. This conversion is done to reduce the number of intensity levels of any frame of the video. The R, G and B components are separated from 24-bit color intensity value of each pixel (i, j) and 8-bit gray level intensity values are calculated. In the present work, color information is not used in the proposed technique to detect the types of vehicles. However, use of a color image leads to increase the computation time of the program and memory requirement. So, the colored input frames are converted into grayscale form in the initial stage. Figure 3 shows the grayscale form of a colored input frame.

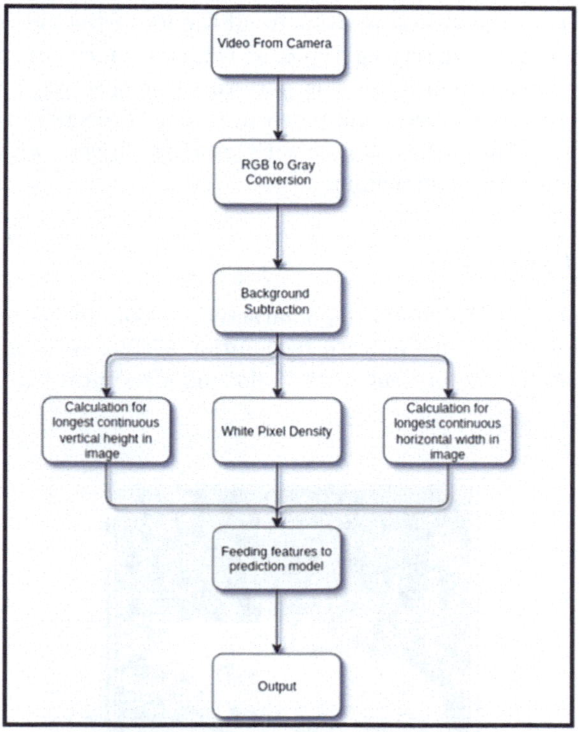

Fig. 2. Detailed framework of the proposed system.

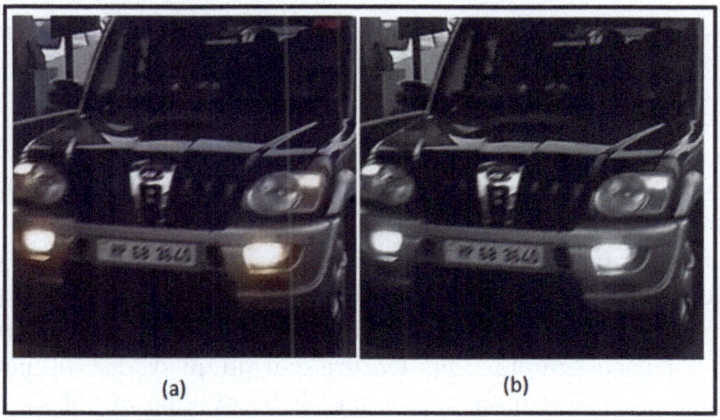

Fig. 3. Conversion of a colored frame into grayscale - (a) Colored form, (b) Gray level form.

4.2 Image Subtraction

The camera used to capture the videos of vehicles arriving at toll plaza was mounted at a fixed position. Thus, the coordinates of pixels associated with the stationary objects of the video are same in all frames of the video. However, the present work focuses on detecting types of vehicles which are moving objects in the video. So, image subtraction is performed at this stage which leads to the cancellation of pixels corresponding to stationary objects in two consecutive frames of a video. Thus, the regions with moving objects which are moving vehicles in this work, are left behind.

4.3 Feature Extraction

Features are unique or discriminating characteristics of any pattern. Extraction of proper features enable one pattern recognition system to assign each pattern to its proper class. In the present work, following structural features of vehicles are extracted:

Fig. 4. After performing image subtraction and white pixel conversion.

White Pixel Density. To extract this feature, initially, all pixels satisfying a certain threshold of gray level intensity are converted into white pixels. The threshold value has been fixed to 200 and it has been decided from experimentation. The other threshold values have also been tried, but the intensity level with value 200 has provided best performance. Next, density of white pixels is calculated for each sample. This feature can uniquely discriminate different classes (heavy weight, medium weight and light weight) of vehicles as volume of the area occupied by these classes of vehicles in associated frames gradually decreases from heavy weight to light weight vehicles. As a consequence, density of white pixels also gradually decreases in relevant frames from heavy weight to light weight vehicles. Figure 4 shows the image of one vehicle after performing image subtraction and white pixel conversion on a gray level image.

Height of Vehicle. This feature calculates the length of the vertical strip with maximum height. Here, vertical strip indicates a continuous block of white pixels going upwards from any pixel in the image. This feature facilitates us to distinguish among heavy weight, medium weight and light weight vehicles through their height.

Width of Vehicle. This feature calculates the length of the horizontal strip with maximum width. Here, horizontal strip indicates a continuous block of white pixels going left or right from any pixel in the image. This feature facilitates us to distinguish among heavy weight, medium weight and light weight vehicles through their width.

5 Classification and Experimental Results

In the present work, the feature values are studied in five different classifier based platforms, namely, Gaussian naive Bayes, Multinomial naive Bayes, Logistic regression, Random forest and SVC. Vehicles are labelled using three different classes—heavy weight, medium weight and light weight. These three types or classes of vehicles are considered because toll taxes vary according to the weight of vehicles.

5.1 Results from Different Classifiers

The performance of the present system has been tested with five different classifiers as mentioned earlier. Among these, Gaussian naive Bayes provides the best classification accuracy. The detailed classification accuracies obtained from five different classifiers are shown in Table 2.

Table 2. Vehicle classification accuracies obtained from various classifiers

Classifier	Accuracy
Gaussian naive Bayes	96.23%
Multinomial naive Bayes	69.23%
Logistic regression	92.30%
Random forest	96.15%
SVC	63.84%

The performance of the present system is also measured using three different performance measurement parameters - *Precision*, *Recall* and *F1-Score*. Figures 5, 6 and 7 show the *Precision*, *Recall* and *F1-Score* respectively.

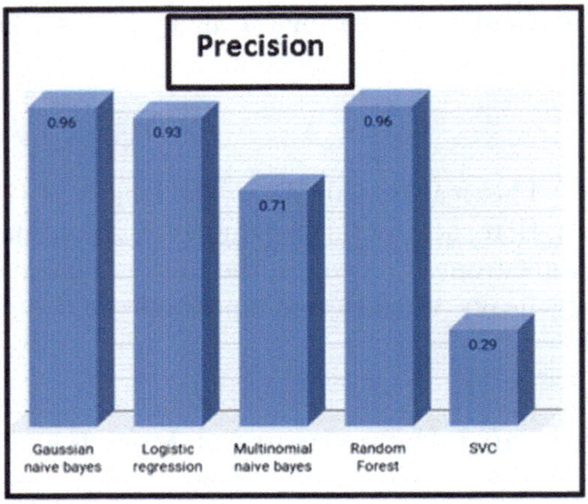

Fig. 5. *Precision* of the present system for various classifiers.

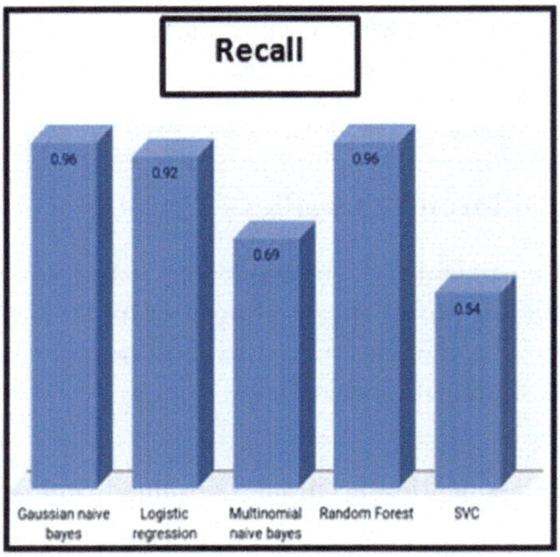

Fig. 6. *Recall* of the present system for various classifiers.

5.2 Comparative Performance Analysis

As most of the existing studies on on-road vehicle tracking have focused on vehicle detection only and very few of them [17,18] have focused on vehicle classification as well, so performance of the present work is compared with the studies worked on vehicle classification only. The comparative performance analysis with these studies is presented in Table 3. As datasets used in these studies are not publicly available, so the performances of the methods presented in these studies have been evaluated using our own dataset used in the present work, to have the comparative performance analysis on the same platform. The proposed structural features help us to get better classification accuracy in comparison to other studies mentioned in Table 3.

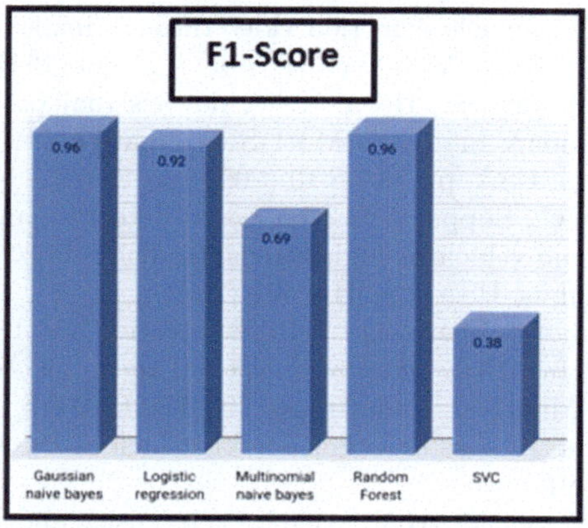

Fig. 7. *F1-Score* of the present system for various classifiers.

Table 3. Comparative performance analysis with few existing studies

Reference	Dataset	Accuracy
Tropartz et al. [17]	Present work	96.11%
Abdelbaki et al. [18]	Present work	95.89%
Proposed method	Present work	96.23%

6 Conclusion and Future Works

This article proposes a system for vehicle detection and classification which offers encouraging classification accuracy. The use of various structural features in machine learning based approach makes this system unique over others. Proposed solution can classify vehicles in real time, but at present the proposed solution has focused to classify them into only three classes. The number of classes of vehicles can be updated as per the requirement. The performance of the present system can be tested by applying deep learning model as a future work. In addition to this, if the present system is combined with number plate recognition system then the new system will record the class of vehicle along with its registration number and refrain from paying at toll plazas by sending the system generated toll bill to vehicle owner's e-mail. Thus, human intervention will not be required at toll plazas. This combined system can also be used for possible video surveillance of vehicles passing by toll plazas as it will recognize the vehicle registration number as well.

References

1. Tai, J.C., Tseng, S.T., Lin, C.P., Song, K.T.: Real-time image tracking for automatic traffic monitoring and enforcement applications. Image Vision Comput. **22**(6), 485–501 (2004)

2. Michalopoulos, P.G.: Vehicle detection video through image processing: the Autoscope system. IEEE Trans. Veh. Technol. **40**(1), 21–29 (1991)
3. Park, S., Kim, T., Kang, S., Heon, K.: A novel signal processing technique for vehicle detection radar. In: IEEE MTT-S International Microwave Symposium Digest, Philadelphia, USA, pp. 607–610 (2003)
4. Wang, C., Thorpe, C., Suppe, A.: Ladar-based detection and tracking of moving objects from a ground vehicle at high speeds. In: IEEE IV2003 Intelligent Vehicles Symposium, Columbus, USA, pp. 416–421 (2003)
5. Chellappa, R., Qian, G., Zheng, Q.: Vehicle detection and tracking using acoustic and video sensors. In: IEEE International Conference on Acoustics, Speech, and Signal Processing, Montreal, Canada, pp. 793–796 (2004)
6. Kuehnle, A.: Symmetry-based recognition for vehicle rears. Pattern Recogn. Lett. **12**(4), 249–258 (1991)
7. Zielke, T., Brauckmann, M., von Seelen, W.: Intensity and edge-based symmetry detection with an application to car-following. CVGIP: Image Underst. **58**(2), 177–190 (1993)
8. Crisman, J., Thorpe, C.: Color vision for road following. In: Cambridge Symposium on Advances in Intelligent Robotics Systems, Boston, USA, pp. 246–249 (1988)
9. Buluswar, S.D., Draper, B.A.: Color machine vision for autonomous vehicles. Eng. Appl. Artif. Intell. **11**(2), 245–256 (1988)
10. Guo, D., Fraichard, T., Xie, M., Laugier, C.: Color modeling by spherical influence field in sensing driving environment. In: IEEE Intelligent Vehicles Symposium 2000, Dearborn, USA, pp. 249–254 (2000)
11. Mori, H., Charkai, N.: Shadow and rhythm as sign patterns of obstacle detection. In: IEEE International Symposium on Industrial Electronics Conference, Budapest, Hungary, pp. 271–277 (1993)
12. Dickmanns, E.D., Behringer, R., Dickmanns, D., Hildebrandt, T., Maurer, M., Thomanek, F., Schiehlen, J.: The seeing passenger car 'VaMoRs-P'. In: Proceedings of the Intelligent Vehicles 1994 Symposium, France, Paris, pp. 24–26 (1994)
13. Bertozzi, M., Broggi, A., Castelluccio, S.: A real-time oriented system for vehicle detection. J. Syst. Archit. **43**(1–5), 317–325 (1997)
14. Matthews, N., An, P., Charnley, D., Harris, C.: Vehicle detection and recognition in greyscale imagery. Control Eng. Pract. **4**(4), 473–479 (1996)
15. Sivaraman, S., ManubhaiTrivedi, M.: A general active-learning framework for on-road vehicle recognition and tracking. IEEE Trans. Intell. Transp. Syst. **11**(2), 267–276 (2010)
16. Chan, Y., Huang, S., Fu, L., Hsiao, P.: Vehicle detection under various lighting conditions by incorporating particle filter. In: IEEE Intelligent Transportation Systems Conference, Seattle, USA, pp. 534–539 (2007)
17. Tropartz, S., Horber, E., Gruner, K.: Experiences and results from vehicle classification using infrared overhead laser sensors at toll plazas in New York City. In: IEEE International Conference on Intelligent Transportation Systems, Tokyo, Japan, pp. 686–691 (1999)
18. Abdelbaki, H.M., Hussain, K., Gelenbe, E.: A laser intensity image based automatic vehicle classification system. In: IEEE Intelligent Transportation Systems, Oakland, USA, pp. 460–465 (2001)
19. Ghosh, R., Thakre, S., Kumar, P.: A vehicle number plate recognition system using region-of-interest based filtering method. In: 2018 Conference on Information and Communication Technology (CICT 2018), Jabalpur, India, pp. 1–6 (2018)

Classification of Chronic Kidney Disease with Genetic Search Intersection Based Feature Selection Technique

Sanat Kumar Sahu[1(✉)] and Prem Kumar Chandrakar[2]

[1] Govt. Kaktiya P.G. College, Jagdalpur, Bastar, CG, India
sanat.kosa1@gmail.com
[2] Mahant Laxmi Narayan Das College, Raipur, CG, India
premchandrakar@gmail.com

Abstract. The objective of this study had reduced the identification time and to get best diagnosis model in Chronic Kidney Disease (CKD). The target was development of the strong and computationally proficient model for classification of CKD. This work had used four classification models like Naive Bayes, Multilayer Perceptron, OneR, Classification and Regression Tree (CART) to classify the CKD data set and compared the classification accuracy with all classifier. Similarly Feature Selection Technique (FST), ranking based namely Chi Squared AttributeEval (CSAE), One-R AttributeEval (ORAE) and Search based namely Genetic Search-J48 (GS-J48), Genetic Search-CART (GS-CART) have used. The contribution of this research work is to recognize and classify the CKD problem and propose a new FST namely Genetic Search Intersection Based Feature Selection Technique (GS-IBFST). The classifier offers higher accuracy with GS-IBFST compare to without FSTs and existing FSTs.

Keywords: Chronic Kidney Disease · Genetic Search Intersection Based Feature Selection Technique · Naive Bayes · Multilayer Perceptron · OneR · Classification and Regression Tree (CART)

1 Introduction and Literature Review

At present, Chronic Kidney Diseases is a very serious problem in the field of medical science and health care. The problem of Chronic Kidney is about 10% of people suffering from the world [1]. The Chronic Kidney Problems (CKD) or Chronic Kidney Disease Grow slowly, which is not detected at the initial stage. Due to this, the disease is not easily detected and a time comes when after months or years, the kidney loses its functionality [2]. Using data processing and Machine Learning Techniques for CKD issues one terribly simple and straightforward methodology will diagnose this disease. With the help of Data Mining CKD can detect at the initial stage. This would be useful for doctors and physician. The CKD patients can be treated in the initial stage and prevented from the effect of this serious problem. Data mining [3, 4] is that the method of finding

N. Nain and S. K. Vipparthi (Eds.): ICIoTCT 2019, AISC 1122, pp. 11–21, 2020.
https://doi.org/10.1007/978-3-030-39875-0_2

patterns in large data sets involving learning strategies, statistics and machine learning at the crossroads of the information system. This is often a necessary method wherever intelligent methods applied to extract knowledge patterns. This is the knowledge base subfield of engineering and computer science. Classification is an important function of data mining in which data is arranged in categories or class. The classification relies on machine learning. Classification is employed within the group of dataset; every item is already classified into outlined class and group. Classification models like Naive Bayes, Multilayer Perceptron, OneR, Classification and Regression Tree (CART) is used to classify the CKD dataset and compared the classification accuracy with all individual classifier. The main task of the feature selection technique [5, 6] is that it removes irrelevant or useless features from the Dataset. In these work uses feature selection techniques evolutionary, ranker algorithm and Proposed GS-IBFST to select the important features the dataset of Chronic Kidney Disease. Genetic Search [7] approach consists in victimization the prediction performance of a given learning machine to judge the relative utility of subsets of features as evolutionary algorithm. Overall, Consistency Subset, Chi Squared [8] AttributeEval (CSAE), One-R [9] AttributeEval (ORAE) are used as ranker algorithms. Feature Selection Techniques on the reduced set of features has increasing benefits. It reduces the number of features in the dataset. It helps to form classification task easier. Then, it enhances the classification accuracy. The authors [10] classify the Chronic Kidney disease dataset used classification algorithms like using Random Forest (RF), Classification and Regression Tree (CART) and Support Vector Machine (SVM) and their ensemble models. The proposed ensemble model with proposed UBFST provides higher accuracy compared to others existing FSTs and every individual classifiers. The authors [2] have used two kind of feature selection strategies, i.e., wrapper and filter approach are accustomed diagnose CKD. Results showed that the Support Vector Machine (SVM) classifier has used filtered subset evaluator with the BFS engine feature selection methodology provides a better accuracy rate (98.5%) within the identification of CKD. The authors [1] have used classifiers Artificial Neural Network, Support Vector Machine, k-Nearest Neighbor, C4.5 and Random Forest for classification of CKD dataset. The random forest (RF) classifier provides highest performances on the classification of CKD. For [11] predictions of the CKD using clinical data. They have used machine learning algorithms like K-nearest neighbors (KNN), support vector machine (SVM), logistic regression (LR), and decision tree classifiers. The outcome shows that SVM classifier gives the highest accuracy, sensitivity after training and testing by the proposed method. For [12] classifying CKD dataset they used classification algorithms like Naive Bayes and Artificial Neural Network (ANN). The results show that Naive Bayes gives higher accuracy than Artificial Neural Network. They have [13] worked to predict kidney disease with completely different machine learning algorithms like Support Vector Machine (SVM), Multilayer Perceptron (MLP), Decision Tree (C4.5), Bayesian Network (BN) and K-Nearest Neighbour (K-NN). The experimental results, was discovered that MLP and C4.5 have the most effective rates. However, compared to with ROC curve, C4.5 appeared the foremost economical.

2 Classification Technique

2.1 Naive Bayes

In machine learning, Naive Bayes [14, 15] classifiers a family of easy "probabilistic classifiers" supported applying Bayes' theorem with sturdy (naive) independence assumptions between the features. Naive Bayes classifiers square measure extremely ascendable, requiring variety of parameters linear within the number of variables (features/predictors) during a learning problem. Maximum-likelihood training will be done by evaluating a closed-form expression, which takes linear time, instead of by overpriced reiterative approximation as used for several different types of classifiers.

2.2 One R

OneR, short for "One Rule", may be a simple, even so correct, classification algorithmic technique that generates one rule for every predictor within the data, so selects the rule with the littlest total error as its "one rule" [16, 17]. To make a rule for a predictor, we have a tendency to construct a frequency table for every predictor against the target. It's been shown that OneR produces rules solely slightly less correct than progressive classification algorithms whereas produced rules that area unit simple for humans to interpret.

2.3 CART

The CART or Classification & Regression Trees [3, 4] methodology was introduced in 1984 by Leo Breiman, Jerome Friedman, Richard Olshen and Charles Stone as an umbrella term to refer to both of the below procedures of decision trees. Classification Trees: where the target variable is categorical and the tree is used to identify the "class" within which a target variable would likely fall into.

Classification Trees: Analysis is once the predicted outcome is that the category to that the data belongs.

Regression tree: Analysis is once the predicted outcome may be thought of a true variety.

2.4 MLP

A multilayer perceptron (MLP) can be a class of feed forward artificial neural network [18]. An MLP consists of a minimum of three layers of nodes [15, 19]. Except for the input nodes, each node can be a cell that uses a nonlinear activation function. MLP utilizes a supervised learning technique referred to as back propagation for work. Its multiple layers and non-linear activation distinguish MLP from a linear perception. It will distinguish data that's not linearly separable.

3 Feature Selection Technique

Feature Selection Technique (FST) [6, 20, 21] reduces the dimensions of the feature space; Remove unnecessary, irrelevant or noise data. This brings instant results for the application: Increasing the performance of fast data processing algorithms, improving data quality and the performance of data mining and increasing the comprehensibility of the mining results. Feature selection techniques [22] have two kinds: Wrappers and Filters. Wrapper evaluates evaluations supported accuracy estimates by learning target algorithms and selects them. On the other hand, the filter uses the general quality of the data and works independently from the learning algorithm.

The feature selections wherever specify the evaluation method and search method.

3.1 Evaluation Methods

Feature Selection selects ‘n’ number of relevant features from the original ‘M’ features [23]. Mainly, feature selections are often thought of as a search downside with some analysis criteria. Subset Evaluation and Single Attribute Evaluation are the two frequently used techniques in feature estimation area. Single Evaluation apply the ranking technique to work out the degree of significance whereas Subset Evaluation uses variation search algorithms to supply the feature subset and the degree of relevance will be measured using different Machine Learning methods (Classifiers, Clustering etc.).

3.2 Search Method

Search algorithms [23, 24] are essential for feature selection for the reason that it afford a way to search for attributes. Single attribute or a set of attributes carried by the algorithm is specified in the form of the feature visualize so that it can measure its eligibility. Search algorithms are essential for search as it is a way to search for properties Search algorithms can start from any no/all all attributes or arbitrary point in space. This closes when the addition/deletion of any remaining features results in reduction in the evaluation. It can also generate a ranking list of attributes by transferring from one side to the other and recording the order in order to record the properties.

4 Data Set

The Chronic Kidney Diseases [25] data set is collected from UCI Machine learning repository. This dataset can be used to predict the chronic kidney disease and it will be collected from the Apollo Hospitals, Tamilnadu, India nearly a pair of months of amount. The CKD dataset having 24 features, 400 instances and 1 class with binary nature (ckd or notckd). The features of data set contain numerical and nominal value.

5 Process Flow and Proposed Planning

5.1 Process Flow

The general structural design of the proposed GS-IBFST and the feature selection techniques is outlined in the Fig. 1. As described in Fig. 1, the core steps of the proposed

GS-IBFST and Feature Selection techniques in approach involve successive steps. The Feature selection techniques are start with the filter and wrapper approach. The filter approach has used two type's algorithms namely Chi Squared AttributeEval (CSAE) and One R AttributeEval (ORAE) have used to Ranked the dataset. The wrapper approach have Genetic Search algorithm with two different learning algorithms like J48 and CART. These Genetic Search-J48 (GS-J48) and The Genetic Search–CART (GS-CART) select the most important feature in the dataset. The GS-IBFST has proposed feature selection technique is explained in proposed planning section in details.

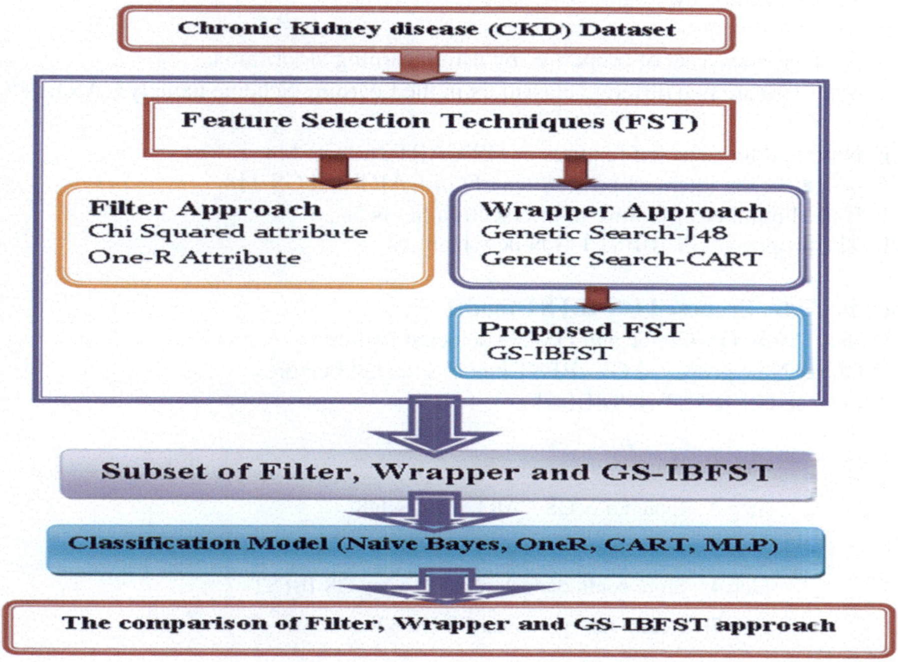

Fig. 1. Process flow of proposed planning.

After getting the selected features FSTs like Filter, Wrapper and GS-IBFST. Classify the dataset with selected feature in the filter, wrapper and GS-IBFST. For which we have used four classified algorithms Naive Bayes, OneR, MLP and CART. Compare the performances of classifiers with existing FSTs and the proposed GS-IBFST.

5.2 Proposed Planning

In this research work, proposed a Genetic Search Intersection Based Feature Selection Techniques (GS-IBFST). For this, we've used 2 completely different classifiers with the Genetic Search methodology. These two FST Genetic Search-CART (GS-CART) and Genetic Search-J48 (GS-J48) selected the features of CKD data set. After this the selected features of Genetic Search method, with the CART and the J48 classifier, the intersection of these two FST GS-CART and GS-J48 selected features has calculated.

The number of features available in the intersection set will be a member of the proposed GS-IBFST and it is a new proposed FST.

Proposed Works as Follows

I. Apply wrapper-based feature selection techniques.

   a. Apply search method as Genetic Search (GS).

II. Select attribute evaluator as WrapperSubsetEval.

   a. Evaluates a set of properties by using learning algorithms.
   b. There are two different classifiers in the Learning Scheme namely CART, J48.

III. Now find the selected features of GS-CART and GS-J48.
IV. Calculate the intersection features of GS-CART and GS-J48.
V. If total number of feature in intersection Set is N.
VI. The proposed GS-IBFST has N new features.

Pseudo Code: Proposed GS-IBFST model

Input: Array GS-CART and GS-J48 selected features.
Output: New proposed GS-IBFST has N Selected Features
Function GS-IBFST (GS-IBFST)

Step 1:	Start
Step 2:	Input array GS-CART and GS-J48
Step 3:	Find common features in to array GS-CART and GS-J48
Step 4:	Store common features in to array GS-IBFST
Step 7:	Return new proposed GS-IBFST

Apply this new features subset GS-IBFST in classification models.
Compare the performance of existing models and proposed GS-IBFST.
The Accuracy of classifier improved by proposed GS-IBFST.

6 Experimental Results

Weka's [26] attribute selection panel where specify the attribute evaluator and search method [24]. Tables 1 and 2 list the choices. Attribute selection is normally done by searching the space of attribute subsets, evaluating each one. This is achieved by combining attribute subset evaluators (AttributeSubsetEvaluator and Single-AttributeEvaluator) in Table 1 with, search methods (Genetic Search and Ranker) in Table 2.

6.1 Evaluation Methods

Table 1 shows the list of feature evaluators employed in this work with its parameter details.

Table 1. Feature evaluation methods and parameter tuning details.

Evaluation algorithm	Name	Parameters tuning	
Attribute Subset Evaluator	*WrapperSubsetEval*	*classifier = "J48"* *fold = 5* *seed = 1* *threshold = 0.01*	Use a classifier plus cross-validation
	WrapperSubsetEval	*classifier = "Simple Cart"* *fold = 5* *seed = 1* *threshold = 0.01*	Use a classifier plus cross-validation
Single Attribute Evaluator	*ChiSquaredAttributeEval*	*binarizeNumericAttributes = false* *missingMerge = True*	Evaluates the worth of an attribute by computing the value of the chi-squared statistic with respect to the class
	OneRAttributeEval	*binarizeNumericAttributes = false* *missingMerge = True*	Use OneR's methodology to evaluate attributes

6.2 Search Methods

Table 2 shows completely different search algorithms employed in our work at the side of the relevant parameter calibration.

The Table 3 Classifier "J48" is used with Attribute Subset Evaluator like Wrapper-SubsetEval and Genetic Search is used as Search Algorithm. The 9 feature selects from

Table 2. Search algorithms and parameter tuning details.

Search algorithm	Parameter tuning	Function
Genetic Search	*crossoverProb = 0.6* *maxGenerations = 20* *mutationProb = 0.033* *PopulationSize = 20* *reportFrequency = 20* *seed = 1* *startSet=*	Performs a search using the simple genetic algorithm described in Goldberg
Ranking Method	*generateRanking = True* *numToSelect = −1* *startSet=* *threshold =−1.7976931348623157E308*	Rank individual attributes (not subsets) according to their evaluation

the CKD dataset. Similarly the classifier: "CART" is used with Attribute Subset Evaluator WrapperSubsetEval and Genetic Search is used as Search Algorithm. The 10 feature selects from the CKD dataset.

Table 3. Genetic search selected features.

Name of FSTs	Selected features	Selected attributes
GS-J48	9	F1, F3, F4, F7, F12, F15, F17, F19, F20
GS-CART	10	F1, F3, F4, F13, F15, F17, F18, F19, F20, F24

The Table 4 ChiSquaredAttributeEval (CSAE) is used as a Single-AttributeEvaluator and Ranker is used as Search Algorithm. Similarly OneRAttributeEval (ORAE) is used as a Single-AttributeEvaluator and Ranker is used as Search Algorithm.

Table 4. Rank of features.

Name of ranked FSTs	Selected features	Selected attributes
CSAE	24	F15, F12, F3, F16, F4, F19, F20, F18, F11, F10, F13, F2, F7, F22, F14, F23, F1, F6, F5, F24, F17, F8, F21, F9
ORAE	24	F15, F16, F18, F12, F3, F4, F6, F14, F19, F13, F20, F11, F10, F2, F17, F1, F7, F9, F8, F21, F23, F24, F22, 5

The Table 5 the common (Intersection) of Genetic Search-J48 and Genetic Search-CART features has been shown has Seven features. This is the proposed FST, which has been named GS-IBFST.

Table 5. Proposed GS-IBFST.

FSTs	Selected features	Selected attributes
GS-IBFST	07	F1, F3, F4, F15, F17, F19, F20

The Table 6 shows Naive Bayes, One R, CART and MLP Classifier accuracy are shown all feature of CKD dataset. After this, the accuracy is shown with the 9 features designated by the GS-J48, and next designated 10 features by GS-CART have shown. Finally, the 7 features of GS-IBFST are shown to be accuracy. This result's clear that every one classifier with GS-IBFST has gives high accuracy.

The Table 7 below shows Chi squared attribute top ranked 7 features, the OneR attribute top ranked 7 features and proposed GS-IBFST has 7 features in CKD dataset.

Table 6. Comparison of classifier accuracy.

Classifiers	All features	GS-J48	GS-CART	GS-IBFST
Naive Bayes	96.25	97.50	96.75	**98.25**
OneR	92.25	92.00	92.00	**92.50**
CART	97.75	97.75	98.00	**98.25**
MLP	97.75	95.75	96.25	**98.00**

Table 7. Features of CKD with different FSTs.

Name of FSTs	Selected features	Selected attributes
CSAE	07	F15, F12, F3, F16, F4, F19, F20
ORAE	07	F15, F16, F18, F12, F3, F4, F6
GS-IBFST	07	F1, F3, F4, F15, F17, F19, F20

In the Table 8, Naive Bayes, One R, CART and MLP Classifier accuracy are shown with all feature of CKD dataset. After this, the accuracy is shown with the 7 features designated by the Chi squared attribute, and next designated 7 features by OneR attribute have shown. Finally, the 7 features of GS-IBFST are shown to be accuracy. This result's clear that every one classifier with GS-IBFST has offer higher accuracy.

Table 8. Comparison of classifier accuracy.

Classifiers	All features	CSAE	ORAE	GS-IBFST
Naive Bayes	96.25	97.50	96.25	**98.25**
OneR	92.25	92.50	92.50	**92.50**
CART	97.75	98.00	98.00	**98.25**
MLP	97.75	96.50	97.75	**98.00**

7 Conclusion and Future Work

Diagnosis of heath condition is incredibly difficult and significant issue in field of medical science. Classification and feature selection techniques play very important role to quick and accurately identification of disease. This helps to diagnosis of disease in easily and perfectly. We have developed robust classification models for classification of CKD. The performances of the classification techniques have enhanced as the numbers of features have reduced. The proposed FST supported the Intersection of the two existing Genetic Search based FST (GS-J48 and GS- CART) known as GS-IBFST. The GS-IBFST has offers higher classification accuracy compared to any or all feature of CKD dataset and existing feature selection techniques. The classification algorithms Naive Bayes, OneR, CART and MLP produces high prediction performance in terms of classification accuracy with propose GS-IBFST. The highest accuracy obtained by Naive Bayes and CART classifiers within the case of GS-IBFST has 98.25 with 7 features.

In future alternative robust and computationally economical model are going to be develop and GS-IBFST is enforced in alternative dataset.

References

1. Subasi, A., Alickovic, E., Kevric, J.: Diagnosis of chronic kidney disease by using random forest. In: 2017 Proceedings of the International Conference on Medical and Biological Engineering, pp. 589–594 (2017)
2. Polat, H., Danaei Mehr, H., Cetin, A.: Diagnosis of chronic kidney disease based on support vector machine by feature selection methods. J. Med. Syst. **41**(4), 55 (2017)
3. Han, J., Kamber, M., Pei, J.: Data Mining: Concepts and Techniques. Elsevier, Amsterdam (2012)
4. Pujari, A.: Data Mining Techniques. University Press, Hyderabad (2013)
5. Jantawan, B., Tsai, C.: A comparison of filter and wrapper approaches with data mining techniques for. Int. J. Innov. Res. Comput. Commun. Eng. **2**, 4501–4508 (2014)
6. Yildirim, P.: Filter based feature selection methods for prediction of risks in hepatitis disease. Int. J. Mach. Learn. Comput. **5**, 258–263 (2015)
7. Hall, M.: Correlation-based feature selection for machine learning. Methodology 21i195-i20, pp. 1–5 (1999)
8. Liu, H., Setiono, R., Science, C., Ridge, K.: Chi2: Feature Selection, pp. 388–391 (1995)
9. Holte, R.C.: Very simple classification rules perform well on most commonly used datasets. Mach. Learn. **11**, 63–91 (1993)
10. Shrivas, A.K., Sahu, S.K., Hota, H.S.: Classification of chronic kidney disease with proposed union based feature selection technique, pp. 503–507 (2018)
11. Charleonnan, A., et al.: Predictive analytics for chronic kidney disease using machine learning techniques. In: 2016 Management and Innovation Technology International Conference, pp. MIT-80–MIT-83 (2016). https://doi.org/10.1109/miticon.2016.8025242
12. Kunwar, V., Chandel, K., Sabitha, A.S., Bansal, A.: Chronic kidney disease analysis using data mining classification. In: 2016 6th International Conference-Cloud System and Big Data Engineering (Confluence), pp. 300–305. IEEE (2016). https://doi.org/10.1109/confluence.2016.7508132
13. Boukenze, B., Haqiq, A., Mousannif, H.: Predicting chronic kidney failure disease using data mining techniques, vol. 397 (2017)
14. Russell, S., Norvig, P.: Artificial Intelligence A Modern Approach. Series in Artificial Intelligence. Prentice Hall, Upper Saddle River (2003)
15. Haykin, S.: Neural Networks and Learning Machines. Pearson Prentice Hall, Upper Saddle River (2008). 936 pLinks 3
16. Nasa, C., Suman, S.: Evaluation of different classification techniques for WEB data. Int. J. Comput. Appl. **52**, 34–40 (2012)
17. Alam, F., Pachauri, S.: Comparative study of J48, Naive Bayes and One-R classification technique for credit card fraud detection using WEKA. Adv. Comput. Sci. Technol. **10**, 1731–1743 (2017)

18. Rumelhart, D.E., Hinton, G.E., Williams, R.J.: Learning internal representations by error propagation (1985)
19. Sivanandam, S.N., Deepa, S.N.: Principles of Soft Computing. Wiley, Hoboken (2014)
20. Novakovic, J., Strbac, P., Bulatovic, D.: Toward optimal feature selection using ranking methods and classification algorithms. Yugosl. J. Oper. Res. **21**, 119–135 (2011)
21. Karegowda, A.G., Manjunath, A., Jayaram, M.: Application of genetic algorithm optimized neural network connection weights for medical diagnosis of PIMA Indians diabetes. Int. J. Soft Comput. **2**, 15–23 (2011)
22. Ashraf, M., Chetty, G., Tran, D.: Feature selection techniques on thyroid, hepatitis, and breast cancer datasets. Int. J. Data Min. Intell. Inf. Technol. Appl. **3**, 1–8 (2013)
23. Arun Kumar, C., Sooraj, M.P., Ramakrishnan, S.: A comparative performance evaluation of supervised feature selection algorithms on microarray datasets. Procedia Comput. Sci. **115**, 209–217 (2017)
24. Witten, I.H., Frank, E., Hall, M.A.: Data Mining. Morgan Kaufmann series in data management systems (2011). https://doi.org/10.1002/1521-3773(20010316)40:6%3c9823::aid-anie9823%3e3.3.co;2-c
25. UCI Machine Learning Repository of machine learning databases (2015). https://archive.ics.uci.edu/ml/datasets/Chronic_Kidney_Disease. Accessed 1 Jan 2016
26. Machine Learning Group at the University of Waikato. https://www.cs.waikato.ac.nz/ml/weka/downloading.html

Analysis of Wormhole Detection Features in Wireless Sensor Networks

Manish Patel[1](✉), Akshai Aggarwal[2], and Nirbhay Chaubey[2]

[1] Smt. S R Patel Engineering College, Unjha, Gujarat, India
it43manish@gmail.com

[2] Gujarat Technological University, Ahmedabad, Gujarat, India

Abstract. Security is very important for resource constrained wireless sensor networks due to their fundamental nature. The possible attacks in wireless sensor networks include sinkhole, wormhole, sybil, selective forwarding, blackhole etc. Wormhole is a gateway of many more attacks. To launch the attack is easy, but to detect it is very hard. For launching the attack, it is not required to know the cryptographic material or protocols used in the network. A malicious node attracts the traffic from one location and tunnels to another location and disturbs the whole routing process. In this paper, we have discussed various features for wormhole attack detection in wireless sensor networks.

Keywords: Wireless sensor networks · Security · Wormhole · Attacker

1 Introduction

For applications such as target tracking, battlefield surveillance and assessment, disaster zones assessment, monitoring tunnels and bridges (civil infrastructure) any compromise of information can have very serious consequences. Providing security in wireless sensor network is a very difficult task due to several constraints. The nodes are resource limited in terms of bandwidth, computation and energy. In such environment, an attacker can easily capture a few nodes. Once the nodes are compromised or captured then the attacker can access the secret material stored on sensor nodes and launch a variety of attacks.

Wormhole attack can be launched by placing one malicious node in part of the network and other malicious node in another location which is far away in other part of the network by creating high speed low latency tunnel.

As shown in Fig. 1, location of malicious node M1 is far away from location of the malicious node M2. First malicious node receives traffic from one part and tunnels it to the second malicious node which replies the traffic to another part of the network. Therefore, routing process in the network is disturbed. As shown in Fig. 1, node Y and Z become one hop neighbors of node W and vice versa. There is no need to know the cryptographic mechanism used in the network and protocols or services offered in the network for launching the attack. The packets pass through the wormhole can propagate faster compared to the normal path.

N. Nain and S. K. Vipparthi (Eds.): ICIoTCT 2019, AISC 1122, pp. 22–29, 2020.
https://doi.org/10.1007/978-3-030-39875-0_3

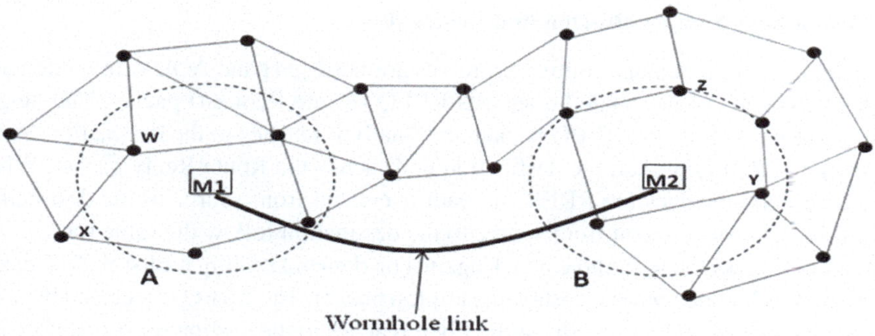

Fig. 1. Wormhole attack

The rest of the paper is organized as follow: Description of wormhole attack is presented in Sect. 2. Section 3 presents wormhole attack taxonomy. Existing wormhole detection techniques are summarizes in Sect. 4. Section 5 presents various wormhole detection features. Finally conclusion and future research challenges are discussed in Sect. 6.

2 Description of Wormhole Attack

2.1 Wormhole Against Periodic Routing Protocols

In distance vector routing algorithm, every node periodically sends its routing table to all its neighbors. Every node receives routing table from all its neighbors and updates its own routing table as per the entries in the neighbor's routing table. The routing table contains the entries of distance from itself to the other nodes in the network. In Fig. 2, node S_9 broadcasts its routing table. All the neighbor nodes of node S_9 hear it. Node S_2 is not a real neighbor but it is connected through wormhole. Therefore, node S_2 also hears it. Node S_2 updates its routing table and makes entry in the routing table that node S_9 is one hop away and the nodes $\{S_8, S_{10}, S_{11}, S_{12}, S_{13}\}$ are two hops away.

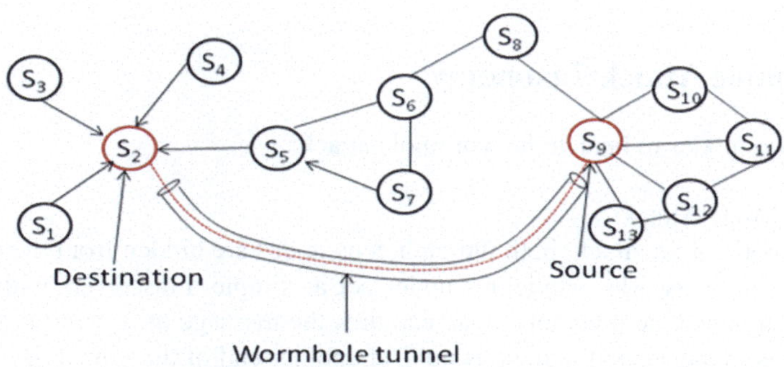

Fig. 2. Wormhole attack against distance vector routing protocol

2.2 Wormhole Against On-Demand Protocols

The examples of on demand routing protocols are DSR [54] and AODV. In on demand routing protocols, source node broadcasts RREQ (Route Request) packet. The neighboring nodes forwards the RREQ packet and finally it reaches to the destination. After receiving the RREQ packet, the destination node sends the Route Reply packet. When the source node receives the RREP, the path is created from source to the destination and all the data packets sent from source to the destination follow the same route.

Figure 3 shows the wormhole attack against on demand routing protocol. To create a path to node S2, node S9 broadcasts route request packet. The RREQ is received by node S8 and it is forwarded to node S6 and finally it reaches to the destination S2. Before the RREQ packet reaches to S2 via the normal path S9-S8-S6-S5-S2, it immediately reaches to node S2 via tunnel. Nodes S2 and S9 are connected through tunnel. Therefore, when S9 broadcasts a RREQ packet, it is captured by the malicious node located at one end of the tunnel and it passes to the malicious node located at another end of the tunnel. The second malicious node replies it to node S2. In this way, path is established between S9 and S2 through tunnel. All the traffic passes from S9 to S2 follows this route via malicious nodes. Malicious nodes located at each end of the tunnel can drop the packet, analyze the traffic or selectively forwards the packets.

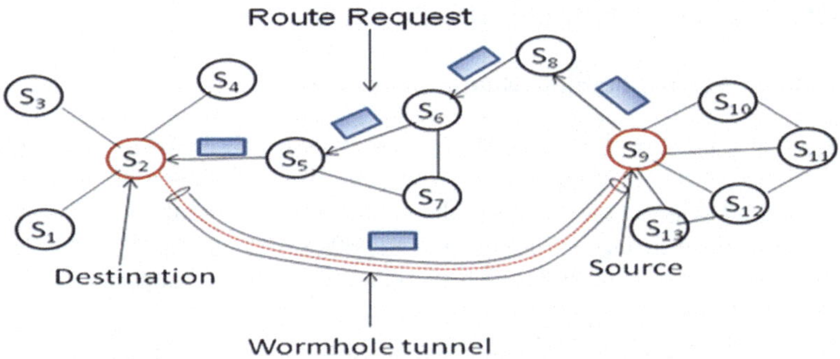

Fig. 3. Wormhole attack against on demand routing protocol

3 Wormhole Attack Taxonomy

There are two modes to launch the wormhole attack:

1. Hidden Mode Attack:
   In this mode of an attack, both the malicious nodes are hidden from the rest of the nodes in the networks. Malicious nodes act as simple transceivers without using their identities. One malicious node captures the message at one end of the tunnel and sends to the second malicious node at another end of the tunnel. By creating a tunnel, malicious nodes capture traffic from one area and reply it to the other area. For hidden mode wormhole attack, the attackers do not require any cryptographic keys.

2. Participation Mode Attack:
   This mode of attack is more powerful. It can be launched by valid cryptographic keys. The malicious nodes participate in the network as legitimate nodes and deliver packets with smaller no. of hop counts. Wormhole attack can be created using encapsulation based method or using out-of band tunnel.

3.1 Wormhole Using Encapsulation

In Fig. 4, node S broadcasts a packet which is received by its neighbor node A and it is also captured by first malicious node. Node S forwards it to node A and the packet reaches to destination node D through the legitimate path S-A-B-C-D. Malicious node M1 forwards the packet to the next node and it reaches to second malicious node M2 which sends the packet to the destination node. Hop count do not increase in between M1 and M2. The path through the malicious nodes has shorter hop count compared to the legitimate path and therefore, the path through the malicious nodes is selected for transmission.

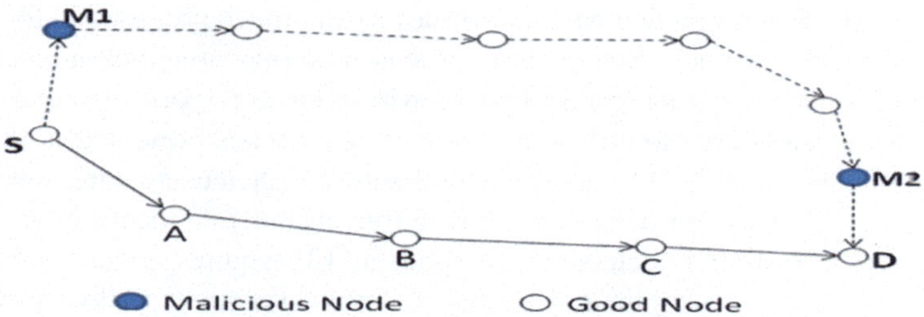

Fig. 4. Wormhole through packet encapsulation

3.2 Wormhole Using Out-of-Band Tunnel

Figure 5 shows wormhole attack using out of band tunnel. The packet broadcasted by source node S is captured by the first malicious node M1. Both the malicious nodes are connected through high speed tunnel. First Malicious node tunnels the packet to the second malicious node. Malicious node M2 sends the packet to the destination D. The legitimate path is S-W-X-Y-Z-D. The time required to reach the packet to the destination via tunnel is less compared to the genuine path. Therefore, the path through the tunnel is selected for transmission. Node M1 attracts the traffic from one part of the network, tunnels to the other node and node M2 replies it to another part of the network.

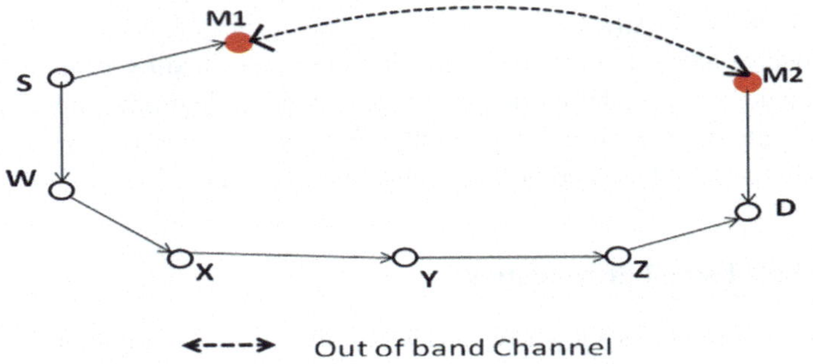

Fig. 5. Wormhole through tunnel

4 Related Work

Geographical leash approach requires loosely synchronized clocks whereas temporal leash approach requires tightly synchronized clocks [1, 2]. Challenge response delay measurement approach proposed in [3] uses distance bounding techniques and one way hash chains. Ranging based neighbor discovery approach proposed in [4] requires microsecond precision clock. Authors have proposed timing based measurement in [5]. Distance consistency based approach proposed in [6, 7] has good performance even when the malicious locators are more than the normal. Approaches presented in [8–10] are based on round trip time. In [11], abnormal nodes are detected using unreasonable rank values. Diameter is used for detection in range free anchor free localization approach [12]. Geographic wormhole detection proposed in [13] requires a pair wise key pre-distribution protocol for wormhole detection. Each node requires radio fingerprinting device in [14]. Each node is equipped with directional antenna for wormhole detection in [15]. Wormhole resistant hybrid technique proposed in [16]. Symmetric key cryptography is used for wormhole detection in [17]. Observer nodes are deployed for monitoring the network in [18]. Passive and real time detection is presented in [19]. Location based keys approach in [20] requires a group of mobile robots having GPS capabilities. An approach presented in [21] is based on network connectivity information. Authors have proposed local connectivity test for wormhole detection in [22]. Neighbor number and all distance tests for wormhole detection were proposed in [23]. Authors have proposed multidimensional scaling visualization based approach in [24–26]. Neighbor node monitoring approach for wormhole detection is presented in [27]. Graph theoretic model in [28] is based on symmetric cryptography. Wormhole resistant secure routing algorithm was proposed in [29]. A mobile application based wormhole detection approach in wireless sensor network is proposed in [30]. Unit disk graph based approach Mobility based beacon wormhole detection is proposed in [31].

5 Detection Features and Comparison

Existing wormhole detection techniques mainly use four detection features: Time, neighborhood, location and hop count. These detection features are described as follow:

(1) Time: Wormhole represents two remote nodes as neighbor nodes. Therefore, the path suffers from wormhole has more average time per hop compared to the path without wormhole. Techniques based on time information require clock synchronization.
(2) Neighborhood: Wormhole attack represents two non neighboring nodes as neighbor nodes. Therefore, by storing the information about neighboring nodes, wormhole can be detected. Detection techniques based on neighborhood store information of one hop neighbors or two hops neighbors. To analyze and store two hops neighbor's data in dense network requires more storage, processing power and memory. Neighbors table is frequently changed in highly mobility based network. Therefore, neighborhood based techniques do not work efficiently for highly mobility based network. It can increase congestion in the network.
(3) Location: Location of node is very important information for detecting wormhole. For knowing the location, existing techniques use GPS (Global Positioning System) device. Directional antenna is also used for getting the relative location of neighboring nodes. The use of GPS and directional antenna decrease the battery timing and increase the cost of the network. For mobility based wireless sensor networks, nodes frequently change their position. Therefore, false positive rate can be increased.
(4) Hop Count: Wormholes attract traffic by presenting shorter route with less no. of hop count. Existing techniques uses hop count feature in collaboration with location or time. The location based techniques require GPS device and time based techniques require clock synchronization. Time value more than normal hop time indicates the presence of abnormal node. Routing delay can be increased using hop count based techniques.

6 Conclusion and Future Research Issues

In this paper, we have presented various detection features that are used for wormhole detection. Techniques based on hop count would be better than other techniques for detecting wormhole. In mobile wireless sensor networks, two far away located nodes can become one hop neighbors after some time. It creates an illusion that wormhole has been launched. Differentiating these legitimate nodes from the malicious nodes is a challenging research issue. Existing detection methods are resource hungry. They require additional hardware device which increases the cost of the node. Software based techniques for wormhole detection has some special assumptions. Detecting wormhole with high accuracy and low overhead is a crucial issue. Distance and time based techniques proposed in the literature for wormhole detection assume that distance or time information cannot be altered. Malicious nodes can alter this information. Therefore, cryptographic schemes are needed to support time and distance based approaches.

References

1. Hu, Y.C., Perrig, A., Johnson, D.B.: Packet leashes: a defense against wormhole attacks in wireless networks. In: IEEE Computer and Communications Societies, vol. 3, pp. 1976–1986. IEEE (2003)

2. Hu, Y.-C., Perrig, A., Johnson, D.B.: Wormhole attacks in wireless networks. IEEE J. Sel. Areas Commun. **24**(2), 370–380 (2006)
3. Capkun, S., Buttyan, L., Hubaux, J.P.: SECTOR: secure tracking of node encounters in multi-hop wireless networks. In: Proceedings of the 1st ACM Workshop on Security of Ad-Hoc and Sensor Networks (SASN 2003), pp. 21–32, October 2003
4. Shokri, R., Poturalski, M.: A practical secure neighbor verification protocol for wireless sensor networks. In: ACM, WiSec 2009, Zurich, Switzerland, 16–18 March 2009
5. Khabbazian, M., Mercier, H., Bhargava, V.K.: Severity analysis and countermeasure for the wormhole attack in wireless ad hoc networks. IEEE Trans. Wirel. Commun. **8**(2), 736–745 (2009)
6. Maheshwari, R., Gao, J., Das, S.R.: Detecting wormhole attacks in wireless networks using connectivity information. In: Proceedings INFOCOM, pp. 107–115 (2007)
7. Chen, H., Lou, W., Sun, X., Wang, Z.: A secure localization approach against wormhole attacks using distance consistency. EURASIP J. Wirel. Commun. Netw. **2010**, 11 (2009)
8. Zhao, Z., Wei, B., Dong, X., Yao, L., Gao, F.: Detecting wormhole attacks in wireless sensor networks with statistical analysis. In: International Conference on Information Engineering (ICIE), pp. 251–254 (2010)
9. Chiu, H.S., Lui, K.S.: DelPHI: wormhole detection mechanism for ad hoc wireless networks. In: 1st IEEE International Symposium on Wireless Pervasive Computing, p. 6 (2006)
10. Qazi, S., Raad, R., Mu, Y., Susilo, W.: Securing DSR against wormhole attacks in multirate ad hoc networks. J. Netw. Comput. Appl. **36**(2), 582–593 (2013)
11. Lai, G.H.: Detection of wormhole attacks on IPv6 mobility-based wireless sensor network. EURASIP J. Wirel. Commun. Netw. **2016**(1), 274 (2016)
12. Xu, Y., Ouyang, Y., Le, Z., Ford, J., Makedon, F.: Analysis of range-free anchor-free localization in a wsn under wormhole attack. In: ACM, MSWiM 2007, Chaina, Greece, 22–26 October 2007
13. Sookhak, M., Akhundzada, A., Sookhak, A., Eslaminejad, M., Gani, A., Khan, M.K., Li, X., Wang, X.: Geographic wormhole detection in wireless sensor networks. J. PLOS ONE. **10**(1) 2015. https://doi.org/10.1371/journal.pone.0115324
14. Rasmussen, K.B., Capkun, S.: Implications of radio fingerprinting on the security of sensor networks. In: Third International Conference on Security and Privacy in Communication Networks and the Workshops, pp. 331–340, September 2007
15. Hu, L., Evans, D.: Using directional antennas to prevent wormhole attacks. In: Network and Distributed System Security Symposium (NDSS), pp. 131–141 (2004)
16. Singh, R., Singh, J., Singh, R.: WRHT: a hybrid technique for detection of wormhole attack in wireless sensor networks. J. Mob. Inf. Syst. **2016**, 13 (2016). Hindawi Publishing Corporation, Article ID 8354930
17. Madria, S., Yin, J.: SeRWA: a secure routing protocol against wormhole attacks in sensor networks. J. Ad Hoc Netw. **7**(6), 1051–1063 (2008)
18. Triki, B., Rekhis, S., Boudriga, N.: Digital investigation of wormhole attacks in wireless sensor networks. In: Eighth IEEE International Symposium on Network Computing and Applications, pp. 179–186 (2009)
19. Lu, L., Hussain, M.J., Luo, G., Han, Z.: Pworm: passive and real-time wormhole detection scheme for WSNs. Int. J. Distrib. Sens. Netw. **2015**(356382), 16 (2015). Article ID 356382
20. Zhang, Y., Liu, W., Lou, W., Fang, Y.: Location-based compromise–tolerant security mechanisms for wireless sensor networks. IEEE J. Sel. Areas Commun. **24**(2), 247–260 (2006)
21. Dong, D., Liu, Y., Li, X. Y., Liao, X., Li, M.: Topological detection on wormholes in wireless ad hoc and sensor networks. In: 17th IEEE International Conference on Network Protocols, pp. 314–323 (2009)

22. Ban, X., Sarkar, R., Gao, J.: Local connectivity tests to identify wormholes in wireless networks. In: ACM, MobiHoc 2011, Paris, France 16–20 May 2011
23. Buttyan, L., Dora, L., Vajda, I.: Statistical wormhole detection in sensor networks. In: SAS 2005, pp. 128–141. Springer (2005)
24. Wang, W., Bhargava, B.: Visualization of wormholes in sensor networks. In: WiSe 2004, Proceeding of the 2004 ACM Workshop on Wireless Security, pp. 51–60. ACM Press (2004)
25. Lu, X., Dong, D., Liao, X.: MDS-based wormhole detection using local topology in wireless sensor networks. Int. J. Distrib. Sensor Netw. **2012**, 9 (2012). Article ID 145702
26. Mukherjee, S., Chattopadhyay, M., Chattopadhyay, S., Kar, P.: Wormhole detection based on ordinal MDS using RTT in wireless sensor network. J. Comput. Netw. Commun. **2016**, 15 (2016). Article ID 3405264
27. Biswas, J., Gupta, A., Singh, D.: WADP: a wormhole attack detection and prevention technique in MANET using modified AODV routing protocol. In: Proceedings of the 9th IEEE International Conference on Industrial and Information Systems, pp. 1–6 (2014)
28. Poovendran, R., Lazos, L.: A graph theoretic framework for preventing the wormhole attack in wireless ad hoc networks. Wireless Netw. **13**, 27–59 (2007)
29. Matam, R., Tripathy, S.: WRSR: wormhole-resistant secure routing for wireless mesh networks. EURASIP J. Wirel. Commun. Netw. **2013**(1), 180 (2013). Springer
30. Tsitsiroudi, N., Sarigiannidis, P., Karapistoli, E., Economides, A.: EyeSim: a mobile application for visual-assisted wormhole attack detection in IoT-enabled WSNs. In: Proceedings of the 9th IFIP Wireless and Mobile Networking Conference, pp. 103–109 (2016)
31. Chen, H., Chen, W., Wang, Z., Li, Y.: Mobile beacon based wormhole attackers detection and positioning in wireless sensor networks. Int. J. Distrib. Sens. Netw. **2014**(1), 10 (2014)

Propagating Minimal Messages Using Multi Set-Cover in Wireless Ad-Hoc Network

Sadia Sharmin(✉)

Bangladesh University of Engineering and Technology, Dhaka, Bangladesh
sadiasharmin.ss@gmail.com

Abstract. Wireless (ad hoc) network is a group of nodes to communicate with each other over a wireless channel to receive and transmit data packets among their neighbors residing in the transmission range. We propose to build a wireless ad hoc network to maintain data transmission in a cost-effective way with minimal data packets. This transmission of data can commence in conditions/circumstances which are challenged and harsh to continue without failure caused by any network, environmental or other factors and ensure fault tolerance in such kind of noisy network. In spite of having vast research experiments in this area, existent studies shows lack effort in establishing both balanced coverage of two hop neighbors of sender and fault tolerance in the network with minimal message propagation at the same time. In this paper, a new algorithm using two hop distant neighbor information is proposed and a special version of set cover problem suited to ad hoc networks is used to output multiple forwarder lists by which, all the uncovered two hop neighbors are covered once. Extensive simulations made in NS2 implementation show that the algorithm provides high delivery rate (88% in average) of packets and transmission of average number of packets is lower than existent Dominant Pruning algorithm. Later we analyzed the reachability aspect of the algorithm by calculating the percentage of nodes (receivers) reached in presence of various fractions of malfunctioning nodes integrated deliberately by us in the network.

Keywords: Wireless ad hoc network · Fault tolerance · Multi set cover · Balanced coverage · Forwarder list · Broadcasting packets

1 Introduction

Among recent research areas of Computer Science, significant interest has been bestowed upon multi hop wireless networks due to it's potential applications in different sectors. Specially with the advent of wireless mobile communication, wireless ad hoc network is undoubtedly one of the best choices to communicate in such places where negligible or no communication infrastructure exists or the existing infrastructure is inconvenient or out of date to use. This kind of networks

N. Nain and S. K. Vipparthi (Eds.): ICIoTCT 2019, AISC 1122, pp. 30–43, 2020.
https://doi.org/10.1007/978-3-030-39875-0_4

is formed by a group of nodes communicating with each other over a wireless channel. No pre-installed infrastructure exists in this type of network which causes communication to be supported by intermediate nodes relaying packets between communicating parties in multiple hops. As a result, each node performs their role as a router to forward traffic from it's source and transmit it to another node where both of the sender and receiver of the packet lies in current node's transmission range [19]. This radio range is fixed for each of the nodes because having large transmission range increases the distance progress of data packets toward their final destinations but energy consumption per transmission is high. On the contrary, a short transmission range uses less energy to forward packets to the next hop, but a large number of hops are required for packets to reach their destinations. Hence, selecting an optimum value of the radio transmission range, wireless multi-hop ad hoc network has emerged. Figure 1 denotes such a network by *Graph Model*, where node S is sending a message and node T is designated to receive it.

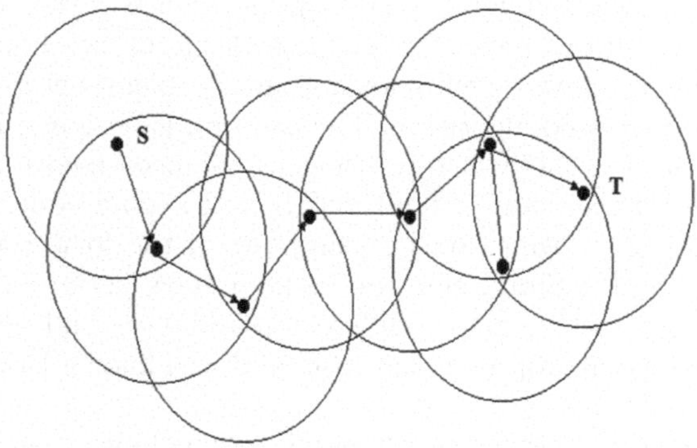

Fig. 1. Wireless ad-hoc multihop network. Circles denote the transmission range of each node.

From the construction of the wireless ad hoc network, it is evident that whether the data transmission mode in that network is unicast or broadcast, intermediate nodes play crucial role for transporting the data packets to subsequent regions of the network. Ad hoc network is inherently deployed in such areas where data transmission can be interrupted by harshness or noise from the environment is prevalent, compounded by the selfishness or shyness of the nodes to transmit the data packet and their malicious behaviors. Example of such failures include link failure, congestion, selfishness of the nodes to forward the packet, limited power supply to the batteries, error in reading and many other sources of faults, listed in [11,16–18]. Thus the intermediate nodes have to develop fault tolerant nature to deal with these advert situations and continue relaying data packets, specially the broadcasting packet, as they are picked up and relayed by multiple nodes unlike the unicasting ones. Flooding can be a straight forward approach to ensure the broadcasting transmission up to the receiver, but relatively huge amount of redundant packets will soon create collision and contention, which in turn results in serious problems like broadcasting storms in MANET [20].

A lot of approaches, listed in the next section, already focused on how to secure this transmission of broadcasting packets. However, finding the optimum number of forwarders to limit redundant transmission still remained NP hard [3]. Again, the number of transmitting data packets is one of the major concerns in such networks as the total number of transmissions is served as the cost criterion for broadcasting [4], so the number of messages being propagated is required to be minimal. Unlike the case of broadcasting, achieving minimal number of message transmission is less challenging in case of unicasting data transmission. Even if the data packet has to be retransmitted in case of failure, the retransmission is made to a specific single neighbor, which makes the transmission highly reliable via the four-way handshake of IEEE. 802.11 [1].

In this paper, three important part of Wireless multi-hop ad hoc network are pointed out which are minimal message propagation, balanced coverage of the target receivers and ensuring fault tolerant performance of the network devising a single solution to all of them. Here the knowledge of two hop distant neighbors for each node and a very well known graph problem, Set Cover guarantees uniqueness of the proposed algorithm. The solution is designed to provide multiple set of k forwarding lists which provides us balanced k-coverage; alternative set(s) of forwarder list in case of fault occurring to some node. It also ensures the minimal message transmission by using any of the forwarder list not containing any faulty nodes. Simulation results later show performance analysis for various performance metrics under different parameters and also performance comparison between our proposed algorithm and well known Dominant Pruning algorithm.

The rest of paper is designed as follows: Sect. 2 includes some previous works related to algorithms for finding minimum forwarder list, providing fault tolerance and coverage of the nodes of wireless networks. Section 3 sheds light onto our proposed algorithm with an example along with the characteristics of our algorithm. Section 4 includes simulation environments with experimental studies for performance comparison and analysis. Section 5 draws conclusion to the paper outlining some future works.

2 Motivation and Background Work

Wireless ad hoc networks have been providing a lot of diverge applications for over last two decades, which made this sector a point of interest from numerous researchers. Their contribution have enriched the literature about the node coverage, fault tolerance, load-balance, power consumption, etc properties of wireless ad hoc networks. But most of these works have shown appreciable reluctance to offer a balanced coverage of the nodes of a multi-hop network with minimal number of message propagation. Again question of combining a balanced coverage of nodes with fault tolerance of the network at the time has also been left unanswered till now. Choosing the appropriate forwarders to broadcast the data packets plays vital role in performance of multi-hop wireless network. One easily fathomable solution to the problem of finding optimal forwarders is to find a minimum dominating set [9], which has been proven Np-hard [5] far ago.

Lim and Kim [3] introduced a greedy based heuristic algorithm, *dominant pruning*. In this algorithm, each node of the network is assumed to have the two hop distant neighbors information and each node maintain a *forwarder list*. This forward list is constructed from the set of the one hop distant neighbors and only the members of this list will forward the data packet to ensure the coverage of all the two hop distant neighbors. As a result, the forwarder list has to be minimum but computing the minimum sized forwarder list is also a NP hard problem [6]. Their suggested greedy approach iteratively chose those nodes which covers maximum number of elements from the uncovered two hop neighbors until all the nodes are being covered. As a result, an unbalanced or non-uniform coverage of the two hop distant neighbors is made by their algorithm, which is literally opposite to our goal. Figure 2 shows such an example of unbalanced coverage.

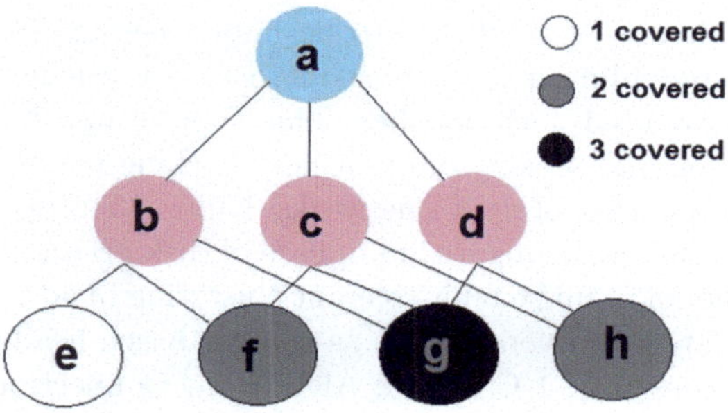

Fig. 2. Non uniform coverage of the two hop distant neighbors. Node e is covered once, Node f and h is covered twice and Node g is covered thrice

The Dominant pruning algorithm had been enhanced including controlled redundancy into retransmission in [1], to pave the way for *Multicover Dominant Pruning (MDP)*. This algorithm constructs the forwarder list of minimum nodes which covers the uncovered two hop neighbors at least m times to make the network more fault tolerant. Their experiments showed that $\mathrm{m} = 2$ is the best choice, for which this algorithm can also be termed as 2-cover *Dominant Pruning algorithm*. They also proposed *Trust-based Dominant Pruning* considering the perceived trustworthiness of nodes. *Total Dominant Pruning (TDP)* and *Partial Dominant Pruning (PDP)* [4] are other two very famous heuristic variant of dominant pruning. With a far more refinement of MDP algorithm because of it's too much redundancy, three better fault-tolerant heuristics named *Multicover Total Dominant Pruning (MTDP)*, *Multicover Partial Dominant Pruning (MPDP)* and *Multicover Improved Dominant Pruning (MIDP)* have been introduced in [2] very recently which utilize 2-hop neighborhood information more effectively to reduce redundant transmissions and improve the reach-ability of nodes in an unreliable ad hoc wireless network. All these kind of dominant pruning algorithm have the similarity of using 2 hop distant neighbors information to provide redundancy into the network for increasing the reachability aspect of ad hoc networks.

Laouiti et al. [7] and Sinha et al. [8] mapped the problem of finding forwarding set to the *classical* set cover problem, assuming no prior knowledge of the nodes. This may sound related to the addressed problem but they introduced the concept of *Multi Point Relay (MPR)* with a number of heuristics that reduce duplicated transmission of the data packet in forwarding broadcasting packets. They efficiently chose the subset of forwarders, called *multipoint relays* from the neighbors. The region of the network covered by these nodes is same as the region covered by all the members of the neighbor set. As a result, the retransmission of the data packets have been reduced substantially. In [10], problem of finding minimum sized forwarding set has been dealt by proposing a number of algorithms with at least three or greater than three approximation ratio using partition based or disk covering algorithms. Wireless networks are inherently noisy where a node can fail due to a lot of factors described in preceding section, hence a wireless network has to be fault tolerant. In [12], fault tolerance is issued by providing a heterogeneous back-up scheme, where one type of resources is substituted with another. This type of approaches are efficient when communication and sensing are mutually backing up each other, which is mainly broadly used in case of sensor networks. Kuhn [13] proposed a distributed algorithm and a probabilistic algorithm in case of disk graph using the standard message passing model to make fault tolerant clustering in ad hoc networks. Various algorithms related to coverage such as Cluster-based head election coverage reservation techniques [14]; 1-Coverage Algorithm, Connection based coverage algorithms, k-Coverage Algorithm, point coverage [15] are available in literature for wireless ad hoc or sensor networks but these algorithms don't aim for the uniform coverage of the nodes or sensors. In [22] Adrian et al. builds a realistic mobility model for Mobile Ad Hoc Networks.

Most of the extent works tried to minimize the packet transmission by finding a minimum sized forwarder lists which in turns to be the generalized case of some NP hard problems like finding connecting dominating set or the set cover problems where different kinds of heuristics are actually applied in most of the cases. These works also neither achieve a balanced coverage of the nodes of the network nor could enhance fault tolerance by the same solution causing increased overhead. Inspired by all such odds, to address the current problem by proposing a single solution which can coalesce all these three aspects of wireless multi-hop ad hoc network is designed.

3 Proposed Algorithm

This section is going to present the proposed algorithm named **Multi Set Cover Algorithm** in this section. At first let us define the classical Set Cover problem.

Let's consider, given a collection of n elements called **Universe** and a number of subsets of the universe, U, say $S = \{S_1, S_2, S_3, \ldots, S_m\}$ Each of the subset, S_i, has an associated cost but for our algorithm, all the subsets will have the same cost. Our target is to find a sub-collection of S, which has the minimum cost and cover all elements of U.

Let's assume node r receives a packet from node s. In the algorithm, the forwarder list, $F(r)$, i.e., the list of the nodes which will rebroadcast the packet, is inserted into the packet header by the node, r. So, if r finds itself into the forwarder list of s, then r will create its own forwarder list, $F(r)$. This list is then inserted into the packet header and only the nodes present in this list will rebroadcast the packet. The packet is transported through the network in this manner. This part of the algorithm is analogous to Dominant Pruning but the aim of **Set Cover Algorithm** is to construct a set of node dis-joint forwarder lists.

It is assumed each nodes know the information of it's two hop distant neighbors, so node r first constructs the set of its all **uncovered** two hop neighbours, U_r, the word 'uncovered' means the just arrived packet has not covered those two hop neighbors yet. This uncovered set of two hop distant neighbors is analogically the Universe of our set cover problem, need to cover. The set of the *Candidate Nodes* for being included into the forwarder list is also defined as $\mathrm{C(s, r)} = \{\text{Neighbors of node r}\} - \{\text{neighbors of node s}\}$.

At first $F(r)$ is set to null and $C(s,r)$ is constructed. Now for each node, i, in the candidate set $C(s,r)$, the subset, S_i, containing the nodes from the set of uncovered nodes, U_r, which are covered by that node, i is created. These collection of subsets, M, are analogically the subsets of the Universe from the Set Cover problem, with which the Universe is needed to be covered.

Now the rest of the algorithm, creating multiple set of forwarders, is pretty straight-forward and works in greedy approach. At first, that node is selected, $w_m \in \mathrm{C(s, r)}$ which covers the maximum nodes from the set of uncovered two hop distant neighbors, U_r and insert that node into the current forwarder list, $F(r)$. This is computed easily by finding which subset (S_m in this case) has the largest number of elements. The node w_m is excluded from the candidate set $C(s,r)$ and the set of nodes, S_m, that is covered by w_m is also excluded from M. Now, the node which covers the maximum nodes from the remaining uncovered two hop distant neighbors is looked for and this new node is enqueued into the current forwarder list, computed similarly as before. Excluding this node from the candidate list and the set of nodes covered by it from M is continued until all the two hop distant neighbors are covered.

Thus one forwarder list is prepared from the candidate list, $C(s,r)$, which covers all the two hop distant neighbors of r. Now the nodes of $F(r)$ are excluded from the candidate list and repeat the whole process once again to find another **new** forwarder list. Clearly the new forwarder list created is node-disjoint from the previous one but each of them individually covers all the two hop distant neighbors of r. The algorithm terminates when anymore node-disjoint set from the candidate set, $C(s,r)$ is found.

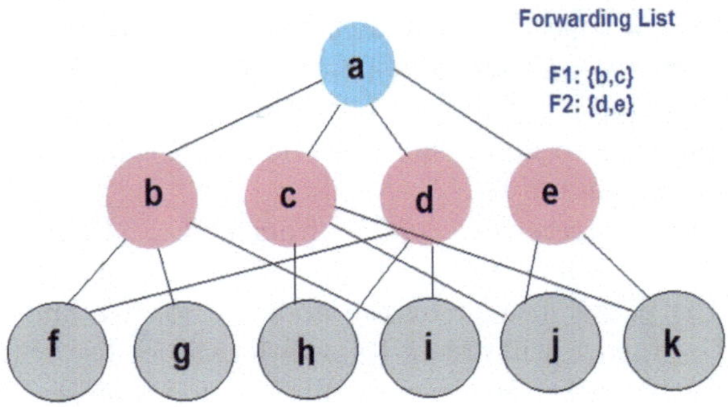

Fig. 3. Illustration of multi set cover algorithm

For better illustration, the algorithm is demonstrated with a random example. Let's consider, the node a received a packet from some of it's neighbor and it has nodes $\{b, c, d, e\}$ in its one hop distance and these 4 nodes actually construct the candidate list, C by whom it's two hop distant neighbors $\{f, g, h, i, j, k\}$ are covered. At first, node, b is selected as it covers $\{\text{f}, \text{g}, \text{i}\}$ and pushed into a forwarder list, F_1 and then node, c is selected as it covers $\{h, j, k\}$ and it is also pushed into F_1. All the uncovered two hop distant nodes got covered and F_1 is returned as one forwarder list. Now $\{b, c\}$ is excluded from the list, C and repeating the similar procedure, another forwarder list $F_2 = \{d, e\}$ is found. So there are two forwarder lists, F_1 and F_2 which covers all the two hop distant neighbors.

Assuming node r has just received just a packet from node s, without loss of any generality. Here, $N(w_i)$ and U_r denotes neighbors of w_i and the uncovered two hop distant neighbors of r respectively.

Multi Set Cover algorithm not only provides a balanced coverage of nodes but also makes a fault tolerant transmission with a minimal number of messages being propagated. This algorithm returns us a list of forwarder lists of which each list covers the two hop distant neighbors once. So, if there are k number of forwarder lists being returned by the algorithm, balanced$k$coverage of second hop neighbors of the sender is found. The algorithm is already fault tolerant because in case of any node being failed to transmit the packet, the corresponding list containing the fault node taking the other available forwarder list can be ignored. Since the output forwarding lists are node disjoint, it is obvious that a faulty node can occur at most one of the forwarding lists, not in two at a time. However, as a worst case complication, in case of presence of at least one faulty node in each forwarding lists will result in the less number of nodes reachable and consequently, all the two hop distant neighbors will not be covered. In that case, initiating the algorithm once again excluding the faulty nodes, another forwarder list to cover the nodes of U_r causing an extra overhead is found. The two hop distant neighbors have been covered with the minimal number of the data packets taking any of the available set from the output list and nominating them as forwarder. Even in case of minimal message propagation, the algorithm

Algorithm 1. Multi Set Cover

```
1: F_r = φ
2: Z = φ
3: L = φ
4:
5: for Each node w_i ∈ C(s,r) do
6:     Create the set S_i = N(w_i) ∩ U_r
7: end for
8: Let M = {S_1, S_2, ..., S_n}
9: Suppose S_m is the set such that,
10: |S_m| = max_{s_i∈M}{|S_i|}
11:
12: if |S_m| = φ then
13:     exit
14: end if
15: F_r = F_r ∪ {w_m}
16:
17: for Each node x ∈ S_m do
18:     Z ← Z ∪ {x}
19: end for
20: M = M − S_m
21: C(s,r) = C(s,r) − {w_m}
22: if Z = U(r) then
23:     save F_r into L
24:     F_r ← φ
25:     go to step 9
26: else
27:     Suppose S_n is the set such that,
28:     |S_n| = max_{s_i∈M}{|S_i ∪ S_m|}
29:
30:     for Each node x ∈ S_n do
31:         Z ← Z ∪ {x}
32:     end for
33:     M = M − S_n
34:     C(s,r) = C(s,r) − {w_n}
35:     go to step 22
36: end if
37: output L
```

can be slightly changed to terminate as soon as the first forwarder list is found, at line no 22 of Algorithm 3. However, to gain this, it has to be ensured that any of the nodes in the chosen forwarder list will not be faulty.

4 Experimental Studies

A detailed simulation model is constructed in NS2 version 2.35 using it's AODV protocol [21] after modifying it to measure the performance of algorithms.

All the experiments are conducted using Intel Core i5 2.30 GHz CPU installed with 4 GB Ram with Ubuntu versions 14.04. The existing flooding protocol in AODV caching packet ids are modified in order to prevent the repeated broadcasting of same packets. In case of upward and downward going packets, the destination address is replaced by broadcast address. Ids of the nodes of forwarding lists are stored to ensure that only the matched nodes will forward the packets and the inbound packets are dropped for the rest of the nodes.

4.1 Simulation Environment and Traffic Generation

The wireless network is modeled varying the node density (number of nodes) and the transmission range of each nodes in an rectangular space area of 625 m by 625 m. Number of nodes in a range of 100 nodes to 500 nodes with an interval of 50 nodes in random position varying the transmission range also of each node from 125 to 225 in an interval of 25 m is simulated. 5 random static scenarios for each simulation is generated and the average of those readings of that specific scenario is taken. CBR traffic broadcasting at a rate of 4 packets of 512 bytes in size per second is produced.

4.2 Performance Metrics and Performance Parameters

In order to examine the effect of our proposed algorithm on the network, *Delivery rate of Data Packets* and *Reach-ability* is considered as performance metrics. The number of *Average Packets Forwarded (APF)* and the number of *Average Packets Received (APR)* are calculated to find the delivery rate of the data packets. For the reach-ability aspect, the number of the *malfunctioning* nodes (nodes that can't/don't transmit packets due to any kind of faults listed previously in Sect. 1) is varied and the percentage of the *Average Nodes Reachable (ANR)* is counted. This aspect indicates that the algorithm is inherently fault tolerant because it continues to transmit packets to the receivers even after having malfunctioning nodes into the network. Each readings were itself an average of 5 different static scenarios to ensure the performance of the algorithm in random network. Note that, all this simulations have been parameterized by varying node density, transmission range and percentage of malfunctioning nodes.

4.3 Performance Analysis

Here the performance of our proposed algorithm is analyzed on basis of performance metrics listed in the preceding section. The data delivery rate is calculated using the APF and APR. First the node density is varied from 100 to 500 in an interval of 50 nodes, as inscribed in Sect. 4.1 and calculated the APF and APR, depicted in Fig. 4. The transmission range is varied similarly from 125 m to 225 m, carried out the similar experiment and showed the results in Fig. 5.

From both figures, APF and APR increases as node density increases, as expected. But they increase more rapidly in case of varying transmission range

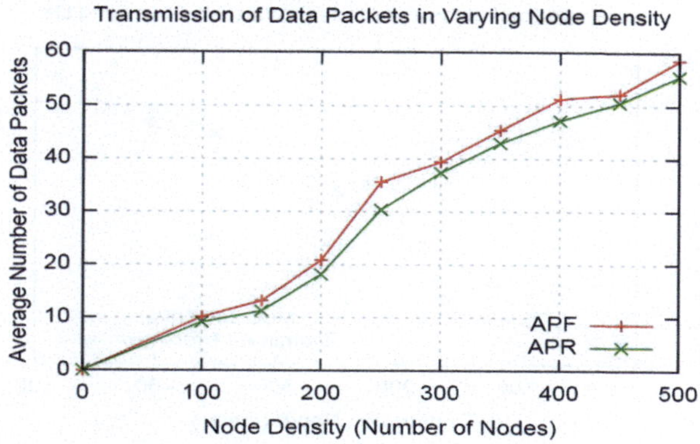

Fig. 4. Effect of data transmission under varying node density

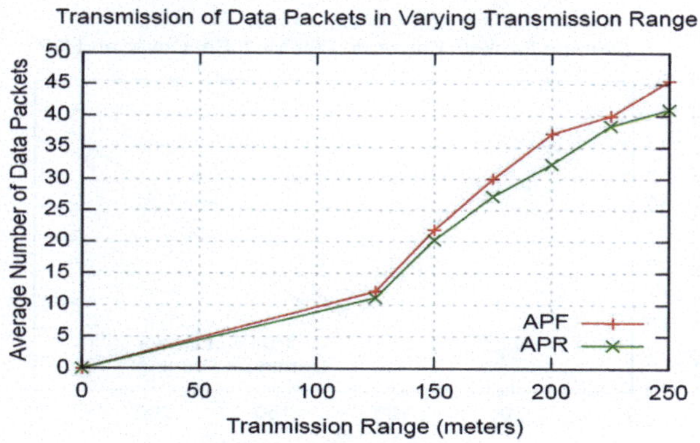

Fig. 5. Effect of data transmission under varying transmission range

than varying node density. The average delivery rates are 88.222% and 88.977% for varying node density and transmission range.

In Figs. 6 and 7, a performance comparison between the existent Dominant Pruning algorithm and the newly proposed Multi Set Cover algorithm under varying Node density and transmission range in shown. To cover the uncovered two hop distant neighbor with minimal number of messages (data packets), only one forwarding set is enough which reduces the APF in both cases as expected. For both of the cases, Multi Set Cover reported less number of average transmitted packets than the number of packets transmitted by Dominant Pruning algorithm.

Now let us compare the reach-ability of our algorithm in case of percentage of malfunctioning nodes included deliberately to some extent into the network. The percentage of malfunction is varied from 0% to 40% and calculated ANR for

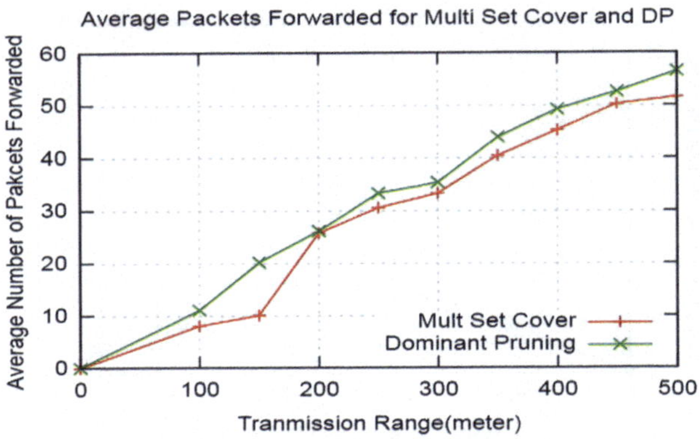

Fig. 6. Average packets forwarded for multi set cover and dominant pruning for varying node density

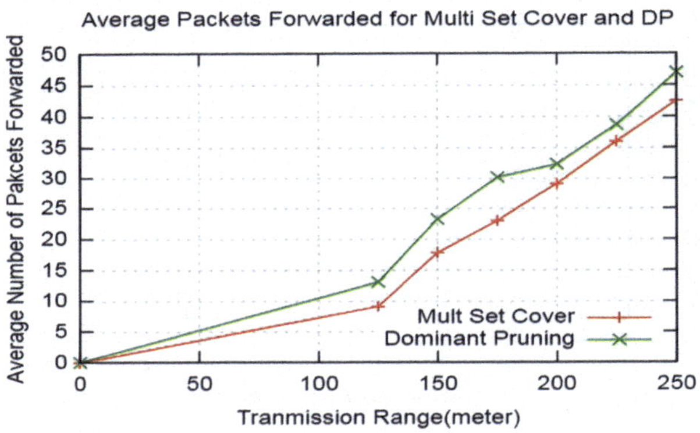

Fig. 7. Average packets forwarded for multi set cover and dominant pruning for varying transmission range

varying Node Density, shown in Fig. 9. The figure reveals that increasing malfunction percentage decreases ANR. At the presence of 40% malfunction nodes, ANR doesn't fall below than 50% in case of 100 nodes. But to fix the percentage of malfunction nodes, if more nodes are augmented in the simulation area, ANR increases more than 65%. Averagely, during 20% of malfunction nodes, the algorithm can reach upto 75% of the receivers. The transmission range is also varied and the same experiment is done. Similarly, ANR remains greater than 75% at 20% of malfunction nodes. Unlike previously, ANR falls below 50% at 40% malfunctioning nodes in case of the nodes having transmission range 125 m. The results are shown in the Fig. 8.

In both of these reach-ability experiments, it is evident that increasing node density or transmission range causes more reach-ability of the receivers, which is expected.

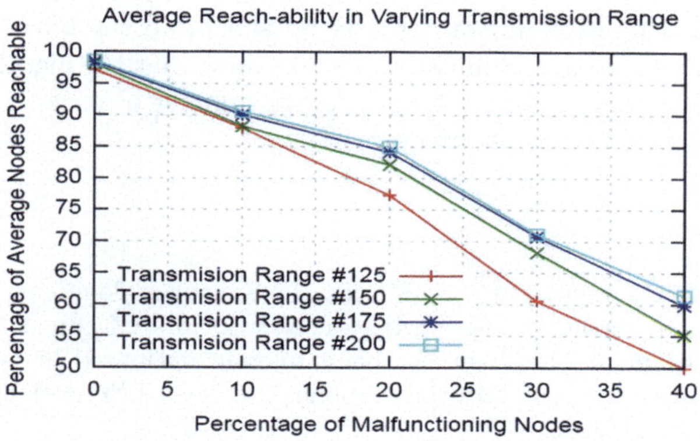

Fig. 8. Effect of reach-ability under varying transmission range

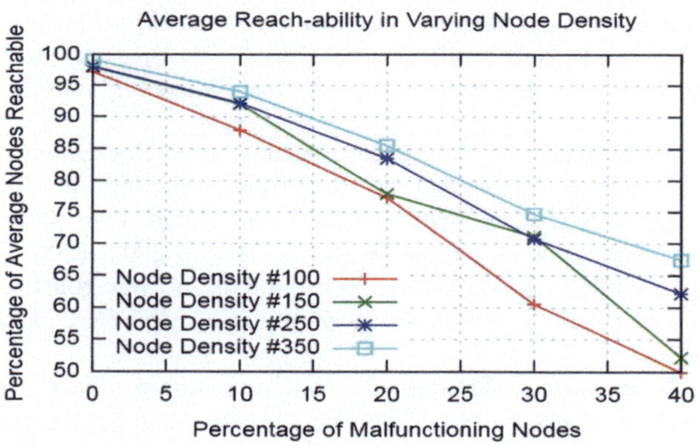

Fig. 9. Effect of reach-ability under varying node density

5 Conclusion and Future Work

The proposed algorithm tries to achieves its goal of minimal propagated packets. In this algorithm, the weight of the subset is kept same. It would be interesting to monitor how the algorithm behaves if some weight is added to the subsets, S_i, because adding weight to a subset would mean that some nodes from the candidate sets are being preferred than some other nodes. This happens when the network learns that at least there are few nodes in the network which are always fail-safe and identifying such nodes will always be beneficial in packet transmission. Here the mobility of the nodes in the network is not considered, so it would be a good side to explore how the algorithm works in case of mobile nodes and what changed might be needed to provide the same coverage and fault tolerance. The experiments are run by generating CBR traffic, which might be enhanced more to visualize the change with other types of data packets. Our algorithm is a special version of set cover problem suited to ad hoc networks is used to output multiple forwarder lists by which, all the uncovered two hop neighbors are covered once. Simulation results show that the algorithm provides high delivery rate and transmission of average number of packets is lower than existent Dominant Pruning algorithm.

References

1. Rahman, A., Gburzynski, P., Kaminska, B.: Enhanced dominant pruning-based broadcasting in untrusted ad-hoc wireless networks. In: ICC, pp. 3389–3394 (2007)
2. Akter, M., Islam, A., Rahman, A.: Fault tolerant optimized broadcast for wireless ad-Hoc networks. In: 2016 International Conference Networking Systems and Security (NSysS) (2016)
3. Lim, H., Kim, C.: Multicast tree construction and flooding in wireless ad hoc networks. In: Proceedings of the 3rd ACM International Workshop on Modeling, Analysis and Simulation of Wireless and Mobile Systems, pp. 61–68. ACM (2000)
4. Peng, W., Lu, X.-C.: On the reduction of broadcast redundancy in mobile ad hoc networks. In: Proceedings of the 1st ACM International Symposium on Mobile Ad Hoc Networking and Computing, pp. 129–130. IEEE Press (2000)
5. Lichtenstein, D.: Planar formulae and their uses. SIAM J. Comput. **11**(2), 329–343 (1982)
6. Even, S.: Graph Algorithms. Computer Science Press, Rockville (1979)
7. Qayyum, A., Viennot, L., Laouiti, A.: Multipoint relaying: an efficient technique for flooding in mobile wireless networks. In: Proceedings of the 35th Annual Hawaii International Conference on System Sciences (HICSS 2002) (2002)
8. Sinha, P., Sivakumar, R., Vaduvur, B.: Enhancing ad hoc routing with dynamic virtual infrastructures. In: Proceedings of the IEEE INFOCOM 2001, pp. 1763–1772 (2001)
9. Alzoubi, K.M., Wan, P.J., Frieder, O.: New distributed algorithm for connected dominating set in wireless ad hoc networks. In: Proceedings HICSS-35 (2002)
10. Calinescu, G., Mandoiu, I., Wan, P.J., Zelikovsky, A.: Selecting forwarding neighbors in wireless ad hoc networks. In: Proceedings of the ACM DIALM 2001, pp. 34–43, December 2001
11. Tipper, D., Dahlberg, T., Shin, H., Charnsripinyo, C.: Providing fault tolerance in wireless access networks. IEEE Commun. Mag. **40**(1), 58–64 (2002)
12. Koushanfar, F., Potkonjak, M., Sangiovanni-Vincentell, A.: Fault tolerance techniques for wireless ad hoc sensor networks. In: 2002 Proceedings of IEEE Sensors, vol. 2. IEEE (2002)
13. Kuhn, F., Moscibroda, T., Wattenhofer, R.: Fault-tolerant clustering in ad hoc and sensor networks. In: 2006 26th IEEE International Conference on Distributed Computing Systems, ICDCS 2006. IEEE (2006)
14. Soro, S., Heinzelman, W.B.: Cluster head election techniques for coverage preservation in wireless sensor networks. Ad Hoc Netw. **7**(5), 955–972 (2009)
15. Liang, J., Liu, M., Kui, X.: A survey of coverage problems in wireless sensor networks. Sens. Transd. **163**(1), 240 (2014)
16. Chen, J., Kher, S., Somani, A.: Distributed fault detection of wireless sensor networks. In: Proceedings of the 2006 Workshop on Dependability Issues in Wireless Ad Hoc Networks and Sensor Networks, pp. 65–72. ACM (2006)
17. Gupta, G., Younis, M.: Fault-tolerant clustering of wireless sensor networks. In: 2003 IEEE Wireless Communications and Networking, WCNC 2003, vol. 3, pp. 1579–1584. IEEE (2003)

18. Paradis, L., Han, Q.: A survey of fault management in wireless sensor networks. J. Netw. Syst. Manag. **15**(2), 171–190 (2007)
19. Deng, J., Han, Y.S., Chen, P., Varshney, P.K.: Optimal transmission range for wireless ad hoc networks based on energy efficiency. IEEE Trans. Commun. **55**(7), 1439–1439 (2007)
20. Tseng, Y.-C., Ni, S.-Y., Chen, Y.-S., Sheu, J.-P.: The broadcast storm problem in a mobile ad hoc network. Wirel. Netw. **8**(2/3), 153–167 (2002)
21. Royer, E.M., Perkins, C.E.: An implementation study of the AODV routing protocol. In: 2000 IEEE Wireless Communications and Networking Conference, WCNC 2000, vol. 3, pp. 1003–1008. IEEE (2000)
22. Pullin, A., Pattinson, C., Kor, A.: Building realistic mobility models for mobile ad hoc networks. Informatics **5**(2), 22 (2018). http://www.mdpi.com/2227-9709/5/2/22

Multi Header Based Ultra Low Power MTCMOS Technique to Reduce NBTI Effect in Combinational Circuit

Anjan Kumar[1(✉)], Shelesh Krishna Saraswat[1], Preeti Agrawal[1], and Shweta Singh[2]

[1] VLSI Design Lab, GLA University, Mathura, India
anjan.kumar@gla.ac.in
[2] ABV-IIITM, Gwalior, India

Abstract. With the advancement of technology, negative bias temperature instability (NBTI) emerges out as a major problem for VLSI circuits. Meanwhile, the leakage power increases dramatically as the supply/threshold voltage continues to scale down. These two issues present extreme unwavering quality issues for CMOS devices. Since both the NBTI and leakage are reliant on input test vector of the circuit however input vector control strategy isn't viable for bigger circuits. Thus in this paper two design is proposed (1) Single header Based Ultra low Power Diode Tri-mode Technique is designed for reducing leakage and delay of the circuit (2) Multi header Based Ultra low Power Diode Tri mode Technique with Body Bias on all sleep pMOS transistor to mitigate NBTI effect. Experimentations are done on 1 bit full adder circuit with the usage of tanner EDA at 90 nm CMOS technology node and supply voltage 1 V. The results reveal that by using first proposed technique leakage current is reduces by 74.93% and 0.217% respectively as compare to Stacking with delay based MTCMOS method and ULP diode based MTCMOS method. First proposed design is also effective in terms of delay. Result shows that delay get reduced by 20.91% and 7.84% as compare to prevalent techniques. Second Proposed design at 50% Duty cycle is very much effective for NBTI effect mitigation. Experimental result shows that Second Proposed design with reduction in duty cycle(from 100% to 50%) the V_{th} shift on pMOS transistor reduces and with this decrement in V_{th} of pMOS transistor NBTI effect on the circuit will get reduced.

Keywords: Leakage reduction · Delay reduction · NBTI mitigation · Forward body biasing · Reverse body biasing

1 Introduction

As technology scales, the increment in leakage current and Negative Bias Temperature Instability (NBTI) are major problems to handle [1, 2]. The Leakage current contributes 50% of the digital circuit operational power. The different types of powers of circuit like standby power, Active power and reliability of the circuit is affected by this. [3, 4]. When the circuit is in the OFF condition or say in standby mode then during this mode

N. Nain and S. K. Vipparthi (Eds.): ICIoTCT 2019, AISC 1122, pp. 44–53, 2020.
https://doi.org/10.1007/978-3-030-39875-0_5

the standby power of circuit is get affected by the leakage current. So it is necessary to minimize leakage in standby mode for which a popular technique Power Gating is used with multi threshold (MTCMOS) transistors [5]. It is having a usage of high threshold sleep transistor which becomes inactive during standby mode as a result of which low Vth circuit gets isolated from supply voltage and ground in standby mode which as a result minimizes the leakage current. [5, 6].

The NBTI effects occurs in the p-type transistors when a logic "0" is used at the gate-to-source terminal, i.e. Vgs of pMOS transistor is negative bias ($V_{gs} = V_{gs}$).This condition is called "stress state", due to which the magnitude of threshold voltage(Vth) of the pMOS transistor increases over time. This increment of V_{th} the performance of the circuit. which shown in Fig. 1. This is An primary thought in today's without limitations, is improve those execution of the circuit.

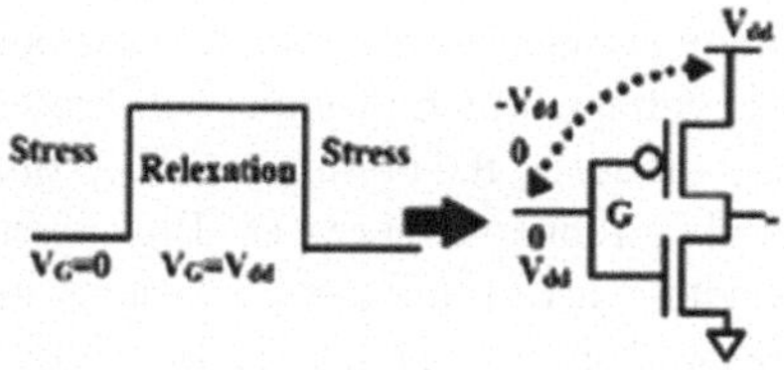

Fig. 1. Stress State on pMOS

After a long time this increment in V_{th} leads to further current degradation between the real Vdd and virtual Vdd line. It degrades the overall circuit performance with time. The Vth shift of pMOS can be controlled by (1) Body biasing and (2) Dividing stress time on parallel pMOS [7].

In this paper different MTCMOS techniques are presented which uses concept of forward body biasing, reverse body biasing, diode connected transistor with parallel pMOS. [8, 9]. Different Standby leakage current, Delay improvement and Negative bias temperature instability aware power gat-ing technique are presented here. A comparative analysis of various power gating design along Vs and V_{bs} shows changes in reduction in delay and leakage

2 Related Techniques

Leakage current and life time of circuit is a most critical issue in VLSI industry So, here we discuss related techniques in terms of leakage reduction and NBTI effect Mitigation.

2.1 Stacking Transistor With Delay Based MTCMOS Circuit

In this technique, three transistor are use at the footer of low logic block in which N_1, N_2 are stacked and P_1 are connected in parallel with biasing, a current flowing through these sleep transistor which is controlled by C_1, The sub threshold leakage current Eq. (1) represents minimization in V_{ds} and the sub threshold leakage current when body to source voltage ($-V_{bs}$) increases (Fig. 2).

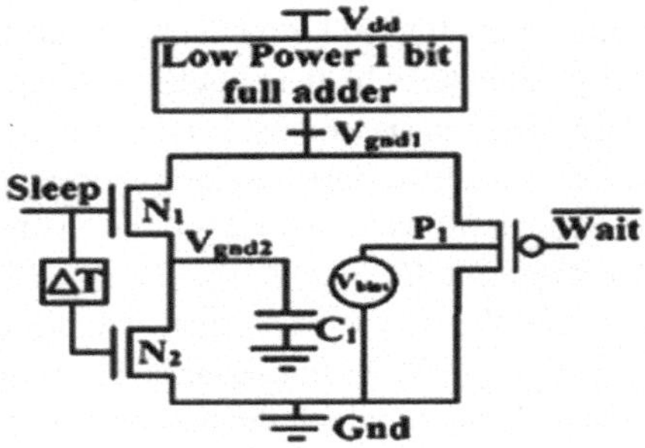

Fig. 2. Stacking transistor with delay based MTCMOS circuit

2.2 Ultra Low Power Diode Based MTCMOS Technique

The Ultra Low Power strategy is functioned as when diode is reverse bias both N_2 and P_2 works with negative V_{gs} voltage which prompts solid reduction of leakage current in equivalence with standard diode. At the point when V_{ds} voltage increased then diode reverse voltage increases and current also increases. The current pass on the peak value and then strongly reduces with V_{gs} of transistor decorously more negative. P_1, P_2, N_1 and N_2 represents high threshold control gated ST's. In this method as comparison with diode connected method, the decrement in leakage current is more effective [11, 13] (Fig. 3).

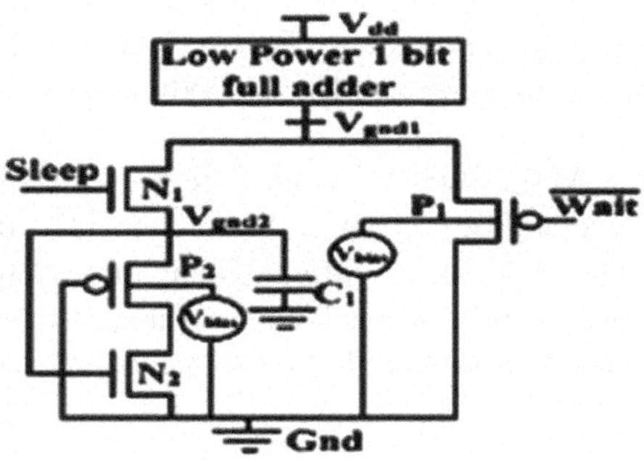

Fig. 3. Ultra low power diode based MTCMOS circuit

3 Proposed Design

3.1 Leakage Aware Single Header Based Ultra Low Power Diode Tri Mode Technique

In first proposed design, high vth transistor P_4 are connected head of the low logic block. N_1, N_2 sleep transistor are stacked with parallely connected P_1, P_2, and P_3 grip transistor. In active mode, P_1, P_2 and P_3 are forward body biased which reduced delay of circuit and in sleep mode, are reverse body biased grip transistor increases the V_{th} of the transistor as a result of which leakage is minimized in the sleep mode. In this design, an additional Header pMOS transistor is added which is helpful to minimizes the circuit's wake up noise (Fig. 4).

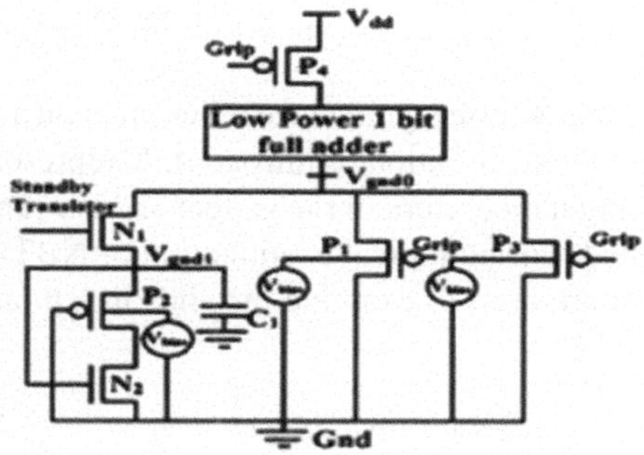

Fig. 4. Single header Based Ultra low Power Diode Tri mode Technique

3.2 NBTI Aware Multiheader Header Based ULP Diode Technique

In proposed design which is shown in Fig. 5 N_1, N_2 sleep transistor are stacked and P_1, P_2, P_3 grip transistor are parallely connected.P_4 and P_5 are high threshold STs which are connected at the head of low logic block, the FBB is used to give the decrease in Vth of the transistor due to which delay can be lowered in dynamic mode and RBB on sleep pMOS transistor in sleep mode to build the Vth of the transistor due to which leakage will be decreased in the sleep mode.

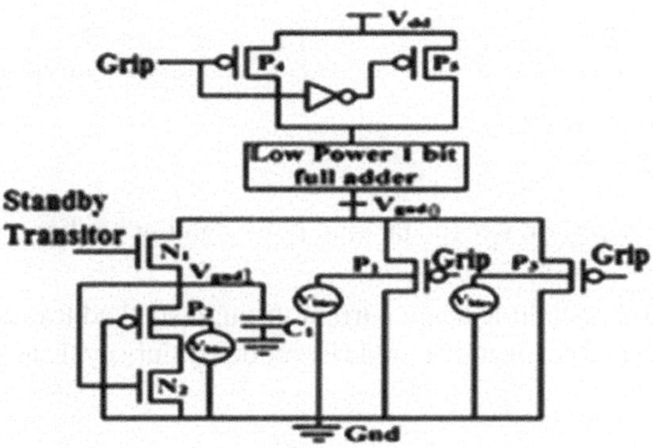

Fig. 5. NBTI Aware Two header Based Ultra low Power Diode Tri mode Technique

To relieve the NBTI issue in the circuit stress time is spilits into half on header transistors P_4 and P_5 of Fig. 5. For this an inverter is utilized among transistors P_4 and P_5, by utilizing this inverter, when transistor P_5 is OFF the transistor P_4 will be ON and when transistor P_4 is OFF the transistor P_5 will be ON, along these lines the aggregate stress time splits over parallel header pMOS transistor. Along these lines stress time on single pMOS transistor decreases. The Eq. (1) shows that stress time decreases the change in V_{th} because of NBTI likewise diminishes and it minimizes the lowering of current because of NBTI. that's why NBTI influenced circuit's aging time can be moved forward.

$$\text{Stress}: V_{gs} = -V_{dd} \rightarrow \Delta Vth \propto k_s e^{-Ea/KT}(t - t_{str})^{1/4.} \quad (1)$$

The two factors Kv and kr whose significance relies on upon a some technical factors (such as channel strain and concentration of nitrogen), K represents Boltzmann steady, t represents nearby working temperature of the gadget and tstr represents time after that NBTI impact get presented. For numerical examination of NBTI impact in the circuit, condition (2), (3) and (4) are used to examine the shift in Vth along with variation in time.

$$\Delta V_{th} = K_v \, \beta^{\frac{1}{4}} (t)^{\frac{1}{4}} + \delta_v \quad (2)$$

Where Δ V_{th} represents the variation in threshold voltage of pMOS (Threshold voltage shift), β is the duty cycle of active sleep pMOS transistor of power gating circuit and the duration after which NBTI aging effect comes into play is represented by t. The relationship between V_{th} and the V_{gs} is represented by Eq. (3) given below

$$K_v \propto \sqrt{C_{ox}(V_{gs} - V_{th})} \exp^{\frac{E_{ox}}{E_o}} \exp^{\frac{-Ea}{kT}} \quad (3)$$

Where V_{th} is the threshold voltage of pMOS transistor and the oxide capacitance is represented by C_{ox} whose value is equal to 0.0135f, V_{gs} represents the gate to source voltage and K is the Boltzmann constant and T is the temperature and Ea = 0.13 eV and $\delta_v = 5$ mv and E0X and E0 = 2 mV are calculated by Eq. (4).

$$E_{ox} = \frac{V_{gs} - V_{th}}{T_{ox}} \quad (4)$$

4 Methodology

Next are the two activities on which stacking power gating configuration works:-

1. Reduction in subthreshold leakage current through stacked transistors.
2. Minimizing NBTI effect in active mode by reducing duty cycle at two header parallel pMOS.

4.1 Method to Reduce Leakage Current in Standby Mode

The leakage in OFF state can be effectively diminished to a degree with the assistance of stacking effect.In this technique all transistors are in off state for reducing the leakage current. By adding a header pMOS grip transistor with reverse body biasing, leakage current is reduced more effectively in sleep mode and delay is also reduces for batter performance of circuit.

4.2 Method to Reduce NBTI Effect

In this method, A circuit shown below is proposed to reduce NBTI effect. Here we are including two (P_4 and P_5) header parallel pMOS grip transistor with forward body biasing for separating stress time (T_{stress}). This infers, for the first half time T_{stress}, P5 is OFF and P4 is ON, while for the other half-time frame, P5 is ON and P4 is OFF. Hence, both P4 and P5 are divide for half time so duty cycle will be half separated on these header transistor. Due to diving this time period of device performance will be improved and NBTI effect will get reduced more effectively.

5 Simulation Results and Comparison

The 90 nm standard CMOS technology along with Tanner EDA was used for simulation of the circuit for 1-bit full adder circuit. (HighjV$_{th}$nMOS = 320 mV, lowV$_{th}$nMOS = 72 mV, HighjV$_{th}$pMOS = −273 mV, lowV$_{th}$pMOS = −56 mV, and V_{DD} = 1 V). Here the effectiveness of the single header based Ultra low Power Diode Tri mode Technique has been applied on 1 bit full adder circuits for the delay and the leakage calculation.

5.1 Leakage Analysis

In sleep mode, we apply a 0 V to nMOS and 1 V to pMOS then all sleep transistor will be in OFF condition. circuit is in sleep mode then we can analysis leakage effectively. All sleep transistor have large threshold voltage for reduction of leakage.

Table 1 represents the comparison between different techniques, by using pMOS transistor with reverse body bias on different techniques which reduces leakage current effectively.

Table 1. Leakage current analysis at different body biasing voltage

Power Gating Technique name	Leakage current in pico amp with Body Bias voltage Vb				
	1.8 V	1.2 V	1 V	0.8 V	0.4 V
Stacking with delay	1087.1	1087.3	1087.5	1087.7	1088.1
ULP Diode based MTCMOS	272.8	272.99	273.18	273.37	273.76
Single header based ULP diode Tri mode	271.78	272.17	272.55	272.93	273.7

Table 2 shows the reduction of leakage for different power gating technique. Leakage depends upon the reverse body biasing variation so with the increment in biasing voltage leakage is reduced in single header based ULP diode tri-mode MTCMOS with reverse body biasing on sleep transistor method least leakage happens which is represented by the graph.

Table 3 demonstrates the relation of leakage current with voltage scaling. Extracted results are tabulated voltage downsize the leakage current is also decreases and with supply voltage increment the Leakage also expands in the circuit.

Table 2. Leakage current result analysis for different power gating techniques with RBB voltage

Power Gating Technique name	Leakage current in pico amp with Reverse Body Bias voltage Vb			Average Leakage Current
	1.4 V	1.2 V	1 V	
Stacking with delay	1087.1	1087.3	1087.5	1087.3
ULP Diode based MTCMOS	272.8	272.99	273.18	272.99
Single header based ULP diode Tri mode	271.78	272.17	272.55	272.26

Table 3. Leakage current analysis for 1 bit full adder for different power gated circuit with change in supply voltage

Power Gating Technique name	Leakage current in pico amp with supply voltage V				
	1.4 V	1.2 V	1 V	0.8 V	0.4 V
Stacking with delay	1157.1	1121.4	1087.5	105.5	920.63
ULP Diode based MTCMOS	143.27	197.13	273.18	366.01	504.18
Single header based ULP diode Tri-mode	287.9	279.94	272.55	95.765	251.67

5.2 Performance Analysis in Terms of Delay

Propagation delay shows the speed of any circuit, it is the combination of routing delay and gate delay in any circuit. In any circuit the duration needed by digital signal to travel from input to the output of the circuit is called propagation delay. Propagation delay may be different for any circuit for different input combination. In any the delay is inversely related to the speed of the circuit so it is a essential factor to increase the speed of the system.

Table 4. Delay result of different power gated circuit with change in bias voltage on pMOS sleep transistor

Power Gating Technique name	Delay in pico sec with Body Bias voltage Vb				
	1.4 V	1.2 V	1 V	0.8 V	0.4 V
Stacking with delay	61.681	51.0487	42.781	35.475	24.328
ULP Diode based MTCMOS	40.32	38.503	35.543	31.46	22.897
Single header based ULP diode Tri-mode	40.704	38.85	35.845	31.801	23.081

Table 4 result analyzes the delay of the distinctive power gated circuit with variation in V_{bs} on pMOS transistor in the circuit. From this table the outcome demonstrates is that speed of the circuit can be increased by using FBB in dynamic mode.

Table 5. Delay result of different power gated circuit with change in forward bias voltage on pMOS grip transistor

Power Gating Technique name	Delay in pico sec with Forward Body Bias voltage		Average Delay(pico sec)
	0.8 V	0.4 V	
Stacking with delay	35.475	24.328	29.72367
ULP Diode based MTCMOS	31.546	22.897	27.17333
Single header based ULP diode Tri-mode	31.801	23.081	27.391

Table 5 clearly shows the batter result in term of delay reduction using proposed single header based ULP diode based MTCMOS technique as compare to other technique.

5.3 NBTI Effect Analysis

The V_{th} shift of the sleep pMOS header transistor along time increases. Result demonstrates that increase in stress time of grip pMOS transistor results increment in V_{th} shift. V_{th} shifting of sleep header transistor improves NBTI impacts in the circuit. Due to increment of NBTI effect reliability of circuit get reduced which is a major factor to improve. For improving this factor we divide Duty cycle(stress or on time) of grip pMOS header transistor.

Table 6. Performance analysis of single Header based ULP diode Design and two header based NBTI aware ULP diode design (With 100% duty cycle)

Time(in sec)	New Vth in (V) after delta Vth shift	Delay in pico sec. with two transistor 100% duty cycle	Delay in pico sec. with single transistor 100% duty cycle
360000	0.273	19.783	19.641
361000	0.335918	19.799	19.453
362000	0.400367	45.101	43.101
363000	0.466353	59.556	50.292
364000	0.533885	233.1	232.1
365000	0.602971	252.02	252.02

Table 6 demonstrates single header based ULP diode tri-mode design and Two header based ULP diode tri-mode method performance analysis with 100% duty cycle(during header sleep transistor). The results demonstrates the time increment which results in the increase in V_{th} shift of pMOS transistor and as a results of which threshold voltage of sleep pMOS transistor and Delay of the circuit likewise increases.

Table 7. Performance analysis of Header based ULP diode Design and NBTI aware Header Based ULP diode design (With 50% duty cycle)

Time in (Sec.)	New Vth in (V) after delta Vth shift	Delay in pico sec. with two transistor 50% duty cycle
360000	0.273	19.485
361000	0.326651	19.507
362000	0.381405	19.532
363000	0.437266	19.563
364000	0.43424	19.601
365000	0.55233	19.652

Table 7 demonstrates the Two header based ULP diode tri-mode design performance analysis at half obligation (50%) on cycle of two header sleep transistor. The results shows the time increment as a result of which V_{th} shift of pMOS transistor also increases and with this increment in V_{th} of pMOS the circuit's delay also increases.

6 Conclusion

In this paper we have designed a circuit which improve leakage as well as delay and also propose a design which is very much effective to mitigate the NBTI effect. The results reveal that by using first proposed technique leakage current is reduces by 74.93% and 0.217% respectively as compare to Stacking with delay based MTCMOS technique and ULP diode based MTCMOS technique. First proposed design is also effective in terms of delay. Result shows that delay get reduced by 20.91% and 7.84% respectively as compare to prevalent techniques. Second Proposed design at 50% Duty cycle is very much effective for NBTI effect mitigation. Experimental result shows that Second Proposed design with reduction in duty cycle(from 100% to 50%) the V_{th} shift on pMOS transistor reduces and with this decrement in V_{th} of pMOS transistor NBTI effect on the circuit will get reduced.

References

1. Narendra, S.G., Chandrakasan, A.P.: Leakage in Nanometer CMOS Technologies. Springer Science & Business Media (2006)
2. Calimera, A., Macii, E., Poncino, M.: Design techniques for nbti-tolerant power-gating architectures. IEEE Trans. Circ. Syst. II: Express Briefs **59**(4), 249–253 (2012)
3. Ravindra, N.M.: International technology roadmap for semiconductors (İTRS). J. Electron. Mater. **30**(12), 1478–1627 (2001). Symposium
4. Pattanaik, M., Agnihotri, S., Varaprashad, M.V.D.L., Arasu, T.A.: Enhanced ground bounce noise reduction in a low leakage 90 nm 1-volt cmos full adder cell. In: 2010 International Symposium on Electronic System Design (ISED), pp. 175–180. IEEE (2010)
5. Deepaksubramanyan, B.S., Nunez, A.: Analysis of subthresh-old leakage reduction in cmos digital circuits. In: 50th Midwest Symposium on Circuits and Systems, 2007. MWSCAS 2007, pp. 1400–1404. IEEE (2007)
6. Roy, S., Liu, D., Singh, J., Um, J., Pan, D.Z.: OSFA: a new paradigm of aging aware gate-sizing for power/performance optimizations under multiple operating conditions. IEEE Trans. Comput.-Aided Des. Integr. Circ. Syst. **35**(10), 1618–1629 (2016)
7. Arya, N., Singh, S., Pattanaik, M.: Temperature insensitive design for power gated circuits. In: 2014 9th International Conference on Industrial and Information Systems (ICIIS), pp. 1–6, December 2014
8. Sharma, S., Kumar, A., Pattanaik, M., Raj, B.: Forward body biased multimode multi-threshold cmos technique for ground bounce noise reduction in static cmos adders. Int. J. Inf. Electron. Eng. **3**(6), 567 (2013)
9. Jiao, H., Kursun, V.: How forward body bias helps to reduce ground bouncing noise and silicon area in mtcmos circuits: divulging the basic mechanism. In: SoC Design Conference (ISOCC), 2010 International, pp. 9–12. IEEE (2010)
10. Calimera, A., Macii, E., Poncino, M.: NBTİ-aware power gating for concurrent leakage and aging optimization. In Proceedings of the 2009 ACM/IEEE İnternational Symposium on Low power Electronics and Design, pp. 127–132. ACM (2009)
11. Ren, P., Liu, C., Wang, R., Gong, N., Liu, J., Wu, H., Huang, R.: New understanding on the single-trap response under nbti stress and the resulted stochastic degradation in nanoscale mosfets. In: 2012 IEEE 11th International Conference on Solid-State and Integrated Circuit Technology, pp. 1–3, October 2012
12. Levacq, D., Dessard, V., Flandre, D.: Low leakage soi cmos static memory cell with ultra-low power diode. IEEE J. Solid-State Circ. **42**(3), 689–702 (2007)

Task Allocation in Distributed Real Time Database Systems in IoT

Shetan Ram Choudhary[1(✉)] and C. K. Jha[2(✉)]

[1] Bioinformatics Centre, Banasthali Vidyapith, P.O. Banasthali Vidyapith, Banasthali 304 022, Rajasthan, India
sr_c2001@hotmail.com

[2] Department of Computer Science, Banasthali Vidyapith, P.O. Banasthali Vidyapith, Vanasthali 304 022, Rajasthan, India
ckjha1@gmail.com

Abstract. In present scenario, distributed and parallel systems in the form of grid, cloud and even cloud based Internet of things (IoT) are cater the needs of demand for computing capacity. Internet of Things (IoT) is a new come up to connect objects/things and therefore transmit information between a variety of entities of the corporeal world or to the control centers where interpret this information. By use of available resources are play very crucial role to ensure systems schedule. In distributed (Real time) database system, data allocation is one of the major problems. It affects the efficiency of the access to the requested data and thereby has large impact on the performance of the whole system. The data allocation involves data splitting, fragment replication, allocation choice to name a few issues. The distributed database system design putting all these factors together into consideration is complex and a Non-deterministic Polynomial (NP) hard. By applying Genetic Algorithm (GA), this work presents a virtual machine (VM) scheduling model to address the job allocation problem aiming to minimize the turnaround time. GA helps to attain a reasonable time for the query execution. The results of experiments have been examined to appraise the efficiency of our approach by comparing with best fit VM scheduling approach.

Keywords: Task allocation · Distributed (real time) database systems · Genetic Algorithm (GA) · Task allocation model · Task scheduling · Turnaround Time (TAT)

1 Introduction

During 1999 at MIT, the Internet of Things (IoT) was introduced for networking of computers, sensors and human itself by cloud computing. Till date, the fundamental use of internet to connect global computers by means of web page. New emerging technology IoT is one step ahead from present internet technology by using wireless network of sensors such as Bluetooth, RFID, Wi-Fi and that latest 4G-LTE. This technology is various applications in data handling and computational needs [37–41].

N. Nain and S. K. Vipparthi (Eds.): ICIoTCT 2019, AISC 1122, pp. 54–68, 2020.
https://doi.org/10.1007/978-3-030-39875-0_6

Cloud computing is adopting significance in the current technology. Due to the advent in the communication technology, any objects can get connected with any other objects. This is the era of where the things speak instead of livings humans. This technology has been termed as Internet of Things (IoT). For IoT to speak, there is a need of powerful essential network infrastructure that is Internet. This is in turn prone to physical threats in the cloud network. IoT is an integrated part of future Internet and can be defined as a dynamic global network infrastructure with self-configuring capabilities based on standard and interoperable communication protocols where physical and virtual "things" have identities, physical attributes, and virtual personalities and use intelligent interfaces, and are seamlessly integrated into the information network [44].

In a study by Choudhary et al. (2013), a single logical operation on the data is called a transaction in the scenario of databases. A RealTime Database (RTDB) is a data storewhose operations execute with predictable response and with application satisfactory levels of logical and temporal consistency of data, as well to appropriate execution of transactions with the ACID (Atomicity, Consistency, Isolation, and Durability) properties [1].

'Grid', term was first coined by Foster and Kesselman(1998) and it great evolution in parallel and distributed computing world [20, 21]. In study by Baruah (2012), these are assigned to run on diverse resources according to their need and are defined as the problem of allocating T number of tasks on M machines in |M||T| ways with the total no. of running states being |T| [6]. It may differ with the number of machines/nodes and the numeral of tasks. According Garey et al. (1979), scheduling becomes an NP complete problem owing to differing objectives and constraints in realizing the same [31].

2 Related Work

In a study by Abraham et al. (2007), the polynomial time algorithms cannot solve scheduling in the cloud being NP hard problem. Thus, to obtain the solutions use meta-heuristic method [4].

In another study by Carretero and Xhafa (2006) suggested that the employ of GA and subsequent to generating initial population by LJFR-SJFR find better results than other heuristics like Min-Min [23]. In the Grid environment, Ant Colony Optimization (ACO) has been used in scheduling in recent years. It shows that in case of ACO, better resources have higher probability of getting selected though it decreases the on the whole performance. So as to obtain better make span, both Ritchie and Levine (2003) joint ACO algorithm with Tabu Search for scheduling dilemma [17].

According to these study of Yarkhan and Dongarra (2002); Abraham et al. (2000) and consideredit avoids local minima and has improved execution time as compared to ad-hoc greedy scheduling through using Simulated Annealing [5, 8]. GA for dependent task scheduling was developed by Falzon and Li (2012) [16].

A study by Yu and Buyya (2006); Wang et al. (1997) proposed that the list scheduling by using priority that are assigned to the tasks and provide a better schedule at a much lower processing time [25, 29]. In another study, Topcuoglu et al. (2002); Yu and Buyya (2007), when abundant resources are there with the duplication based algorithm having higher computation complexity, cluster algorithms assigned the grouped task to the same cluster and are used [12, 19].

In another study, a hybrid GA for the job shop scheduling problem to minimize makespan where Smallest Position Value (SPV) rule is used to convert a real number vector to a job repetition representation this work was proposed by Liu and Xi (2006) [27]. In study by Martino and Mililotti (2002) proposed that GA to solve the scheduling problem to maximize the throughput [30]. Gonçalves et al. (2002) used hybrid GA for job shop scheduling problem [24]. For grid environment a Hierarchical Load Balancing Algorithm (HLBA), it is proposed by Lee et al. (2011) [36].

A study indicated that a new GA using markov decision making process in order that it converges faster. Constraints of budget and minimum execution time as their objectives it is used by Yu and Buyya (2006) [26].

In another study, Dowdy & Foster (1982) considered on comparative model of the file assignment problem [28]. A study by Iordache et al. (2007) suggested that SAGA a decentralized solution for job/task scheduling in heterogeneous environments which shows important convergence when it takesexpeditesince objective function [18].

According to Zhou and Ping (2012) suggested that adaptive GA for minimizing make span and load balancing by considering roulette wheel selection method and find better solution than max-min [34]. Enhanced selection method combining rank and roulette wheel was conducted by Ramachandram et al. (2009) [7].

According to Kumar et at. (2017) that the integration of SCM and IoT proposes an efficient way of scheduling the data on the IoT network. However, there are a number of issues that should be considered while adapting the same [43].

2.1 Task Allocation in Distributed Computing System (DCS)

A set of modules is called a task and module (or task) executes on only one processing node and can communicate with other modules of the task if required as inter module communication (IMC). Therefore, a task is a part of a program or a program in execution at similar or dissimilar locations. In distributed system and distributed software design, task allocation is a necessary stage. A task calls the software application and is a put of cooperate modules in DCS.

In another study, Raj et al. (2013) extracted that Genetic Algorithm (GA) is an evolutionary technique which is popular among the researchers due to its simplicity, wide use in many areas and effective results [32]. GA works on the basis of finding the best solution in a population and then refining the results over generations till we get the satisfactory solution. Thus, selection of the appropriate chromosomes in the given population and the rate at which the population is mutated becomes very important. Roulette wheel selection method is applied in this work to study the behaviour of the scheduling model under various test conditions.

In a study by Imtiaz et al. (1995) [22] for task assignment in distributed computing systems a technique based on the problem space genetic algorithm (PSGA) has been reported.

2.1.1 Criteria of Task Allocation

Distributed database environment has proposed by Cornell and Yu (1989) [15]. Under distributed database environment site assignment of relations is a sensitive/crucial matter. Development of new methodology for assigning relations and query descriptions.

Task allocation algorithms for distributed computer system and these are explain the execution time of the different modules and processing nodes, it was reported by Shen et al. (1985) and Vidyarthi et al. (2001) [11, 14].

2.2 Task Allocation in Distributed Database System (DDBS)

For user query, Distributed database (DDB) system is cluster of distributed computers and it is configured database. A DBMS is managing the user request and their solutions. The fundamental unit of database systems are transactions. Each transaction reflects the basic mechanism for operations for logical solution. DDB based upon allocation of data and operations effectively.

The problems related to data and operation allocation and solved properly. The retrieval and processing of all data depends on optimal allocation, while processing depends on data sets and their data allocation. For data and operations allocation problem is solved by proper DDBS design approach.

3 Problem Statement

This section details the proposed model while providing an insight into the related details.

3.1 Issues in DDBS

The research in DDBS requirements deliberation in the subsequent areas indicated here. Better task allocation model is current research area for efficient DDBS progress

(I) Data fragmentation: Under DDBS, data fragmentation and their allocation is critical matter. Vertical and horizontal fragmentations are application dependent. Arrival frequency transaction is dependent upon fragmentation's set of node.
(II) Data replication: Under DDBS network, the issues of concern for data are duplicated data location, amount of data replicate and how to replicate.
(III) Cost and performance: While data ease of use is a similarly important factor to be considered which will give more reliable, DDB systems Data and operation allocation during distributed transaction execution generally the cost and performance is considered.
(IV) Reallocation of data: In partitioned database, data reallocation must be taken into account when the access pattern of datachange frequently, which is a minor deviation from traditional DDBS (static).
(V) Changing workloads: As overloaded sites have processors memory capacity limitation associated with these, whilst allocating tasks and data to nodes of network the changing workloads should be measured.
(VI) Unit Cost of transaction: By in view of the cost of transaction in that particular concerned path schedule in DDBS, the site of computer network is decided for data allocation.

3.2 Task Assignment Issues

3.2.1 Issues in DCS

The scope and horizon of task allocation fluctuates, it is depending upon the applications of DCS. The following include tasks space, processors and links as:

(I) Task inter dependency: The DCS must think about execution of modules of a range of unrelated (independent) tasks also. Since modules from same task show interdependency but may not depend on the modules of the other tasks thus that process idleness is negligible during execution.
(II) Task preference and task exclusion: The task preference matrix indicates that sure tasks can only be executed in the specific processor. It is represented by m x n matrix. Certain pair of tasks must not be assigned to the same processors. These pairs of tasks are said to be mutually exclusive and are represented by E matrix and $E_{ij} = 0, 1$. In heterogeneous DCS the different amount of execution time specify different degrees of preference of module on the processors.
(III) Task redundancy: It may be provided for system consistency. Multiple copies of a task exist. i.e. if task (i) has redundancy of three, new tasks $i + 1$ and $i + 2$ are added to the original set of tasks and have similar properties as (i) and additional restriction is added such that tasks $i, i + 1, i + 2$ cannot be assigned to the same processor.
(IV) Channel and processor's load: IMC (intermodular communication) plays a essential role and is uniformly important as the speed and capability of the processing nodes while assigning modules of multiple tasks to the processing nodes. Thus during task allocation model development the load on the channels must be considered which are used for inter module communications. Heavily loaded nodes may lead to dead lock or even starvation at times.
(V) Message passing: The processors use non identical communication links for message communication. That is identical amount of message if transmitted through different communication links will require different amounts of transmission time, which specify different degrees of preference of the message on the links.
(VI) Unit value for cost function: The sum of processing cost and IPC is cost function in task allocation that is in fact dissimilar in measurement units inherently. Thus weights (W_{in} equation underneath) should be assigned to these two cost factors to reflect the cruciality of the casts in particular application. Cost function will help in balancing IPC and processor cost and finally led to load balance in that application.

$$\text{Total Processing Cost} = \sum\sum\left(WQ_{ik}X_{ik} + \sum\sum\left(C_{ij} * d_{kl}\right)X_{ik}X_{jl}\right) \quad (1)$$

3.3 Task Allocation Tools

3.3.1 Tools in DCS

To address the task allocation problem the Tools used to attain can be catagorized as:

(I) Use task dependant information as heuristic to diminish search efforts and search space. (II) Use precedence relationship as parameter in heuristic algorithm for task allocation, it is described in study by Wesley Chu et al. (1987) [35]. (III) Maintaining global table (GT) showing that all the processors are being utilized properly for execution. This further leads to load balancing. It is reported by Deo Prakash Vidyarthi et al. (2001) [14]. (IV) Using redundancy (S.W./H.W.) not only increases the system reliability but also improves response time of the system. (V) To use clusters (task and processors) to trim down the communication penality, it is used by Deo Prakash Vidyarthi et al. (2004) [13]. (VI) By Using concurrency (by including interference costs of modules) and parallelism to increase speed of assignment thus leading to minimize turn around time.

3.3.2 It is agreed upon in literature that data's fragmentation, allocation, replication and reallocation are of main concern with the following concerning issues:

1. Load balance: The load on processors and channel affects the task execution speed which finally decides system throughput in DCS. Changing access patterns leading to changing workloads on nodes of network are thus significant that sometime they lead to reallocation of data to have acceptable system throughput in DDES.
2. Unit cost: The value of unit cost for cost function needs weight(W) for normalization since of their different units inherently in DCS. The unit cost value of transaction helps in deciding the lane of schedule by selecting low cost nodes for data allocation in DDBS.

3.4 Task Allocation Models

3.4.1 Developing DCS Allocation Model

Any DCS model to be developed should comprise of the following properties:

1. Processor's load balancing; 2. Minimizing the IPC cost and 3. Increasing the throughput of the system/minimizing the turnaround time of processor.

The details of processor in the network include: (i) Network topology (architecture of processors'connections) which includes interprocessor's distance and (ii) Processor's attributes like H.W. associated may be storage space and speed.

The attribute of task details may include: (i) Task size; (ii) Task dependency (task graph (TG) represents it), the sequence decides successors and predecessors; (iii) Task execution (in terms of execution time matrix); (iv) Coupling factor between tasks(unit data transfer between tasks) and (v) Task communication (in mxn matrixform).

Objective cost function is application and environment dependant.

Total processing cost = IPC(Inter processor communication) cost + Processing cost

Objective function development

$$\text{Total processing cost } = \sum_{i}\sum_{j}\left(\underbrace{\mathbf{WQ_{ik}X_{ik}}}_{\text{Processing cost}} + \underbrace{\sum_{i}\sum_{j}(\mathbf{C_{ij}} * \mathbf{D_{kl}})\mathbf{X_{ik}X_{jl}}}_{\text{IPCcost}}\right) \quad (2)$$

Objective function for DCS

Q_{ik} = Processing cost of task i on processor k; C_{ij} = No. of data units transferred from task i to j;

D_{kl} = distance related communication cost for one unit data trans. from processor k to l;
X_{ik} = assignment variable task i is assigned to processor k (0, 1);

Objective function for task allocation model may be cost model and may include other parameters similar to Precedence Relation (PR) among tasks or it may be common probability formula in case of reliabilitybeing main objective of task allocation. In case of task allocations considering load balanceas main goal. The load on a processor is the amalgamation of all the execution and communication costs associated with tasks assigned to it (Fig. 1).

Task Allocation problem and Goals

Tasks (attributes and details)

Processors (attributes and details)

{

Calculate

Cost function Formulation (IPC andProcess Execution cost)

Task Allocation using MinimizingCost function and load balance

Iterative algorithm for minimum cost solution

Model evaluation for reliability and solution time and cost wise

}

Fig. 1. Developing DCS Task allocation model (Ma et al. 1982) [10]

Objective function is f(v) = h(v) + g(v) in mainly of distributed systems task allocation; where h(v) is lower bound from current node and g(v) is search path from source node to goal node.

In optimal task assignment goal is to minimize the process turnaround time which can be calculated as

Turnaround time = execution time for module + time for IPC cost.

$$\mathbf{Tp\,(A) = t_p^e(A) + t_p^c(A)}$$
$$\mathbf{P = processor;\ A = task\ assigned}$$

Task scheduling in Cloud computing can be single objective or multi-objective according to the requirements and the constraints considered. In this paper, we consider the problem as single objective optimization problem of minimizing the Turnaround Time (TAT). Here, DAG based tasks comprising of sub-modules (which can be database related operations or computational tasks) with their precedence constraints are considered to be mapped on the appropriate resources. The model uses centralized scheduler for task scheduling. Each node has attributes like processing speed, time allocated to a task. Further, each node is considered to have a single processor. Though, practically a node may have more than one processor.

The objective function designed should consider the node attributes as well as task characteristics for appropriate task placement. There are many other concerns such as memory capacity of the node but have been avoided in this work.

In a study by Choudhary et al. (2018), Singh et al. (2015) and Raza et al. (2009), distributed scheduling policy is used when task is submitted at any node while the suitability of any given node being evaluated by the dispatcher which is a part of the cloud scheduler (CS) [2, 9, 42]:

(a) No. of resources/nodes available at any instant of time. (b) Previously allocated workload on each node. (c) Time to finish for the previously allocated jobs. (d) Nature of job in terms of its resource requirements. (e) Processing speed of a task on a node. It can be calculated as:

$$\text{Proc\_Speed} = \text{Noins}/\text{CF}; \quad (3)$$

Here, Noins = no. of instructions in a job/transaction module under consideration, and CF = clock frequency of a node/processor. This information is used in selecting the best resources for a particular taskmodule. The chromosome holds the node number and index of the chromosome gives the task module that are executing on that node. Consider a scenario having 8 machines and 10 tasks. A typical task allocation schedule

(a)

2	1	5	6	8	2	5	4	1	3

Chromosome 1

8	7	4	4	4	5	4	4	3	4

Chromosome 2

(b)

Machine 1	Task2	Task9
Machine 2	Task1	Task6
Machine 3	Task10	
Machine 4	Task8	
Machine 5	Task3	Task7
Machine 6	Task4	
Machine 7		
Machine 8	Task5	

Fig. 2. (a). Chromosomes. (b). Task/Machine Dispatch

as GA chromosome is represented as shown in Fig. 2(a) and (b) shows how the jobs are dispatched on the machine for the same.

Different arrangements of nodes at different index in chromosome results in different TAT values. Using GA, thescheduler tends to converge to the best solution offering the minimum TAT. In another study by Singh et al. (2015) and Raza et al. (2009), the fitness function for the evaluation of the TAT is expressed as follows [24] [Krishan Veer Singh and Zahid Raza, 2015; Zahid Raza and D. P. Vidyarthi, 2009] [2, 9]:

$$NE_{ikl} = E_{ijk} + \sum_{(h=1)}^{(i-1)} max(B_{ihk} * D_{kr}) + RT_{kl} \quad (4)$$

where, NEC_{ikl} = node execution cost for processing a module m_i of a taskT_j on a module P_k at level l; E_{ijk} = processing speed of a module m_i of a taskT_j on a module P_k.
B_{ihk} = inter-module communication cost between module m_i and m_h of taskT_k.
D_{kr} = hamming distance between nodes P_k and P_r of a cluster. This corresponds to the number of links traversed between two nodes P_k and P_r.
RT_{kl} = previous workload by the previous assigned job on node P_k at level l.

Further, RT_{kl} can be calculated as

$$RT_{kl} = max\ (NEC_{ikl}) \forall\ k = 1, 2, 3, \ldots, k \quad (5)$$

This value gets updated with max value of NEC_{ikl} among all node values which are at same level l in the DAG.

The Turnaround Time (TAT_i) for a chromosome can then be calculated as

$$TAT_i = max(NEC_{ikl}) \quad (6)$$

Objective Function: It is to minimize the cost of Retrieval + Update + Storage (R + U+S). Objective function selected to this problem is minimization of system operating cost including communication cost, disk IO, CPU processing and storage costs, it is performed by their study by March et al. (1995) [33].

$$\text{MiniCost} = \sum_k \sum_j f(k, j) \sum_m (COM(k, j, m) + IO(k, j, m) + CPU(k, j, m)) + \sum_t STO(t) \quad (7)$$

Here, f(k, j) is frequency of execution of query k at node j; COM (k, j, m) is communication cost for m step of query k at node j; IO (k, j, m) is input and output cost of disk for m step of query k at node j; CPU (k, j, m) is CPU processing cost for m step of query k at node j; and STO(t) is node t's storage cost

Constraints: Two dissimilar types of constraints are intrinsic problem constraints and resource constraints. Resource constraints are CPU processing and storage capacity of nodes and communication capacity of each link. Which are limited

Intrinsic problem constraints include:

$\Sigma_t X_{it} \geq 1$ all file fragments l = 1, 2 …… must be stored at one or more nodes. $Z_{kit} \leq X_{it}$ for all the queries k for all file fragments cannot be accessed from a node unless it is stored at that node.
$\Sigma_t Y_{kmt} = 1$ all steps m of all queries must be processed at some node.

After developing such mathematical formulation an algorithm can be developed to solve it. One such algorithm was tested, that is a genetic algorithm and found an efficient solution of data and operation allocation in realistic amount of time.

3.5 Task Allocation Algorithms

3.5.1 A work of task assignment in DCS by Harold Stone and Bokhariproposes network-flow algorithms on dual processor system only. This was improved later by considering more than the said number of processors to give a flavour of multi processor scheduling environment. An efficient algorithm for the task assignment problems is always desired aiming for best possible solution within acceptable CPU times.

The Algorithm

GA (Task)
{
Submit the DAG based task
Estimate the processing speed E for the task
Generate and randomly populate the chromosomes *// Size varies as per the experiment*
Do
{
Perform selection *// Roulette wheel selection and Rank selection alternatively*
Perform crossover *// Single point crossover is used, crossover point is randomly chosen.*
Calculate NEC_{ikl} and TAT_i offered by each chromosome
//as per eq. 4 & 6
// Best chromosomes are preserved using elitist approach
Perform mutation *// For every 5th and 10th generation*
// On 25 and 50% of population as per the case
Calculate NEC_{ikl} and TAT_i offered by each chromosome
}
While (Termination condition is not met)
}

4 Experimental Study

Simulation experiments were conducted to observe the allocation of tasks to the servers and results are compared with Genetic Algorithm (GA) and Best-Fit (BF), First-Fit (FF), Worst-Fit (WF). The experiments were conducted on Intel(R) Core(TM) i5 processor @3.20 GHz, 4GBRAM platform version using Matlab R2015a (58.5.197613) 64-bit (win64). Problem instances were randomly generated with the above configuration where the instances are the resource requirements of Tasks and Databases.

4.1 Performance Parameters Settings

The values of the proposed model parameters set in the simulation model are given in Table 1.

Table 1. Proposed model parameters and their assumed values

Sr. No.	Parameters	Description	Range
1	No. of Resources	No_Resources	5–20
2	Clock Frequency Sites	CF (MHz)	10–20
3	Time to finish Previous Load	RT_{kl} (in Time Unit Seconds)	2–5
4	Distance between Two Hops	D_{kr} (No. of Hop Count)	2–5
5	No. of Module in a Task	No_Module	5–20
6	No. of Instructions in a Module	NoN (MIs)	100–500
7	Intermodule Communication	B_{ihk}	2–5
8	Size of Chromosome	Size_Chromosomes	5–20
9	Population Size	No_Pop	100
10	No. of Generation	No_Generation	100
11	Crossover Considered during Process	Single point	--
12	Mutation Performed after (Generation)	MU	5, 10
13	Population Selected for	--	25%, 50%
14	Roulette Wheel Selection	RW_Selection	--

There are big data centres having a pool of server nodes in a cloud environment. Since the infrastructure is fully virtualized, all the applications run on Virtual machines (VMs) and these VMs are deployed on physical servers. In present scenario, only two dimensions of a VM are measured i.e. CPU capacity and memory size. The CPU utilization of a server is estimated as the sum of CPU utilization of all the VMs that are running on that server. The memory utilization can also be calculated in the same manner. For example if there are two VMs running on a server with (30%, 35%) and (35%, 45%) utilizations where (30%, 35%) represents the pair of CPU and memory requests of first VM and (35%, 45%) represents the pair of CPU and memory requests of the second VM then the total CPU and memory utilization of that server would be (65%, 80%) being a sum of the vectors.

In study by Kumar et al. (2015) VMs containing their corresponding CPU and memory utilization requirements are generated randomly. In each case, the number of servers are equal to the number of VMs for supporting the worst case scenario which corresponds to only one VM is running on a single server [3].

5 Results

Experiments 1: TAT Vs Generations with Mutation applied after 5th Generation

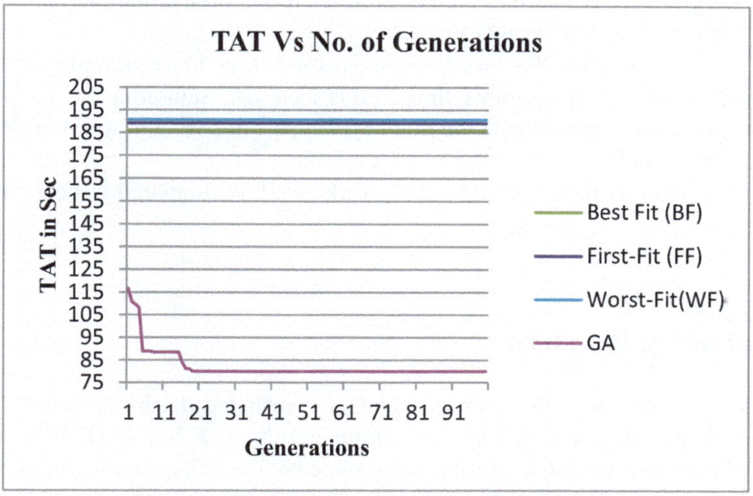

Fig. 3. TAT v/s Generations with Mutation applied after 5th Generation

Experiments 2: TAT Vs Generations with Mutation applied after 10th Generation

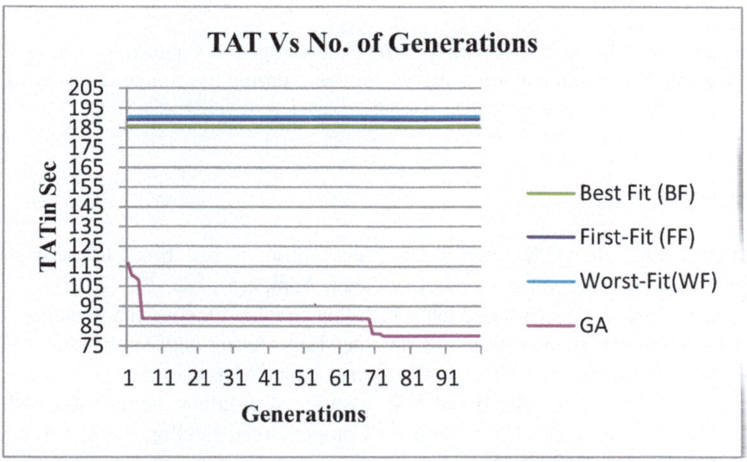

Fig. 4. TAT v/s Generations with Mutation applied after 10th Generation

Observations:

1. The TAT of the jobs/queries submitted is seen reducing over generations (Figs. 3 and 4).
2. From the graph it has been observed that further, it exhibits a little slower convergence of 5th and 10th generation in GA.
3. In 5th generation, after 20 generations, the result is seen to be saturating and in 10th generation, after 70 generations, the result is seen to be saturating.
4. The results are compared with the first fit (FF), best fit (BF) and worst fit (WF) VM placement strategies.
5. It is observed that the proposed model works well as compared to peers in all the cases.

6 Concluding Remarks

This paper presents task allocation problem of distributed database system (DDBS) design research area. The distributed computing systems (DCS) and DDB systems are taken anywhere task allocation appears to be same but basically is quite different which is due to their basic difference which lies in the type of joint resources in each case. Task allocation is found to be more complex in DDBS as compared to DCS. This work presents a task/query allocation model based on GA to effectively minimize the turnaround time of the batch of jobs submitted. The results are compared with the first fit, best fit and worst fit VM placement strategies. It is observed that the proposed model works well as compared to peers in all cases with the saturation emerging after nearly twenty and seventy generations.

Acknowledgement. The authors would like to acknowledge the suggestions and contributions made by Dr. Zahid Raza with numerous discussions held during the course of this work.

References

1. Choudhary, S.R., Jha, C.K.: Performance evaluation of real time database systems in distributed environment. Int. J. Comput. Technol. Appl. **4**(5), 785–792 (2013)
2. Singh, K.V., Raza, Z.: A GA based job scheduling strategy for computational grid. In: International Conference on Advances in Computer Engineering and Applications (ICACEA), pp. 29–34, IMS Engineering College, Ghaziabad, India, March 2015
3. Kumar, D., Raza, Z.: A PSO based VM resource scheduling model for cloud computing. In: IEEE International Conference on Computational Intelligence & Communication Technology, pp. 213–219 (2015)
4. Abraham, A., Carretero, J., Xhafa, F.: Genetic algorithm based schedulers for grid computing systems. Int. J. Innov. Comput. Inf. Control **3**(5), 1053–1071 (2007)
5. Abraham, A., Buyya, R., Nath, B.: Nature's heuristics for scheduling jobs on computational grids. In: The 8th IEEE International Conference on Advanced Computing and Communications (ADCOM 2000), India (2000)

6. Baruah, A.: A GA approach to static task scheduling in grid based systems. Int. J. Comput. Sci. Eng. (IJCSE) **4**(01), 54 (2012)
7. Ramachandram, A.J.S., Al Jadaan, O., Abdulal, W.: An improved rank-based genetic algorithm with limited iterations for grid scheduling. In: 2009 IEEE Symposium on Industrial Electronics and Applications (ISIEA2009), pp. 215–220, Malaysia, Kuala Lumpur, October 2009
8. Yarkhan, A., Dongarra, J.: Experiments with scheduling using simulated annealing in a grid environment. In: 3rd International Workshop on Grid Computing (GRID2002), pp. 232–242 (2002)
9. Raza, Z., Vidyarthi, D.P.: GA based scheduling model for computational grid to minimize turnaround time. Int. J. Grid High Perform. Comput. **I**(IV), 70–90 (2009)
10. Ma, P.-Y.R., Lee, E.Y.S., Tsuchiya, M.: A task allocation model for distributed computing systems. IEEE Trans. Comput. **C-31**(1), January 1982
11. Shen, C.-C., Tsai, W.-H.: A graph matching approach to optimal task assignment in distributed computing systems using a minimax criteria. IEEE Trans. Comput. **C-34**(3), 197–203 (1985)
12. Yu, D.J., Buyya, R.: Workflow scheduling algorithms for grid computing. Technical report, GRIDSTR-2007-10, Grid Computing and Distributed Systems Laboratory, The University of Melbourne, Australia (2007)
13. Vidyarthi, D.P., Tripathi, A.K., Sarkar, B.K.: Cluster based task allocation in distributed systems. In: Proceedings of 18th International Parallel and Distributed Processing Symposium. IEEE (2004)
14. Vidyarthi, D.P., Tripathi, A.K., Sarkar, B.K.: Multiple task management in distributed computing system. J. CSI **31**(1), 19–25 (2001)
15. Cornell, D.W., Yu, P.S.: On optimal site assignment for relations in the distributed database environment. IEEE Trans. Softw. Eng. **5**(8), 1004–1009 (1989)
16. Falzon, G., Li, M.: Enhancing genetic algorithms for dependent job scheduling in grid computing environments. J. Supercomput. **62**(1), 290–314 (2012). Springer
17. Ritchie, G., Levine, J.: A fast, effective local search for scheduling independent jobs in heterogeneous computing environments. Technical report, Centre for Intelligent Systems and their Applications, School of Informatics, University of Edinburgh (2003)
18. Iordache, G.V., Boboila, M.S., Pop, F., Stratan, C., Cristea, V.: A decentralized strategy for genetic scheduling in heterogeneous environments. Multiagent Grid Syst. **3**(4), 355–367 (2007)
19. Topcuoglu, H., Hariri, S., Wu, M.Y.: Performance-effective and lowcomplexity task scheduling for heterogeneous computing. IEEE Trans. Parallel Distrib. Syst. **13**(3), 260–274 (2002)
20. Foster, I., Kesselman, C.: The grid - blueprint for a new computing infrastructure. Morgan Kaufmann Publishers (1998)
21. Foster, I.: What is the Grid? A three point checklist (2002)
22. Ahmad, I., Dhodhi, M.K., Ghafoor, A.: Task Assignment in Distributed Computing Systems, pp. 49–53. IEEE (1995)
23. Carretero, J., Xhafa, F.: Using genetic algorithms for scheduling jobs in large scale grid applications. J. Technol. Econ. Dev. Res. J. Vilnius Gediminas Technical University **12**(1), 11–17 (2006)
24. Gonçalves, J.F., de M. Mendes, J.J., Resende, M.G.C.: A Hybrid Genetic Algorithm for the Job Shop Scheduling Problem, AT&T Labs Research Technical Report TD-5EAL6J, September 2002
25. Yu, J., Buyya, R.: Scheduling scientific workflow applications with deadline and budget constraints using genetic algorithms. Sci Program **14**, 217–230 (2006)

26. Yu, J., Buyya, R.: A budget constrained scheduling of workflow applications on utility grids using genetic algorithms. In: Proceedings of the 15th IEEE International Symposium on High Performance Distributed Computing. IEEE CS Press, Paris (2006)
27. Liu, L., Xi, Y.: A hybrid genetic algorithm for job shop scheduling problem to minimize makespan. In: Proceedings of the Sixth World Congress on Intelligent Control and Automation, pp. 3709–3713 (2006)
28. Dowdy, L.W., Foster, D.V.: Comparative model of the file assignment problem. ACM Comput. Surv. **2**, 287–314 (1982)
29. Wang, L., Siegel, H.J., Chowdhury, V.R., Maciejewski, A.: Task matching and scheduling in heterogeneous computing environments using a genetic-algorithm-based approach. J. Parallel Distrib. Comput. **47**(1), 8–22 (1997)
30. Mililotti, M., Martino, V.D.: Scheduling in a Grid computing environment using Genetic Algorithms, 0-7695-1573-8/02/ (C) IEEE (2002)
31. Garey, M.R. Johnson, D.S.: Computers and Intractability - A Guide to the Theory of NP Completeness. W.H. Freeman and Co. (1979)
32. Raj, J.S., Thomas, R.M.: Genetic based scheduling in grid systems: a survey. In: Computer Communication and Informatics (ICCCI), International Conference on IEEE (2013)
33. March, S.T., Rho, S.: Allocating data and operations to nodes in DDB design. IEEE Trans. Knowl. Data Eng. **7**(2), 305–317 (1995)
34. Zhou, W., Bu, Y.P.: An adaptive genetic algorithm for the grid scheduling problem. In: Control and Decision Conference (CCDC), pp. 730–734 (2012)
35. Chu, W.W., Lan, L.M.-T.: Task allocation and precedence relations for distributed real time systems. IEEE Trans. Comput. **36**(6), 667–679 (1987)
36. Lee, Y.H., Leu, S., Chang, R.S.: Improving job scheduling algorithms in a grid environment. Future Gen. Comput. Syst. **27**(8), 991–998 (2011)
37. Hwang, K., Dongarra, J., Fox, G.: Distributed and Cloud Computing: From Parallel Processing to the Internet of Things. Elsevier Pvt. Ltd, Singapore (2012). ISBN 978-0-12-385880-1
38. Karimi, K., Atkinson, G.: What the Internet of Things (IoT) Needs to Become a Reality. http://www.eetimes.com/document.asp?doc_id=1280077
39. Gubbi, J., Buyya, R., Marusic, S., Palaniswami, M.: Internet of Things (IoT): a vision, architectural elements, and future directions. Fut. Gen. Comput. Syst. **29**(7), 1645–1660 (2013)
40. Atzori, L., Iera, A., Morabito, G.: The Internet of Things: a survey. Int. J. Comput. Telecommun. Netw. **54**(15), 2787–2805 (2010)
41. Zanella, A., Bui, N., Vangelista, L., Zorzi, M.: Internet of Things for smart cities. IEEE Internet Things J. **1**(1), 22–32 (2014)
42. Choudhary, S.R., Jha, C.K.: Task (Transaction) allocation in distributed real time database systems in cloud computing. Int. J. Trend Res. Dev. (IJTRD) **05**(01), 160–167 (2018)
43. Kumar, S., Raza, Z.: Internet of Things: possibilities and challenges. Int. J. Syst. Service-Oriented Eng. **7**(3), 32–52 (2017)
44. Petrolo, R., Loscri, V., Mitton, N.: Cyber-physical objects as key elements for a smart cyber-city. In: Management of Cyber Physical Objects in the Future Internet of Things, pp. 31–49. Springer International Publishing (2016)

Optimization of Block Diagonalization for MU-MIMO Downlink System Using PSO

Archana Doneriya(✉), Manish Panchal, and Jaya Dipti Lal

Shri G. S. Institute of Technology and Science, Indore 452003, MP, India
archanadoneriya25@gmail.com, hellopanchal@gmail.com,
jayadiptilal@yahoo.co.in

Abstract. Multiuser Multiple-Input Multiple-Output (MU-MIMO) system is to serve a number of users simultaneously. Here the aim is performance enhancement of an MU-MIMO system in terms of capacity and bit error rate (BER). To enhance the bit error rate (BER) performance of the system particle swarm optimizer (PSO) combined with Block Diagonalization (BD) precoding technique. PSO algorithm employing on BER function of BD to minimizing BER of the system. The main advantage of PSO is that in case of the performance index cannot be formulated by simple equations it can find out the solution. Simulation results show that PSO-BD can achieve significantly superior BER performance than precoding technique over the different fading channel environment like Rayleigh, Rician, and Nakagami.

Keywords: MU-MIMO · BER · Fading channel · Linear precoder · PSO

1 Introduction

Now a day's demands increase for high data rate transmission and capacity improvement so to achieve capacity improvement MIMO attract enormous interest [2,6]. Space dimension is exploited for scheduling the multiple transmissions for multiple users when the base station (BS) and the mobile station (MS) both equipped with multiple antennas. Spatial division multiple access (SDMA) is realized by MU-MIMO when a BS communicates with multiple MS in the same time slot and frequency band. By using multiple-diversity MU-MIMO system can improve the capacity. Co-channel interference (CCI) prevents the performance improvement of the MU-MIMO system.

The transmitter can design a code to suppress the interference if it has knowledge of interference in advance. Many linear precoders are emerged, like channel inversion techniques (ZF, MMSE) [6] and the block diagonalization (BD) technique [2]. ZF technique can suppress CCI completely but it increases amplification of noise. MMSE technique provides a trade-off between CCI and noise and gives better performance than ZF, but still, performance is not good. During BD

N. Nain and S. K. Vipparthi (Eds.): ICIoTCT 2019, AISC 1122, pp. 69–76, 2020.
https://doi.org/10.1007/978-3-030-39875-0_7

technique [2] MU-MIMO channel decomposes into multiple parallel SU-MIMO channels to eliminate CCI completely by using null space. Noise is suppressed efficiently by the generated null space vector so BD is better than channel inversion techniques. The factors which affect the pre-processing MU-MIMO performance are precoding gain, noise, and CCI. A linear precoding method is proposed in [10] to maximize the rate. Because of the computation of optimized function is to complex, here try to solve the optimal linear precoding in an easy way.

James Kennedy and Russel C. Eberhart have proposed PSO algorithm first in 1995 motivated with the fish schooling and bird flocking behavior [3]. PSO is invoked [4] because of the limited ability of DE algorithm [5]. Implementation of PSO is easy and simple and a few parameter adjustments needed so PSO is used to optimization of minimum BER. Like Genetic Algorithm (GA) global search is achieved by the PSO through competition and collaboration. PSO gets optimal global solution with the particle position adjustment.

An optimal MU-MIMO linear precoding scheme proposed for solving the MIMO processing problem with the adoption of PSO optimization. To measure the optimal linear precoding problem establish a simplified function first. Then, according to the simplified function employ the PSO algorithm to search the optimal linear precoding vector. Proposed scheme outperform BD and channel inversion methods and obtains desire system capacity with minimizing BER.

In this paper first, describe the system model of downlink MU-MIMO. Then discuss channel inversion precoders and Block Diagonalization. Following this learn about PSO optimization and then fading channel. Finally, show the BER performance in simulation result then conclusion.

2 System Model of MU-MIMO

A MIMO created by using antenna array at transmitter and receiver side to improve system performance. Consider the MU-MIMO downlink channel to model. Here over the fading channel H all N_t transmitting antennas at BS communicate simultaneously to K user. Where $K \in (1, 2, \ldots.., K)$ has N_k receiving antennas. Channel between BS and the Kth user modeled by $N_t \times N_K$ channel matrix H_k.

Transmitting symbol vector of Kth user is $S_k \in N_k \times 1$ (Fig. 1).

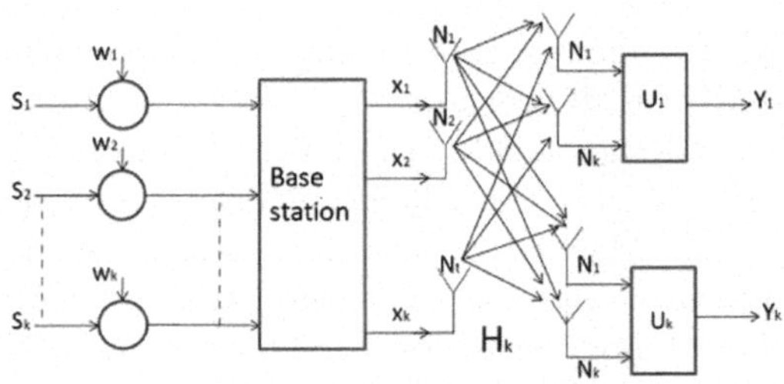

Fig. 1. System model of Multiuser - MIMO

Precoding matrix $W_k \in N_k \times N_t$ to transform data vector S_k to $N_t \times 1$ transmitting vector $(W_k \times S_k)$. Now signal receiving vector for K^{th} user is [2]

$$Y_k = H_k W_k S_k + H_k \sum_{k \neq i} W_i S_i + n_k \tag{1}$$

n_k is the noise vector and $n_{\text{k,i}}$ noise vector component are i.i.d. with mean zero and σ^2 variance and $i = 1, 2, \ldots, NK$. $H_k W_k S_k$ is desired signal and $H_k \sum_{k \neq i} W_i S_i$ is the interference both are received by K^{th} user.

3 Precoding Techniques

3.1 Channel Inversion Precoding Techniques

Due to the multipath channel fading interference occurred in the transmitted signal for eliminating this interference Channel Inversion Precoding techniques Zero Forcing (ZF) and minimum mean square error (MMSE) are used in MU-MIMO. The ZF is the simplest technique and has the lowest computational complexity. By taking the pseudo-inverse of channel matrix (H) is the precoding matrix (W) interference suppresses. Mathematically W is given by [6]

$$W = H^H (HH^H)^{-1} \tag{2}$$

ZF neglected channel noise effect by which noise inhancement is main disadvantage, noise is minimizing by MMSE channel inversion technique with taking noise effect on the channel and reduces mean square error (MSE) of symbols of the transmitted signal which considers interference process as well as noise [6].

$$W = H^H (HH^H + \frac{\sigma}{ea}.I)^{-1} \tag{3}$$

Here $\frac{\sigma}{ea}.I$ is a factor of noise power function. ea and σ Is denotes transmitted signal power and variance of noise respectively. MMSE perform better than ZF.

3.2 Block Diagonalization (BD)

First, all symbol combined and then transmit from all transmitter. Then to cancel the interference $\bar{H}_k W_j = 0, \forall k \neq j$ where W_j is precoding matrix. In the BD algorithm, W_1 and W_2 are calculated to obtain the above conditions [2].

$$\bar{H}_K = [H_1^T, \ldots\ldots.., H_{\text{K}^{\text{th}}-1}^T, H_{\text{K}^{\text{th}}+1}^T, \ldots.., H_K^T]^T \tag{4}$$

$\bar{h} \in ((K-1)N_r \times N_t)$ here channel matrix H_K is excluded. Let the singular value decomposition (SVD) of $\bar{H}_K$ [2]

$$\bar{H}_K = \bar{U}_K \bar{D}_K \bar{V}_K^H \tag{5}$$

$$\bar{H}_K = \bar{U}_K [\bar{D}_K^n \quad 0_{(\text{K}-1)\text{N}_\text{r} \times \text{N}_\text{t}}][\bar{V}_K^n \quad \bar{V}_K^z]^H \tag{6}$$

Where $\bar{D}_K$ is non-negative singular value diagonal matrix, $\bar{V}_K^n, \bar{V}_K^z$ denote right non-zero value singular matrix and zero value right singular matrix respectively and $\bar{U}_K$ left singular value matrix.

To not transferring the signal to all user except K^{th} User, use $\bar{V}_1^z$ to pre-cancel the inter-user interference by existing following relation between $\bar{H}_K$ *and* $\bar{V}_K^z$ [2].

$$H_1\bar{V}_K^z = ... = H_{\mathrm{K^{th}}-1}\bar{V}_K^z = ... = H_{\mathrm{K^{th}}+1}\bar{V}_K^z = ... = H_K\bar{V}_K^z = 0_{\mathrm{N_r}\times(\mathrm{N_t}-(\mathrm{K}-1)\mathrm{N_r}} \tag{7}$$

With $W_K = \bar{V}_K^z$ BD realization is done and total transmit weight is [2]

$$W_K^{\mathrm{BD}} = \bar{V}_K^n\bar{V}_K^z \tag{8}$$

The channel matrix $\bar{H}_K = H_K\bar{V}_K^z$ is regarded as that of Single user MIMO by using $\bar{V}_K^z$ for K^{th} user. In BD algorithm, the Eigenmode transmission beamforming (EM-BF) is employed for the matrix $\bar{H}_K$.

4 Particle Swarm Optimization (PSO)

PSO based on two concepts swarms intelligence and field of transformative figuring. Swarm has many particles; Particles are able to interact with each other through competition and cooperation a search space traversed and the fitness value of each particle evaluated with the fitness function optimization. For the solving problem of higher complex optimization, PSO technique is a very effective powerful optimization algorithm. Flow chart shown in below Fig. 2.

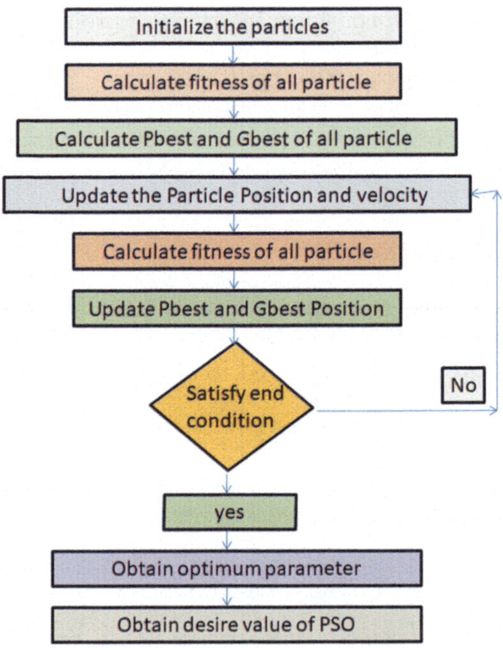

Fig. 2. Flow chart of PSO

4.1 PSO Optimized BD

Following steps describe algorithm for proposed work PSO-BD which is based on standard PSO algorithm.
1 - Initialize particles or algorithm parameter with its coordinator position Vector $x_i = [x_{i1}, x_{i2} \ldots ..x_{iD}]^T$ and velocity vector $v_i = [v_{i1}, v_{i2} \ldots ..v_{iD}]^T$
2 - Consider BER as Fitness function.
3 - Identify the personal best value (P_{best}) by comparing particle's current value with its previous value and global best value g_{best} by comparing current value of all particles. Fitness function is BER so here we take lesser value.
4 - Particle i update its location and velocity according to below equations [1].

$$v_i(t+1) = w.v_i(t) + \varphi_1 c_1[P_{best}(t) - x_i(t)] + \varphi_2 c_2[g_{best}(t) - x_i(t)] \quad (9)$$

$$x_i(t+1) = x_i(t) + v_i(t+1) \quad (10)$$

where t and w current iteration number and inertia factor respectively and φ_1 and φ_2 are the uniformly distributed random numbers from 0 to 1. Learning factors are c_1 and c_2 are usually Content $c_1 = c_2 = 2$.
5 - For the simulation purpose here we take 10 no. of iterations. Steps are repeats from 2nd to 4th until reaching maximum no. of iteration.

5 Fading Channel Model

In wireless communication, multipath propagated signal received by the receiver so because of scattering signal get the various fading effect on it. If there is a line of sight path of signal Multipath propagated signals modeled as Rician fading distribution otherwise modeled as Nakagami and Rayleigh fading distributions with the heavy scattering.

The single element of channel gain matrix can be written as

$$h_{ij} = a + jb = r.exp(j\varphi) \quad (11)$$

Where a,b$\in \mathcal{N}\left(0, \frac{1}{\sqrt{2}}\right)$ are random variables.

The probability density function (PDF) for Rayleigh fading channel [7]

$$P_{RA}(r) = \frac{r}{\sigma^2} \exp\left(\frac{-r^2}{2\sigma^2}\right), r \geq 0 \quad (12)$$

PDF of the Rician fading channel distribution [8].

$$P_{RC}(r) = \frac{r}{\sigma^2} \exp\left(\frac{-r^2 + A^2}{2\sigma^2}\right) J_0\left(\frac{rA}{\sigma^2}\right), r \geq 0 \quad (13)$$

Variable A is the strength of the dominant signal component and channel amplitude r $\geq$ 0, J() is the modified 0th order Bessel function of the first kind.

Nakagami-m distribution is used to model the fading channel and activities calculation when the Rayleigh fading distribution fails then the. PDF of Nakagami-m fading channel characterize as gamma distribution [9].

$$P_N(r) = \frac{2}{\Gamma(m)} \left(\frac{\mathrm{m}}{\Omega}\right)^m (r^{2\mathrm{m}-1}) \exp\left(\frac{-mr^2}{\Omega^2}\right) \tag{14}$$

If m = 0 Nakagami become Rayleigh and if m−1 Nakagami become Rician.

6 Simulation Result

Here MATLAB used for simulation purpose. Here get BER performance of PSO-BD techniques over the different type of fading channels for MU-MIMO system. A fading channel may be Rayleigh fading channel, Rician fading channel or Nakagami fading channels. And here results consist the BER performance of Particle swarm optimized BD (block diagonalization) and comparison of BER performance of different precoding techniques with the optimization technique. Simulation achieve with the considering single cell for K = 2 users and 4 transmitting antennas at a base station, Furthermore, we take N_r = 2 receiving antennas for each user and the data length is 200 and with taking 10 no. of iteration. It is essential to note that the Monte-Carlo simulation method has been applied to obtain the results. Here graphs were plotted between the BER and SNR for getting the MU-MIMO system performance.

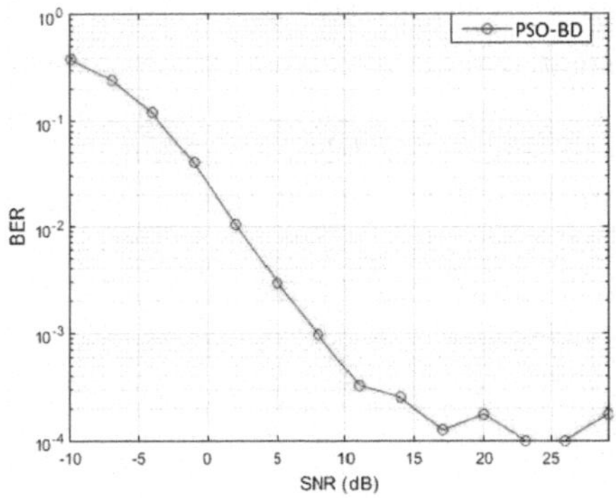

Fig. 3. BER performance of PSO-BD over Rayleigh fading channel

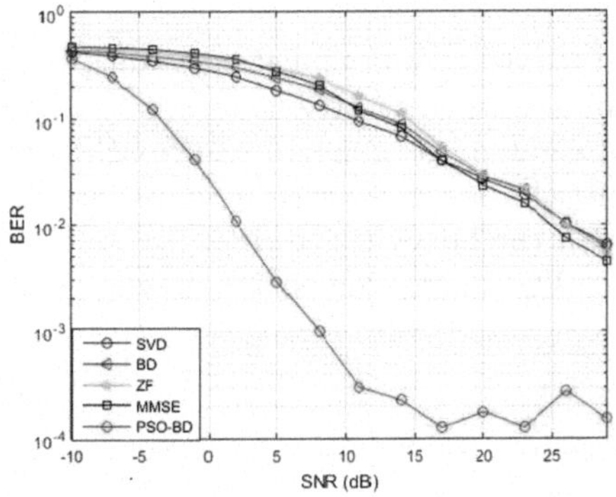

Fig. 4. precoders BER performance over Rayleigh fading channel

Figures 3 and 4 shows the BER performance of ZF, BD, SVD, MMSE and PSO-BD precoders over Rayleigh fading channel. Figure 3 shows the graph of BER with SNR When the BD is optimized by using the PSO technique. Figure 4 shows the comparison between precoders BER performance comparison. Here we can see that minimum BER can obtain with the PSO optimization of BD.

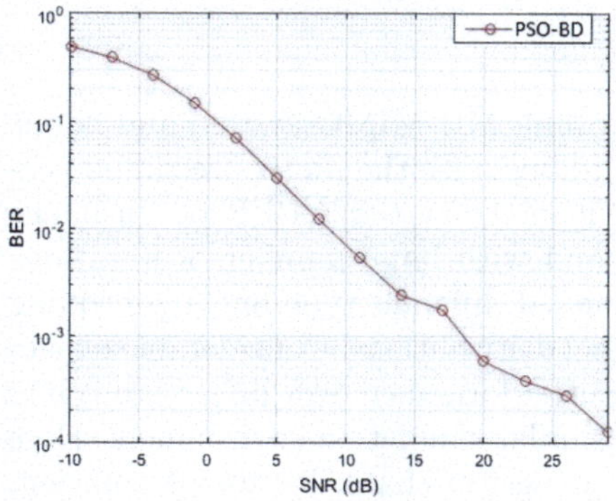

Fig. 5. BER performance of PSO-BD over Rician fading channel

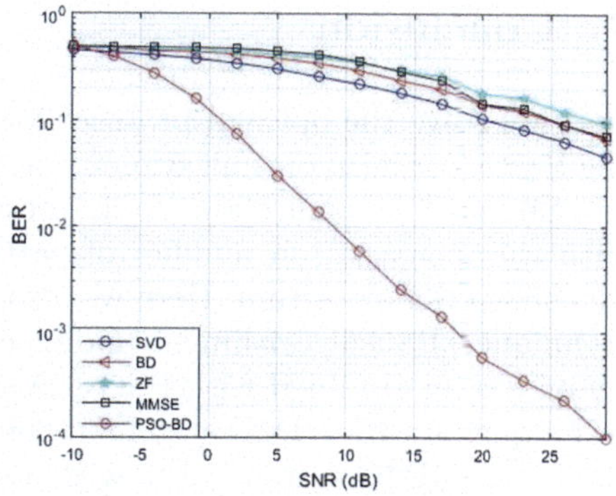

Fig. 6. precoders BER performance over Rician fading channel

Figures 5 and 6 shows that the precoders error rate performance over a Rician fading channel. There is an assumption that there is a LOS or direct path for signal transmission, getting a better result than Nakagami fading channel through MU-MIMO. BER performance of PSO-BD shown in Fig. 5. Comparison of BER performance of all precoders and Particle Swarm Optimized BD shown in Fig. 6.

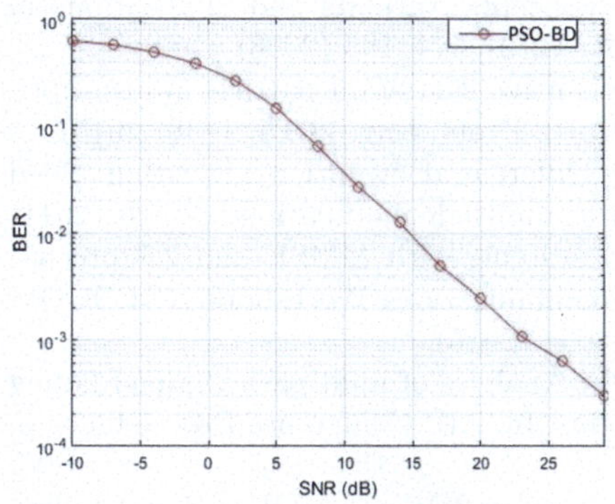

Fig. 7. BER performance of PSO-BD over Nakagami fading channel

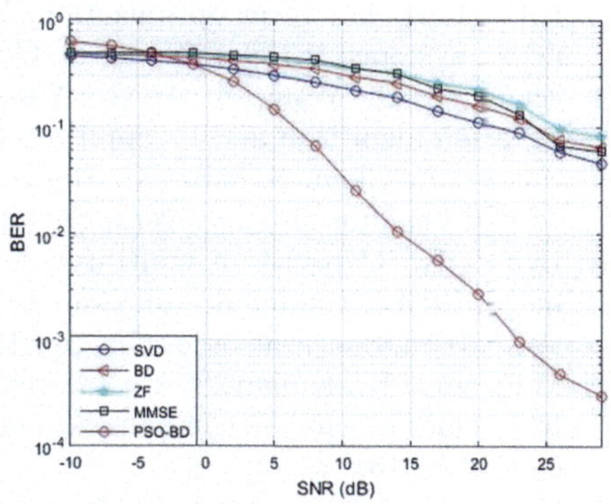

Fig. 8. Precoders BER performance over Nakagami fading channel

Figures 7 and 8 convey BER versus SNR of linear precoders over Nakagami fading channel with m = 0.5. Figure 7 convey the PSO optimized BD performance and Fig. 8 convey the BER performance of Precoders and PSO-BD comparison in MIMO. Here also get that PSO-BD performs better than other precoding techniques.

7 Conclusion

Here we solve the optimal problem for downlink MU-MIMO system and adopt the Particle swarm optimization algorithm to optimize the linear precoding vector. Simulation result show the improvement in the system's BER performance with the increasing in number of particle and with large number of iterations by using PSO algorithm. Here we also discussed different type linear precoding techniques, PSO algorithm and fading channel distributions. Considerable capacity of system obtained with PSO-BD and outperformance over linear precoding technique for MU-MIMO system under the similar simulation environment. Here single cell MIMO system taken with assuming perfect channel state information, in the future it can be develope for multicell MIMO system by taking imperfect channel state information using Non-linear precoding techniques.

References

1. Zhang, R., Song, R.: A linear precoding strategy based on particle swarm optimization in multicell cooperative transmission. J. Electron. (China) **28**(1), 15 (2011)
2. Crâşmariu, V., Arvinte, M., Enescu, A., Ciochină, S.: Performance analysis of the singular value decomposition with block-diagonalization precoding in multi-user massive MIMO systems. In: 12th IEEE International Symposium on Electronics and Telecommunications (ISETC), Timisoara, pp. 71–74 (2016)
3. del Valle, Y., Venayagamoorthy, G.K., Mohagheghi, S., Hernandez, J., Harley, R.G.: Particle swarm optimization: basic concepts, variants and applications in power systems. IEEE Trans. Evol. Comput. **12**(2), 171–195 (2008)
4. Zhang, Y.-D., Wang, S., Ji, G.: A comprehensive survey on particle swarm optimization algorithm and its applications. Math. Probl. Eng. **2015**, 1–38 (2015)
5. Zhang, J., Chen, S., Mu, X., Hanzo, L.: Differential evolution algorithm aided MBER beamforming receiver for quadrature amplitude modulation systems (2011)
6. Lyu, T.K.: Capacity of multi-user MIMO systems with MMSE and ZF precoding. In: IEEE Conference on Computer Communications Workshops (INFOCOM WKSHPS), San Francisco, CA, pp. 1083–1084 (2016)
7. Mahender, K., Kumar, T.A., Ramesh, K.S.: Analysis of multipath channel fading techniques in wireless communication systems. In: AIP Conference Proceedings, p. 020050 (2018)
8. Wongtrairat, W., Supnithi, P.: New simple form for PDF and MGF of Rician fading distribution. In: 2011 International Symposium on Intelligent Signal Processing and Communications Systems (ISPACS), Chiang Mai, pp. 1–4 (2011)
9. Popa, S., Draghiciu, N., Reiz, R.: Fading types in wireless communications systems. J. Electr. Electron. Eng. **1**(1), 232 (2008)
10. Stojnic, M., Vikalo, H., Hassibi, B.: Rate maximization in multi-antenna broadcast channels with linear preprocessing. IEEE Trans. Wirel. Commun. **5**(9), 2338–2342 (2006)

Simulation Analysis of DDoS Attack in IoT Environment

Vikash Kumar[1(✉)], Vivek Kumar[1], Ditipriya Sinha[1], and Ayan Kumar Das[2]

[1] National Institute of Technology Patna, Patna 800005, Bihar, India
vika96snz@gmail.com, vivek5796@gmail.com,
ditipriya.cse@nitp.ac.in

[2] Birla Institute of Technology, Mesra, Patna Campus, Patna 800014, Bihar, India
das.ayan@bitmesra.ac.in

Abstract. Now a day, Internet of Things (IoT) has touched almost every corner of human and unimaginably affect our life by its applications. Resources and environment are being more susceptible to security threats like Virus, DoS/DDoS, Ransomware, Spyware, IP Spoofing, etc. To consider security services and IoT devices capabilities, low power and processing constraints, response rate, this paper has proposed a Decision Tree-Based IDS for IoT environment to prevent intra and inter network from DoS/DDoS attacks. In this paper, the analysis is done in two ways- (a) Power consumption and (b) Attack Detection. The experiments are conducted in the Cooja simulator pre-installed in Contiki operating system within the virtual machine. From attack detection mode it is concluded that C5 Decision Tree-Based IDS model shows high accuracy with low false alarm rate (FAR). Whereas, from power consumption mode it is observed that the simulated network suffers from high-power consumption and around three times more CPU power and two-time Listening Power consumption during attack as compare to their normal behavior.

Keywords: Internet of Thing · Intrusion Detection System · Monitor node · Decision tree · CPU power · LPM power · Listen power

1 Introduction

Today the rapid growth of internet evolves several Internet of Things (IoT) applications. Now a day, IoT devices and its unimaginably applications (e.g. Air-pollution Control System, Street Lightening System, Waste Management System to name a few [1]) highly affected our life. With time, the dependency on IoT devices rapidly grew in the field of services such as Home automation, Traffic Monitoring System, healthcare monitoring devices [2] and many more. The security services like traditional Firewalls, are not very much efficient to prevent from today's attackers as these days the IP/Port filtering and blocking are no longer effective. Unfortunately, till now, many of IoT devices, thought to be safe, are still susceptible and vulnerable for outside world [3, 4]. The "Mirai" botnet [3] attack has drawn the vulnerabilities of IoT system and brought IoT threats in

N. Nain and S. K. Vipparthi (Eds.): ICIoTCT 2019, AISC 1122, pp. 77–87, 2020.
https://doi.org/10.1007/978-3-030-39875-0_8

limelight. Aside from the traditional security services, the Intrusion Detection System (IDS) is being developed but it has their own limitation [5]. To consider security services and IoT devices capabilities, low power and processing constraints, response rate, this paper has proposed a Decision Tree-Based IDS [6–8] for IoT environment to prevent network from DoS/DDoS [9] attacks. This proposed model can be used to achieve the high amount of security in IoT environments. This paper describes a secure simulated IoT network, where the IDS is deployed over some resourceful sensor nodes that protect the other nodes from the both intra network attack and inter network attack. In addition to the usual IoT sensors and network elements, proposed model consists of IDS based monitor, border-gateway and malicious node. Figure 1, depicts the proposed IDS model in the simulated environment where two type of nodes are depicted such as- (a) Normal nodes (b) Attacker nodes and, (c) Monitor nodes. Normal nodes represent the victim node on which attack is performed and the Monitor node is implemented using the Decision Tree-Based IDS [6–8] and take care of the devices lies within their frequency range and monitor and control the network traffic within the network. The model uses multiple monitor nodes which has several advantages. Cooja tool [10] has been used to simulate the proposed model. Cooja is a most widely used simulator for the simulation of IoT environment which has variety of sensors such as Sky mote, Z1 mote, Mica mote, etc. Here, we have analyzed the power consumption of sensor nodes during attack and compare it with the normal IoT sensor network data. The analysis shows that when the attack is occurred, the network suffers from very high-power consumption as compare to their normal behavior. In this situation the network suffers from around three time more CPU power and two-time Listen Power consumption. This proposed work also captures the data from simulated environment during DoS attack and evaluates the performance of IDS on C5 model.

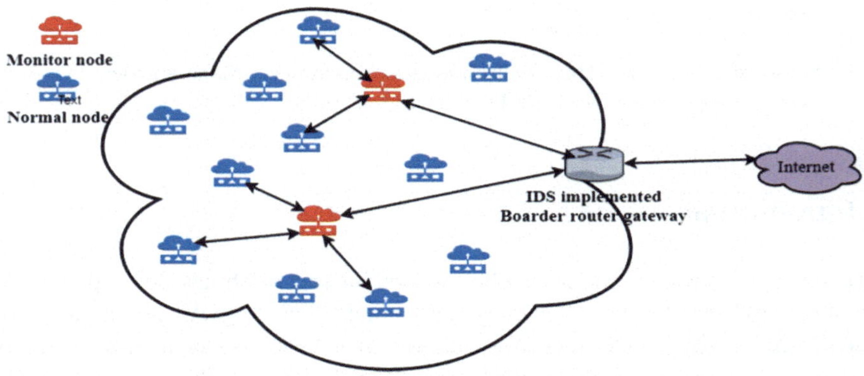

Fig. 1. Proposed sensor-based IDS model in IoT environment

Rest of the paper is organized as: Sect. 2 review some related work. Section 3 describes the proposed model. Experimental setups are presented in Sect. 4. Result analysis is discussed in Sect. 5. Finally, Sect. 6 concludes the work followed by future scope.

2 Review Work

In this section, several recent works on IDS for IoT environment has been reviewed. The review is done based on several factors like detection technique, energy consumption etc.

Moustafa et al. [11] have performed an analysis of different datasets (UNSW-NB15 [11], KDD99 [12]) including the recent one to evaluate the performance of different ML techniques. In similar way, Koroniotis et al. [7] have investigated several ML techniques in order to find the most suited for attack detection. Based on the analysis they concluded that the Decision Tree based model can outperform other ML techniques. [13] have proposed a model called *Dendron* which uses the decision tree model for detecting any attack. Diaz et al. [13] presented a methodology for security analysis in WSN using attacker modeling. The analysis is made based on the power consumption and node's software execution time. They concluded that the power consumption and software execution time can be utilized to detect different attacks.

Shreenivas et al. [5] proposed an approach which is an extension of SVELTE [14] technique to improve the security in 6LoWPAN to other protocols: RPL, IPv6, UDP and IEEE 802.15.4. In addition, to improve the detection rate they proposed a different IDS model that uses the Expected Transmissions metric and a geographical approach to detect attack in the network. The Contiki OS and Cooja tool is used to simulate their whole work. Mehare et al. [15] proposed an IDS architecture to detect DDoS attack along with attacker location and simulated it using Cooja simulator. Their architecture is developed in ebbits network. Eastman et al. [9] proposed an approach to determine if the packet loss and power consumption can be useful parameter in detecting the attacks on IoT environment. To address this, he implemented several threat models and collected the power consumption data using power trace plugin in Contiki. Their analysis concludes that packet loss can be used to detect the sinkhole attack whilst power consumption parameter can be useful to detect DoS attack.

Loulianou et al. [4] proposed a signature-based IDS design, which includes centralized and distributed IDS model. Their implementation considered two variants of DoS attack namely Hello Flood attack and Version attack in the simulated IoT environment in Cooja. However, their work only concerns with the design part of such stated high-level architecture IDS and their result states only about the impact of attack on certain IoT devices and Power consumption.

3 Proposed Model

Several recent works on IDS have shown that the Decision Tree-Based model is efficient for attack detection models. Here, we have proposed a decision tree based IDS for an IOT environment and demonstrated in simulation environment. In this paper, the analysis is done in two ways- (a) Power consumption and, (b) Attack Detection. The main constraint of the IoT devices is power, processing and response rate. Most of the IoT devices has very low power, low processing speed and slower response rate due to high amount of network traffic and the analysis of whole network simultaneously is a bit difficult task for one node which can be overcome through using multiple monitor node. Another advantage of monitor node based distributed IDS model is that it reduces the constrain of a traditional IDS system by distributing the tasks at monitor node level and each monitor node is responsible for IoT security equally. The proposed IDS is also deployed in the boarder-router gateway which ensures the security from inter-network communication.

From Fig. 1, it is observed that the model consists of several IoT sensors, two IDS based sensor node (monitor node) and one IDS based boarder-router gateway node. The gateway sensor is responsible for detecting attacks in inter network communication. Thus the model ensures high security to IoT environment. Flow diagram for the proposed work is shown in Fig. 2.

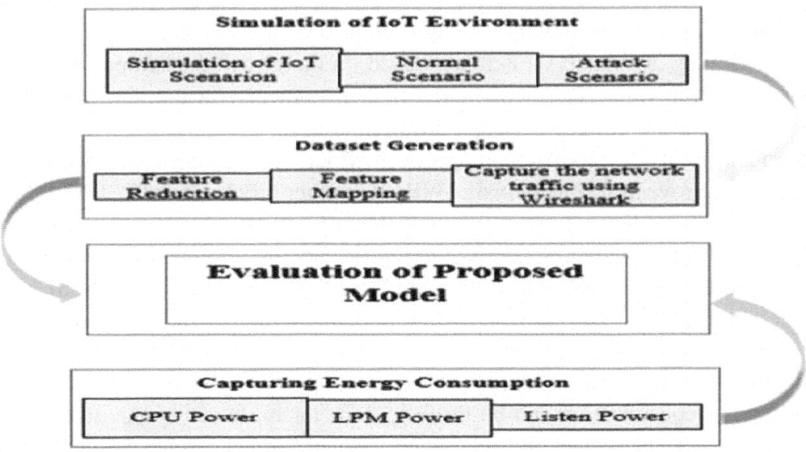

Fig. 2. Flow diagram for the proposed work

3.1 Dataset Capturing

In this section we discuss about the dataset which is created using the simulated environment in Cooja by performing two different DoS attack which are-

(a) Hello Flood attack [4]: It is intended to make network resources or services unavailable by generating a large network traffic. This is achieved by broadcasting DIS message consistently to other nodes and forces it to send DIO message as reply.
(b) Version attack [4]: It is performed by generating high version number DIO control packets that make other nodes suffer new DAG tree each time. As a result loss of data packets and network congestion are occurred due to control message overheads. In turns it makes network topology inconsistent and inefficient.

Both attacks have same motive to make node unavailable for any services. Thus our aim is to achieve the DoS attack through the simulation. The total time of running the simulation environment is 6 min 40 s, whilst the simulation is in steady mode around after 1 min. After that the traffic is started to capture the data for non-attack environment and then for the attack environment by performing the same using Wireshark packet sniffing tool plug-in itself in Contiki OS. While running the attack simulation model 2 sensor nodes started acting maliciously remaining condition kept same. All the captured data are saved and exported in .csv file for dataset preparation.

3.2 Dataset Mapping and Description

This section discusses the feature mapping of the captured dataset and describes the features of the dataset as well. [16] analyzed a set of features of UNSW-NB15 which are most important for attack detection. Hence, the captured data are mapped to those important features which ensures high contribution of all chosen features for classifying the attacks. The developed IDS model is trained and tested on the mapped dataset created at NIT Patna during the experiment. This dataset contains total 11 features with DoS and Normal categories. It contains 14686 records with 7168 DoS attack and 7518 normal data records.

The dataset description and a sample of dataset is presented in the Table 1 and Table 2 respectively. Table 3 shows the distribution of training and testing dataset used to evaluate the proposed IDS model.

Table 1. Dataset description

SI No.	Features	Types	Feature description
1	proto	Nominal	Transmission protocols used
2	src_dst_bytes	Integer	Number of source to destination transaction bytes
3	dst_src_bytes	Integer	Number of destination to source transaction bytes
4	src_dst_ttl	Integer	Source to destination time to live value in milli second
5	dst_pkt_loss	Integer	Number of destination packet dropped or loss
6	dst_inpkt_arrt	Integer	Destination interpacket arrival time (in milli second)
7	nct_dst_lt	Integer	Number of connection of same destination (IP/Physical) address in last 100 records
8	nct_src_dstport_lt	Integer	Number of connections of same source IP and destinations port address in last 100 records
9	nct_dst_src_port_lt	Integer	Number of connections of same destinations IP and source port address in last 100 records
10	nct_dst_src_lt	Integer	Total number of connections with same destination and source IP in last 100 records
11	atk_catg	Nominal	Attack category (DoS and Normal)

Table 2. A sample of created dataset

proto	src_dst_bytes	dst_src_bytes	src_dst_ttl	dst_pkt_loss	dst_inpkt_arrt	nct_dst_lt	nct_src_dstport_lt	nct_dst_src_port_lt	nct_dst_src_lt	atk_catg
igmp	56	56	1	0	0.024646	0	0	0	0	DoS
udp	568	304	31	0	1.277667	3	1	1	3	normal
udp	92	92	64	0	0.007935	0	2	0	0	DoS

Table 3. Distribution of dataset

Instances in final dataset	Instances in training set		Instances in testing set	
14686	Normal	DoS	Normal	DoS
	4895	4300	3262	2229
Total	9195		5491	

4 Experimental Setup

Here, we discuss the experimental setup of IoT network simulator, capturing the network traffic and evaluation based on developed decision-tree based IDS model. The experiments are conducted in the Cooja simulator pre-installed in Contiki operating system [10]. Sky mote sensor type is used to design the IoT environment due to its compatibility and number of sensors supports.

A simulated network is designed with 15 sensor node including 2 malicious nodes shown in Fig. 3. Further the scenario could be scalable as per need and application. The malicious nodes of simulation environment performs DoS attack in the network. Sensor nodes deployed can communicate to each other within the network through IDS based monitor node. Whereas, any communication outside the network is done through IDS based network-gateway node. The ***Collect View*** in Cooja tool is applied to collect the power consumption in network. To implement the malicious node and control its traffic flow we change the normal behavior of sensor nodes such that they can generate the desired attack by tuning the parameters of malicious nodes. Malicious node reduces the package latency timer i.e. trickle timer set to 0 s from 60 s, which results rapid increase in network traffic and the channel is flooded by DIS and DIO control messages. Thus it increases the more power consumption as well as CPU consumption at the infected sensors in the network.

The network traffic is captured with the help of Wireshark during the normal network behavior as well as attack mode. After capturing the network traffic, the feature mapping is used to prepare the dataset using python program. The important factor while creating the dataset is the feature mapping. Further the features which has high information gain

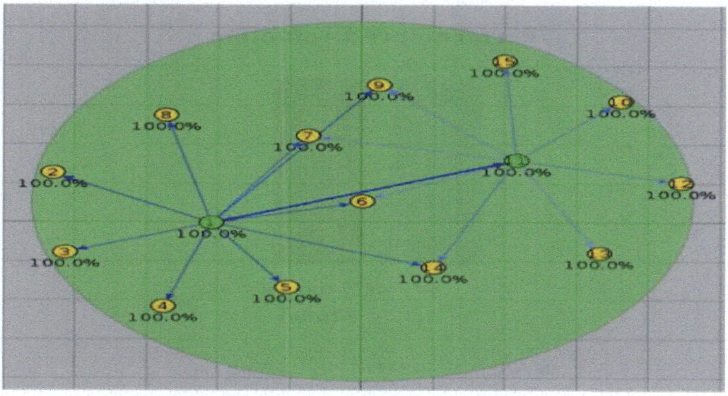

Fig. 3. Experimental setup for the proposed work

and useful for the IDS are captured during the normal network behavior as well as during attack behavior. In this way the dataset is created and further divided in 60:40 ratios to train our C5 Decision Tree-Based IDS model [16] that is used to test our proposed model.

5 Result and Analysis

In this section we have presented the analysis, based on two major approaches: (a) Based on Attack Detection and (b) Power Consumption approach. The both the approaches are discussed below.

5.1 Based on Attack Detection

In Attack Detection approach, we have used the C5 Decision Tree-Based IDS model [16] to test and measure the accuracy and other performance parameters [8]. Confusion matrix of the proposed model for detection of attack with precision and recall of respective categories is given in Table 4.

Table 4. Confusion matrix for the proposed IDS model

	DoS	Normal	Recall (%)
Dos	3108	154	95.3
Normal	36	2832	98.74
Precision (%)	98.85	94.84	

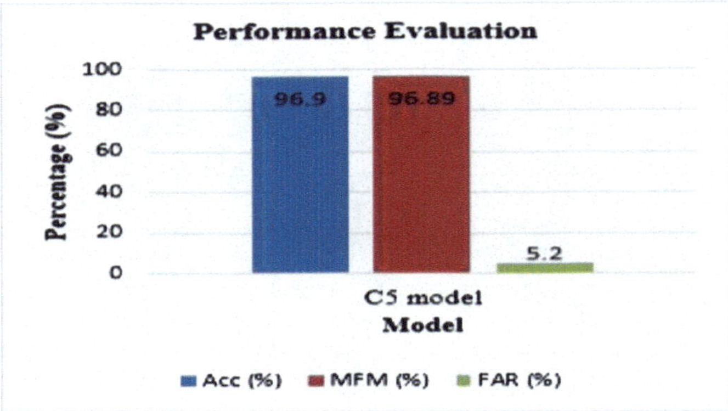

Fig. 4. Performance analysis of IDS on C5 model

Figure 3 shows the value of different metrics [8] which is used to evaluate the proposed IDS model. It can be observed that the accuracy of the model is high which is 99.9% and false alarm rate (FAR) is 5.2% (Fig. 4).

5.2 Based on Power Consumption Techniques

Whenever an intruder tries to access unauthorizedly any of the IoT devices or starts to attack on any of the devices or performing any type of malicious behavior which results in rapid increase in power consumption of IoT devices. Due to the high-power consumption at infected IoT devices the power dissipation increases which further results to unavailability of services. It leads to harm the IoT devices or environment. At this stage intruder may steal confidential information, monitor the traffic and make the service unavailable. Basically, in this approach we are more concerned with power. To show this, different power consumption data such as CPU power, LPM power and Listen power is captured using ***Collect View*** in Cooja.

Tables 5 and 6 contain the captured CPU and listening power consumption data (in mW) for the 15 nodes including one malicious node during normal condition as well as during attack condition respectively (* denotes malicious node). CPU power is the energy that spent to process each node tasks. On the basis of the power consumptions, the network devices are analyzed. Network is suspected to any type of malicious activities, if any node consumes more power than usual, otherwise they treated as normal behavior.

Observing the below result, it is concluded that the network data in attack has almost three-time CPU power and two times Listen power consumption as compared to normal condition network. So, during attack IoT network devices suffer from high power consumption. As a result, the low power IoT devices become unavailable. This also makes the services unavailable and leads to compromise the security of IoT environment. Figures 5 and 6 shows, before and after CPU and Listening power consumption distribution of the environment.

Table 5. Normal condition network power consumption data (in mW)

Node Sl. No.	CPU power	LPM power	Listen power
1*	0	0	0
2	0.336	0.153	0.411
3	0.393	0.152	0.403
4	0.343	0.153	0.396
5	0.346	0.153	0.383
6	0.334	0.153	0.392
7	0.345	0.153	0.389
8	0.365	0.152	0.444
9	0.336	0.153	0.401
10	0.356	0.153	0.377
11*	0	0	0
12	0.346	0.152	0.385
13	0.341	0.153	0.409
14	0.392	0.153	0.396
15	0.355	0.152	0.418
Average	0.329	0.152	0.400

Table 6. Attack condition network power consumption data (in mW)

Node Sl. No.	CPU power	LPM power	Listen power
1*	0	0	0
2	0.935	0.153	0.685
3	0.964	0.152	0.672
4	0.970	0.152	0.660
5	0.895	0.153	0.638
6	0.943	0.153	0.653
7	0.898	0.153	0.648
8	0.993	0.152	0.740
9	0.964	0.153	0.668
10	0.925	0.152	0.628
11*	0	0	0
12	0.957	0.152	0.667
13	0.899	0.153	0.650
14	0.925	0.152	0.721
15	0.918	0.152	0.598
Average	0.937	0.152	0.571

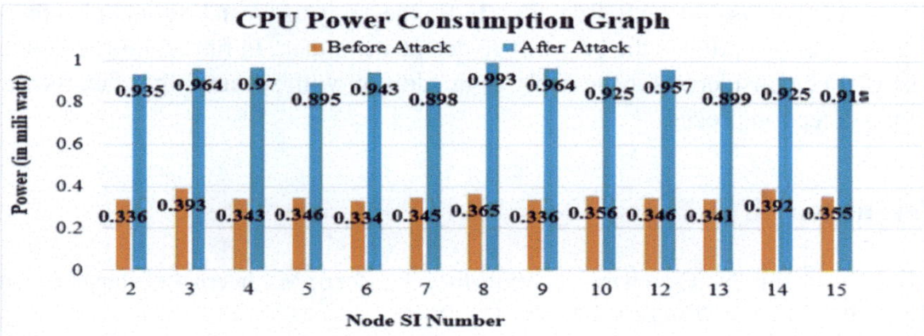

Fig. 5. Comparison of CPU power consumption graph before and after attack

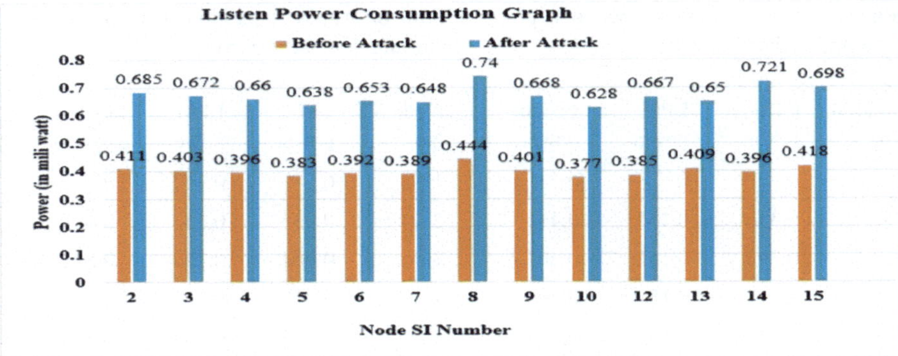

Fig. 6. Comparison for listen power consumption graph before and after attack

6 Conclusions

This paper analyzed the simulated IoT environment using IDS for 3 types of motes (sensors) which are- C5-Based boarder-router gateway node, second one is normal node in which few acts as the monitor node and third is malicious node, which acts as an attacker and is deployed to make the resources unavailable and consume the power of normal nodes. The simulation is done into two modes (i.e. Attack detection mode and Power consumption mode). Attack detection mode concludes that C5 Decision Tree-Based IDS model shows 99.9% accuracy with 5.2% FAR. On the other hand from power consumption mode it is observed that the simulated network suffers from high-power consumption and around three times more CPU power and two-time Listening Power consumption during attack as compare to their normal behavior. This will affect the IoT devices and make disable the IoT devices. From the analysis it can be concluded that the CPU power consumption is the best way to detect any attack. In future, more number of attacks will be considered for the analysis in addition with several other factors which affect the detection process.

References

1. Zanella, A., Bui, N., Castellani, A., Vangelista, L., Zorzi, M.: Internet of things for smart cities. IEEE Internet Things J. **1**(1), 22–32 (2014)
2. Bellavista, P., Cardone, G., Corradi, A., Foschini, L.: Convergence of MANET and WSN in IoT urban scenarios. IEEE Sens. J. **13**(10), 3558–3567 (2013)
3. The Mirai botnet explained: How teen scammers and CCTV cameras almost brought down the internet. https://www.csoonline.com/article/3258748/security/the-mirai-botnet-explained-how-teen-scammers-and-cctv-cameras-almost-brought-down-the-internet.html
4. Loulianou, P., Vasilakis, V., Moscholios, I., Logothetis, M.: A signature-based intrusion detection system for the internet of things. In: Information and Communication Technology Form (2018)
5. Shreenivas, D., Raza, S., Voigt, T.: Intrusion detection in the RPL-connected 6LoWPAN networks. In: Proceedings of the 3rd ACM International Workshop on IoT Privacy, Trust, and Security, pp. 31–38. ACM, April 2017

6. Moustafa, N., Slay, J.: The evaluation of Network anomaly detection systems: statistical analysis of the UNSW-NB15 data set and the comparison with the KDD99 data set. Inf. Secur. J. A Glob. Perspect. **25**(1–3), 18–31 (2016)
7. Koroniotis, N., Moustafa, N., Sitnikova, E., Slay, J.: Towards developing network forensic mechanism for Botnet Activities in the IoT based on machine learning techniques. In: International Conference on Mobile Networks and Management, pp. 30–44. Springer, Cham (2017)
8. Papamartzivanos, D., Mármol, F.G., Kambourakis, G.: Dendron: genetic trees driven rule induction for network intrusion detection systems. Future Gener. Comput. Syst. **79**, 558–574 (2018)
9. Eastman, D., Kumar, S.A.: A simulation study to detect attacks on internet of things. In: 2017 IEEE 15th International Dependable, Autonomic and Secure Computing, 15th International Conference on Pervasive Intelligence & Computing, 3rd International Conference on Big Data Intelligence and Computing and Cyber Science and Technology Congress (DASC/PiCom/DataCom/CyberSciTech), pp. 645–650. IEEE, November 2017
10. Cooja Simulator – Contiki. https://anrg.usc.edu/contiki/index.php/Cooja_Simulator
11. Moustafa, N., Slay, J.: UNSW-NB15: a comprehensive data set for network intrusion detection systems (UNSW-NB15 network data set). In: Military Communications and Information Systems Conference, pp. 1–6 (2015)
12. KDD 99 data set. http://kdd.ics.uci.edu/databases/kddcup99/kddcup99.html
13. Diaz, A., Sanchez, P.: Simulation of attacks for security in wireless sensor network. Sensors **16**(11) (2016)
14. Raza, S., Wallgren, L., Voigt, T.: SVELTE: real-time intrusion detection in the Internet of Things. Ad Hoc Netw. **11**(8), 2661–2674 (2013)
15. Mehare, T.M., Bhosale, S.: Design and development of intrusion detection system for internet of things. Int. J. Innovative Res. Comput. Commun. Eng. **5**(7), 13469–13475 (2017)
16. Kumar, V., Das, K.A., Sinha, D.: Statistical analysis of the UNSW-NB15 dataset for intrusion detection. In: 1st International Conference on Computational Intelligence in Pattern Recognition. Springer (2019, in press)

Blockchain Application Framework for Priority Metric Based Academic Record Repository

Ajita Banerjee, Arman Singhal, Lokesh Gujral, and Kavita Choudhary(✉)

JK Lakshmipat University, Jaipur, India
kavitachoudhary@jklu.edu.in

Abstract. Blockchain as the name suggests can be virtually visualized as an endless chain of blocks. The complex algorithms and computational power required to process the transactions for inserting blocks into the chain laid to the basic structure. Blockchain functionally work on consent of all, removing the requirement of middle man authority. This paper proposed a Blockchain fabric where an academic track record can be managed for every Indian Citizen after getting verified through Aadhaar Number. The resultant information generated by the proposed fabric would be immutable, decentralized and on distributed ledger. This is beneficial for all stakeholders like, government recruitment portal, institutes, organizations, etc. All electronic academic records, summary, vitals, transcripts, certifications etc. are mapped (Aadhaar UID) and are kept on Blockchain fabric, which can be accessed anytime with no loss of data (fully secured). In the proposed framework, all Stakeholders can filter the data as per skill set in demand. Job information and opportunities can reach to beneficiary directly with minimum efforts.

Keywords: Blockchain · Decentralized · Distributed · Smart contract

1 Introduction

It can be defined as a distributed database that maintains records in the form of blocks. It is a decentralized structure that maintains a linked chain of blocks being added continuously. Each of the block records a transaction that has taken place. It stores a timestamp to mark the time of transaction and a link to denote the prior block. As the timestamp can't be changed, blocks can't be modified. Hence the design enforces trustworthiness in a decentralized structure. So, no one has all authority to make decisions. But the most important advantage of Blockchain is the solution to double spending. This revolutionary approach will make every system economically stronger by cutting the cost of maintenance. Initially Blockchain came as a platform to support the mining of Bitcoin the famous digital currency. As proposed by Satoshi Nakamoto in late 2008, the core component of Blockchain is peer to peer technology combined with encryption. The whole structure is governed by a set of rules that each node has to abide by. Though from the time of inception, it is applied in many types of applications on experimental basis, however It is seen that there are used cased where Blockchain has opened a new

N. Nain and S. K. Vipparthi (Eds.): ICIoTCT 2019, AISC 1122, pp. 88–94, 2020.
https://doi.org/10.1007/978-3-030-39875-0_9

dimension for business which is discussed in literature survey section. The issue of trust in absence of a central authority is addressed in a firmer way. Its adoption in government and private sectors is leading to innovation of new business models and also is opening new research opportunities. This paper is focused to investigate the potential of Blockchain in developing academic record repository to make recruitment process transparent and efficient. As there is no secure and distributed plat-form for skill tracking and managing is available in India, this paper will propose a model to have a non-centralized repository for all academic reports as an evidence of certain skill acquirement, which could be furthered to rate the skill sets on the basis of priority. This proposed model will be used by various Government and non-government organizations while hiring employees.

2 Literature Survey

In the paper "Enigma: Decentralized Computation Platform with Guaranteed Privacy", Zyskind et al. [1] has attacked the issue of protection of personal data. So a lightweight architecture for blockchain is conceptualized to improvise the strength. In this paper a method is proposed where an external blockchain is used only to store the references to actual data hence responsible for light-weight processing whereas the actual data is stored off-chain. This model is based on a highly optimized version of secure multi-party computation. In "Towards a More Democratic Mining in Bitcoins" by Paul et al. [2] a new scheme is proposed to mine Bitcoin in an energy efficient technique. This technique is suggested as an alternative way to the existing Proof-of-Work scheme where some extra bytes are added to the present block header so that the timestamp is utilized more effectively in the hash generation. This paper basically introduces a new defense mechanism cyber-attack. Although many mechanisms such as distributed consensus and cryptography are there to prevent Blockchain from common cyber-attacks, however attacks like 51% attack of 14th June 2014, which is peculiar to a Blockchain, is a matter of concern. As discussed by Underwood [3] in his paper, immutability is a prime reason behind the acceptance of Blockchain in the finance sector. However, this assumption of security by the organizations adopting it initially hasn't been same after 51% attack. Yli-Huumo et al. [4], authenticates that in a case like this a single node can control all other nodes and can smartly manipulate the records during attack. Lim et al. [5], describes that attacks like DDoS and hacking of private accounts are also other common breaches in security in a Blockchain. This analysis is supported by Atzori [6], which says that as the nodes have access to all the data, which may pose threat to confidentiality and privacy. According to Kakavand et al. [7] and also Peters and Panayi [8], Blockchain can be used as a boon in application areas that involve smart contracts or financial asset settlements and even in areas which involve financial transactions. In today's competitive economy, new innovative ways to reduce transaction costs can surely be explored by the use of Blockchain technology. One of such innovation chain.com a start-up venture of NASDAQ aiming for private equity exchange is introduced by Crosby et al. [9]. This paper also discussed about the applications which are non-financial like the electronic voting system proposed by Bitcoin foundation to enforce transparency. Other applications like management of digital ownership and notary services based on Blockchain are also discussed in the paper. According to Kim and Laskowski [10], the application of Blockchain in supply chain management is described in elaboration. It is said that the consensus-based ledger system is quite useful for product provenance as the origin and processes can be tracked easily in a Blockchain. In the paper "A maturity model for blockchain adoption" by Wang et al. [11], a five-stage maturity model is proposed which has four dimensions for blockchain. For adoption of the same model a progressive procedure is also proposed to guide step by step. In the paper, "Are Blockchains Immune to All Malicious Attacks" by Xu [12], various types of malicious activities and fraud that can be prevented by blockchain technology are identified. Even the attacks that are vulnerable for Blockchain are also discussed. This paper has recommendations for measures that could be taken for security of blockchains. Sun et al. [13] has discussed an altogether new perspective of Blockchain that is its contribution in making smart cities in future. Zhu and Zhou [14], has analysed some important issues of China's equity crowdfunding. This proposal will be helpful in reducing the cost in equity registration or operations related to them. From this valuable information can be generated to understand the crowdfunding market in a more efficient way.

3 Proposed Framework of Academic Repository

3.1 Academic Repository: Block Diagram

To give transparency and automation in the recruitment process of any organization the proposed blockchain network provides a trustworthy platform for Institutions, Companies, Students and Job Seekers. The proposed framework discusses how the processes, protocols, role of miners, setting of priorities are integrated altogether. National Institutes and individuals will be able to store academic and additional skills on the platform. They will upload all the achievements and skills of them respectively. Miners from recognized institutes will verify the authenticity of the documents. An individual can update additional skills on the platform. This will be verified by Miners from Domain Specific Communities where they can validate the skill. The proposed framework will generate a Resume of the person with all the verified skills on the platform as well as it is helpful for any organization (Government Sector) to keep track of skills in individuals after Aadhaar verification (UID). Now if an organization wants to hire people on a particular set of skills, they can simply upload the skills and search for worthy people. They can send offer letters to all these people on a single click. Every process till the final job confirmation can be verified by Verifiers at all stages. First categorization of miners is from pool of recognized and approved organizations. Second categorization of miners is from domain specific communities. This community comprises similarly skilled peers who are responsible for the verifying process of additional skills added by individuals (Fig. 1).

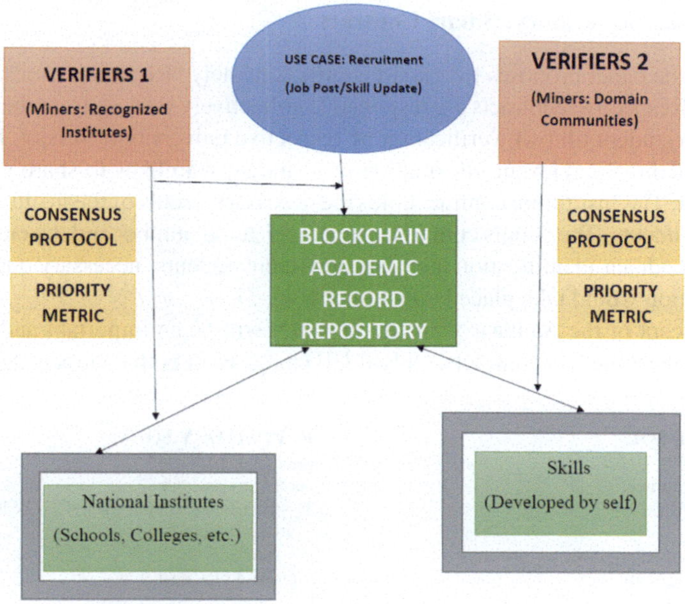

Fig. 1. Block diagram for Academic Record Repository

3.2 Academic Repository: Class Diagram

See Fig. 2.

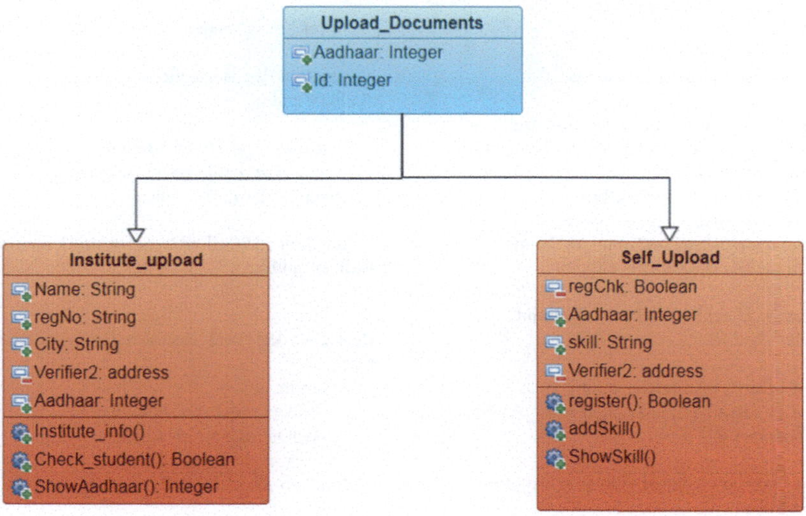

Fig. 2. Class diagram for Blockchain academic repository

3.3 Academic Repository: Smart Contract

Henceforth the paper proposes two smart contracts namely "Repository.sol" and "Institution.sol". Respective contracts are responsible collectively starting from the process of registering a student till own verification by respective university or school. Repository contract contains the relevant information of a student who likes to share skills on an online portal. The Institution contract holds the necessary details of the institution and its respective students. The details comprise name, registered number and the city. With the help of the Aadhaar identification they can store their students' necessary qualifications and verification would take place by institutes (Fig. 3).

The concept of the Aadhaar verification (UID) will be an important and necessary aspect to connect the two contracts. Only if UID matches in both contracts then it would

INSTITUTE.SOL

```
pragma solidity ^0.5.1;
contract institution{
   struct InstiInfo
   {
      string name;
      string regNum;
      string city;
   }
   struct stuId{
      uint256 Aadhaar;
   }
   stuId[] stud_identity;
   address ver1;
   uint counter=0;
   uint public uid;
   mapping(address =>InstiInfo) Institute;
   constructor(uint numberOfSt)public
   {
      ver1=msg.sender;
      stud_identity.length=numberOfSt;
   }
   function inst_infor(address ofInst,string
memory _name,string memory _regNum,string
memory _city)public{
      if(msg.sender!=ver1)return;
      Institute[ofInst].name=_name;
      Institute[ofInst].regNum=_regNum;
      Institute[ofInst].city=_city;
   }
   function checkFor(address ofInst,uint256
_Aadhaar)public
   {
      if(ofInst!=msg.sender)return;
      stud_identity[counter].Aadhaar=_Aadhaar;
      counter++;
   }
   function showAdhaar(address ofInst,uint
place)public{
      uid=stud_identity[place].Aadhaar;
   }
}
```

REPOSITORY.SOL

```
pragma solidity ^0.5.1;
contract repository
{  string public output_skill="No skill";
   struct student{
      uint id;
      bool regCheck;
   }
   struct skillSet
   {
      string skill;
      uint id;
   }
   mapping(address => student) smembers;
   mapping(address => skillSet) skillOption;
    address ver2;
    constructor()public
    {
       ver2=msg.sender;
    }
    function Register(address stu,uint
Aadhaar)public
    {
       if(msg.sender!=ver2)return;
       smembers[stu].regCheck=true;
    smembers[stu].id=Aadhaar;
    }
    function addSkill(address stu,string memory
skill_stu)public
    {

if(msg.sender!=ver2||smembers[stu].regCheck!=
true)
       return;
       skillOption[stu].skill=skill_stu;
       skillOption[stu].id=smembers[stu].id;
    }
    function showSkill(address stu)public
    {
       output_skill=skillOption[stu].skill;

    }
}
```

Fig. 3. Smart contracts

be possible to retrieve the qualifications/academic record of any student. It serves as a vigilant method to provide the abstraction and transparency of any students' details who wishes so.

3.4 Academic Repository: Output

This is the output for the repository for students. The smart contract is deployed using an account address. Following contract contains the details of any student. Register function registers a student using his/her Aadhaar number. addSkill adds the skill set of the particular student. output_skill displays the skill set of that student (Fig. 4).

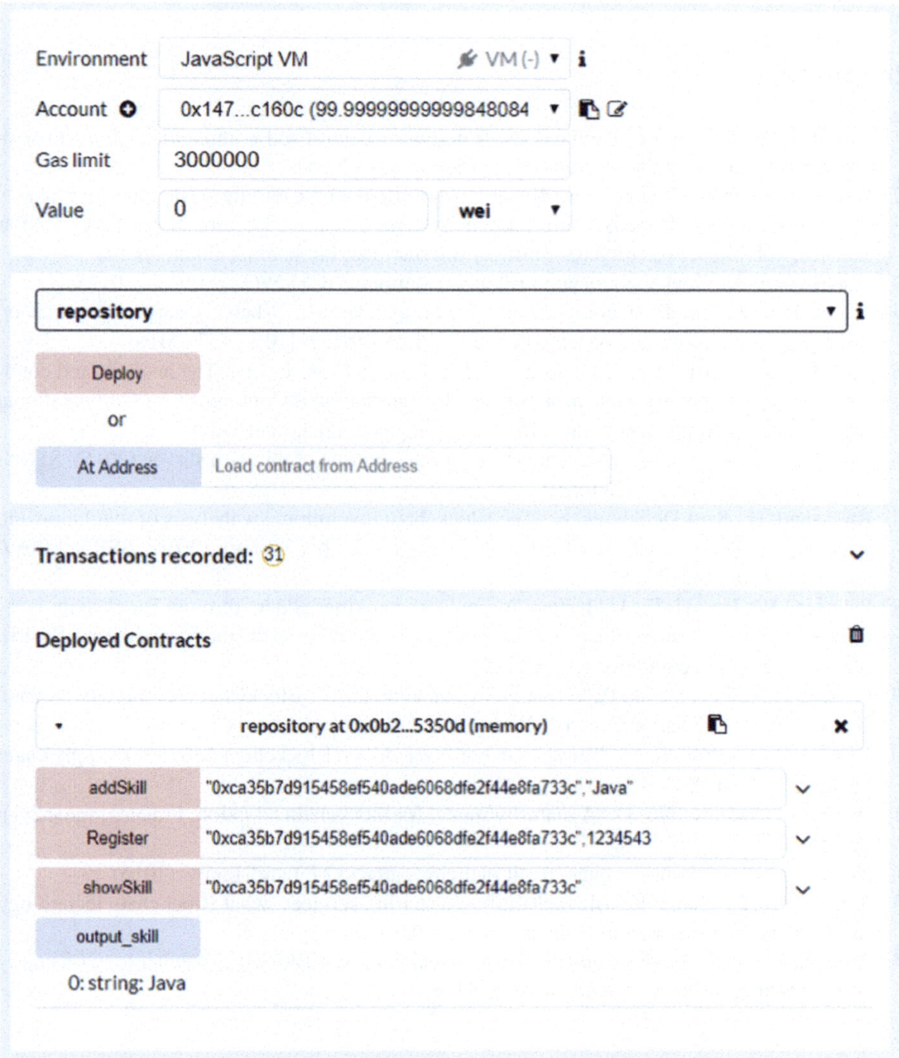

Fig. 4. Academic Record Repository

4 Conclusion and Future Scope

The block and class diagrams discussed in the above sections provide a gateway for several institutions and other companies to hire required skill employees. Government Recruitment portal can directly forward the job-related openings to concerned skill set available across India with single click. Verification Process can be managed where Proof of work and Proof of stake would be applied. Government and other stakeholders can filter the data as per requirements and can deduce the "Skill Metric" of India. Job information and opportunities can reach to beneficiary directly with no misleading/misguidance at all. The proposed Blockchain framework can be further extended by integration of artificial intelligence and machine learning. This welcomes variety of high-volume data analytic solutions for various applications.

References

1. Zyskind, G., Nathan, O., Pentland, A.: Enigma: decentralized computation platform with guaranteed privacy. arXiv preprint arXiv:1506.03471 (2015)
2. Paul, G., Sarkar, P., Mukherjee, S.: Towards a more democratic mining in Bitcoins. In: Prakash, A., Shyamasundar, R. (eds.) Information Systems Security. Lecture Notes in Computer Science, vol. 8880, pp. 185–203. Springer International Publishing, Cham (2014)
3. Underwood, S.: Blockchain beyond Bitcoin. Commun. ACM **59**(11), 15–17 (2016)
4. Yli-Huumo, J., Ko, D., Choi, S., Park, S., Smolander, K.: Where is current research on blockchain technology – a systematic review. PLoS ONE **11**(10), 1–27 (2016)
5. Lim, I.K., Kim, Y.H., Lee, J.G., Lee, J.P., Nam-Gung, H., Lee, J.K.: The analysis and countermeasures on security breach of Bitcoin. In: International Conference on Computational Science and Its Applications, pp. 720–732. Springer International Publishing (2014)
6. Atzori, M.: Blockchain-based architectures for the internet of things: a survey (2016). SSRN: https://ssrn.com/abstract=2846810
7. Kakavand, H., Kost De Sevres, N.: The Blockchain revolution: an analysis of regulation and technology related to distributed ledger technologies. Luther Systems (2016). SSRN: https://ssrn.com/abstract=2849251
8. Peters, G.W., Panayi, E.: Understanding modern banking ledgers through Blockchain technologies: future of transaction processing and smart contracts on the internet of money (2015). SSRN: https://ssrn.com/abstract=2692487
9. Crosby, M.A., Pattanayak, P., Verma, S., Kalyanaraman, V.: BlockChain technology: beyond Bitcoin. Appl. Innovation **2**, 6–10 (2016)
10. Kim, H.M., Laskowski, M.: Towards an ontology-driven Blockchain design for supply chain provenance (2016). CoRRabs/1610.02922
11. Wang, H., Chen, K., Xu, D.: A maturity model for blockchain adoption. Financ. Innov. **2**(1), 12 (2016)
12. Xu, J.: Are Blockchains immune to all malicious attacks? Financ. Innov. (2016)
13. Sun, J., Yan, J., Zhang, K.: Blockchain-based sharing services: what Blockchain technology can contribute to smart cities. Financ. Innov. (2016)
14. Zhu, H., Zhou, Z.: Analysis and outlook of applications of Blockchain technology on equity crowdfunding in China. Financ. Innov. (2016)

Design of Attribute Based Authenticated Group Key Agreement Protocol Without Pairing

Reshu Verma(✉) and Abhimanyu Kumar

National Institute of Technology, Uttarakhand, Srinagar 246174, India
{reshu.verma.cse17,abhi.kumar}@nituk.ac.in

Abstract. The paper present a multi-party attribute based authenticated key agreement protocol without using bilinear pairing. Attribute based key agreement is needed in verity of applications wherever a fine grained access control is required like design of access control for cloud storage, military applications, broadcast encryption etc. Such applications usually need to establish, a secure session among the participants without knowing their actual identities for subsequent cryptographic (symmetric key) operations. Instead of identities, the participants are authenticates based on a descriptive set of attributes (called as access policy). As per the literature-survey the protocols based on attribute based authentication presented so far requires bilinear pairing computations. Since the implementation of pairing is much costlier than ordinary scalar point multiplication over elliptic curve. Therefore pairing free protocols over attribute-based authentication is much desirable especially in resource constraint network. The present paper employs elliptic curve arithmetic along with Lagrange interpolation instead of bilinear pairing computations. The authentication process of the protocol is information theoretically secure while the confidentiality of the shared key relies on the intractability of Elliptic Curve Discrete Logarithm Problem (ECDLP).

Keywords: Attribute Based Encryption · ECC · Shamir secret sharing scheme · Lagrange interpolation · Broadcast encryption

1 Introduction

There are several vulnerability to the applications, which uses wireless medium for communication such as eavesdropping, impersonation, etc. which may be overcomes by cryptographic protocols. Symmetric key cryptosystem is more suited for large size messages but it needs a common key to establish among the communicating parties. A key establishment protocol allows two (or more) number of participants to establish a common secret key that would be uses in symmetric cryptographic operations. However, in some of the scenario one can often need to establish a secure session between the participants based on some

N. Nain and S. K. Vipparthi (Eds.): ICIoTCT 2019, AISC 1122, pp. 95–104, 2020.
https://doi.org/10.1007/978-3-030-39875-0_10

specified set of attributes instead of their actual identity. In this case the participant's authentication is depends on the fulfilment of certain access control policy defined over a set of attributes.

The present paper proposes an attribute based key agreement protocol where every participant can agree on a common secret key if and only if it individually holds sufficient set of attribute to satisfy the required access policy. Thus, the actual identities of the participants are need not be disclosed hence the protocol also provides the user anonymity features by default. As per the literature survey all existing attribute based protocols depends upon the computation of bilinear pairing in their operations. While the proposed protocol is, free from bilinear pairing and uses elliptic curve arithmetic, that is faster and simpler than pairing. The protocol also satisfies all security attributes necessary for the attribute based key establishment.

In the proposed protocol Initiator define the access policy based on necessary set of attributes and share it publicly. The protocol also needs an offline trusted authority called Key Generation Center (KGC) responsible for verifying the attributes of participants and computing attribute secret value for both the participants. The participant's authentication is done with the help of attribute secret value given by the KGC and lagrange interpolation of points.

The rest of the paper is organized as follows. In Sect. 2, related works on key agreement protocol specially in attribute-based setting are discussed. The preliminaries related to proposed work have been addressed in Sect. 3. The Sect. 4 proposes the protocol, while Sects. 5 and 6 provides the performance comparison and security analysis of the proposed protocol respectively followed by the conclusion in Sect. 7.

2 Related Work

Research on developing of key agreement protocols is continues since the development of Diffiee-Hellman protocol in 1976. Smart [1] presented the first two-party identity-based key agreement protocol (ID-KA) which uses the IBE scheme [2]. After that many identity based key establishment protocols are designed like [3–5] which has proved their security in different models. Later Sahai et al. [6] introduced the concept of Attribute Based Encryption (ABE) scheme as an special type of identity based cryptosystem, where the identity can be viewed as the set of descriptive attributes. Later et al. [7] extends the idea of ABE and introduces its two variants Key Policy Attribute-Based Encryption (KP-ABE) and Cipher Text Policy Attribute-Based Encryption (CP-ABE). From the literature, it is revealed that few key exchange protocols [8–11] under attribute-based authentication are available. Moreover, it is also noticed that two party attribute based key agreement protocols are also available, for e.g., [8,9]. In these protocols, two parties use attribute information of the other party instead of the traditional public key to generate the session key shared between them. These protocols uses Lagrange interpolation with bilinear pairing and generates private key corresponding to their descriptive attributes (functioned as his/her public key). Recently, some protocols [12–14] are introduced based on CP-ABE scheme with efficient attribute revocation, which resists the user collusion attack and supports fine-grained access control. Some existing research proposals are available on attribute based group key management [15,16] for cloud computing mechanism, however, it does not focus on key establishment schemes.

An attribute based authenticated key agreement protocol is proposed here without requiring the computation of billinear pairing.

3 Preliminaries

In this section we define the preliminaries required for our proposed protocol.

3.1 Elliptic Curve Group

An elliptic curve E over a prime field F_p is denoted by E/F_p and defined by the equation $Y^2 = x^3 + ax + b$ where $a, b \in F_p$ An elliptic curve should be non singular and fulfil the following condition $\delta = (4a^3 + 27b^2)\ mod\ p \neq 0$ The points on E/F_p together with an extra point O called the point at infinity form a group based on the binary operation '+' called point addition operation defined between the two points on elliptic curve.

$$G = \{(x, y) \in E/F_p : x, y \in F_p\} \cup O$$

Adding an elliptic curve point to itself on a number of times are called as scalar point multiplication as $k.P = kP = P + P + ... + P$ (k times) where P is an elliptic curve point and k is an scalar quantity.

The detail description of Elliptic Curve Cryptography (ECC) can be found in various articles including [17–19].

3.2 Shamir Secret Sharing

Shamir's Secret Sharing is an algorithm in cryptography is introduced by Adi Shamir. It is a form of secret sharing, where a secret s is divided into n shares and shared among the shareholders in such a way that t or more that t shares are able to reconstruct the secret s but with less than t parts it cannot reconstruct the secret [20,21]. Shamir secret sharing scheme is based on polynomial interpolation.

4 Proposed Protocol

The proposed protocol is described in the subsequent subsections. Assume there is a trusted authority whose responsibility is to verify the attributes claims by the participants and provides the secrets corresponding to that. The set $A = \{a_1, a_2, ..., a_m\}$ assumed as universal set of attributes and every participant have a subset of this.

4.1 Setup Phase

On taking a security parameter $k \in Z^+$ as input KGC choose a k bit prime p and determine the following tuple and publish it as public parameter

$$\{F_p, E/F_p, G, P, H, P_{pub}\}$$

where F_p is a prime field, E/F_p is an elliptic curve defined over F_p and P is a base point of E/F_p.

H is a hash cryptographic function defined as:

$$H : \{0,1\}^* \rightarrow Z_p^*$$

G represent an elliptic curve group as discussed in Sect. 3.

$P_{pub} = s.P$ where $s \in_R Z_p^*$ and keep s as master private key.

4.2 Attribute Secret Generation

- KGC calculates h_j for all the universal set of attributes $A = \{a_1, a_2, ..., a_m\}$ where m is the number of attributes in A as:

$$h_j = H(a_j)$$

  for $1 \leq j \leq m$.
- Every user U_i who wants to participates in the secure session must show their list of attributes to the KGC for validation.
- KGC first verify each of the attributes claimed by the user. The user needs to somehow prove that they actually holding the claimed attribute in front of KGC.
- After validating the attributes KGC chooses a random number r_i and calculate $R_i = r_i P$ for every user U_i.
- KGC chooses a random number r_{in} and compute $R_{in} = r_{in} P$ for initiator.
- Now KGC computes the secret value for each verified attributes of every user as:
  - KGC computes $S_j = \frac{s.h_j \bmod p}{r_i}$ for each attribute belonging to user U_i where $1 \leq j \leq m_i$ and m_i is the number of attributes verified by KGC belonging to user U_i.
  - KGC delivers $\{S_j \forall 1 \leq j \leq m_i\}$ and R_i where $1 \leq i \leq n$ to the corresponding users U_i via a secure channel.
  - Every user can validate S_j received from KGC as $h_j.P_{pub} = S_j.R_i$ where $1 \leq i \leq n$.

4.3 Key Agreement Phase

The key agreement among the participants will be done as follows:

- Initiator define the access policy over the set of attributes for key establishment among n participants as follows: $\{A_1 = (a_1 \wedge a_2 \wedge a_3)\} \vee \{A_2 = (a_1 \wedge a_2 \wedge a_4 \wedge a_5)\} \vee \{...\} \vee \{A_w = (a_{w-2} \wedge a_{w-4} \wedge a_{w-1} \wedge a_w)\}$ where w is the number of set of attributes and make it publicly available.
- The index of attributes in the set $A_t; 1 \leq t \leq w$ are numbered from 1 to num_t where num_t is the number of attributes in the set A_t. So, points for constructing the polynomial will be $\{(1, S_1), (1, S_2), (1, S_3)\}$ for set A_1 and $\{(1, S_1), (1, S_2), (1, S_4), (1, S_5)\}$ for set A_2 where $\{(S_1, S_2, ..., S_n)\}$ is the attribute secret value received from KGC. In the similar way points for constructing the polynomial is made for other sets of attributes.
- Initiator construct polynomials P_t for each set of attributes A_t and compute

$$P_t(0) = \frac{a_t}{r_{in}}.$$

- Polynomial of degree $num_t - 1$ is to be constructed with each set of attributes where num_t is the number of attributes present in the set of attributes.
- Multiply $P_t(0)$ with R_{in} received from KGC and get $P_t(0).R_in = \frac{a_t}{r_{in}} r_{in} P = a_t P$.
- Choose a random number $x_{in} \in Z_p^*$ and compute $x_{in}P$.
- Now encrypt the packet C_t for each set of attributes A_t with the x coordinate of symmetric key a_tP as

$$C_t = Encrypt(a_tP + x_{in}P); 1 \leq t \leq w$$

  and broadcast this packet.
- On receiving the encrypted packets from Initiator, participants chooses the set of attributes for which they are satisfying and compute the polynomial P_t' with the attribute secret value S_j received from KGC for that set of attributes.
- Compute $P_t'(0) = \frac{a_t'}{r_i}$ where $P_t'(0)$ is the polynomial constructed by other participants and multiply it with R_i as

$$P_t'(0).R_i = \frac{a_t'}{r_i} r_i P = a_t' P$$

  where R_i is received from KGC.
- If $a_t'P == a_tP$, then they can decrypt the packet C_t belonging to the set of attributes A_t with the symmetric key a_tP.
- Compute inverse of the point a_tP and add it with decrypted packet C_t as $(a_tP + x_1P + (-a_tP) = x_{in}P$ and get $x_{in}P$.
- Choose a random number $x_i \in Z_p^*$ and multiply it with base point as x_iP.
- Now encrypt the point x_iP with symmetric key $x_{in}P$ and broadcast this packet.
- On receiving the above encrypted packet, the participants who satisfies the access policy will able to get the point $x_{in}P$ so that they will be able to decrypt the above packet and compute the session key as follows: $K = x_{in} + x_1 + x_2 + ... + x_n$.

5 Performance Comparison

As per the literature survey made the proposed work is the first attribute based group key agreement Protocol without using bilinear pairing. However, this section also gives comparison of the proposed protocol with some other existing group key establishment protocols in the Table 1. Based on the comparison given in Table 1 one can observe that proposed work have significant performance over the existing protocol and fulfill the necessary security features.

Table 1. Performance comparison with the existing protocols

Protocol	Security parameter	Operations required	Computation cost	Attribute based authentication
Wang et al. [8]	Bellare-Rogaway key agreement model	Bilinear pairing, lagrange interpolation	High	Yes
Wang et al. [9]	Bellare-Rogaway key agreement model	Bilinear pairing, lagrange interpolation	High	Yes
Li et al. [22]	aMSE-DDH	Bilinear maps	High	Yes
Tian et al. [23]	CDHP, DBDHP	Multilinear mapping, bilinear mapping, polynomial interpolation	High	Yes
Proposed	ECDLP	Polynomial interpolation, ECPM	Low	Yes

The Notations used in Table 1
ECPM: Elliptic Curve Scalar Point Multiplication
aMSE-DDH: Augmented multisequence of exponents decisional Diffie-Hellman
ECDLP: Elliptic Curve Discrete Logarithmic Problem
DHKE: Diffie Hellman Key Exchange
CDHP: Computational Diffie Hellman Problem
DBDHP: Decisional Bilinear Diffie Hellman Problem

The overall computation cost of the proposed protocol is considered as low because as compare to other protocol it is free from bilinear pairing, more hashing requirement for the participants, modular exponentiations, etc. This protocol only require some scalar point multiplications over the elliptic curve which is a lighter operation as compared to the bilinear pairing, modular exponentiation, etc.

6 Security Analysis

This section explains that our proposed protocol is secure against different kind of attacks.

6.1 Correctness

An attribute-based key agreement protocol is correct if all the users are honest and they deliver all messages honestly. Every users having sufficient attributes to satisfy the access policy should be authenticated by KGC. Correctness of the protocol is rely on the authentication mechanism used. The entire process can be visualize as follows:

Initiator construct the polynomial P_t for each set of attributes A_t of degree $num_i -1$ where num_t is the number of attributes in the set of attributes A_t by interpolation of points. Let the following num_t points for the set A_t:

$$(x_1, \frac{y_1}{r_i}), (x_2, \frac{y_2}{r_i}), ..., (x_{num_t}, \frac{y_{num_t}}{r_i})$$

U_i have to calculate a polynomial P'_t by interpolation of above points only as follows

$$\frac{y_1}{r_i}(\frac{x-x_2}{x_1-x_2} \cdot\cdot \quad ... \quad \cdot \frac{x-x_{num_t}}{x_1-x_{num_t}})$$
$$+ \frac{y_2}{r_i}(\frac{x-x_1}{x_2-x_1} \cdot \frac{x-x_3}{x_2-x_3} \cdot \quad ... \quad \cdot \frac{x-x_{num_t}}{x_2-x_{num_t}})$$
$$\vdots$$
$$+ \frac{y_t}{r_i}(\frac{x-x_1}{x_t-x_1} \cdot \frac{x-x_3}{x_t-x_2} \cdot \quad ... \quad \cdot \frac{x-x_{num_t-1}}{x_{num_t}-x_{num_t-1}})$$

Since the participant wants only the constant term of above expression i.e $P'_t(0)$ it can be calculated as:

$$\begin{aligned} P'_i(0) = \; & \frac{y_1}{r_i}(\frac{x_2}{x_1-x_2} \cdot \frac{x_3}{x_1-x_3} \cdot \quad ... \quad \cdot \frac{x_{num_t}}{x_1-x_{num_t}}) \\ & + \frac{y_2}{r_i}(\frac{x_1}{x_2-x_1} \cdot \frac{x_3}{x_2-x_3} \cdot \quad ... \quad \cdot \frac{x_{num_t}}{x_2-x_{num_t}}) \\ & \vdots \\ & + \frac{y_{num_t}}{r_i}(\frac{x_1}{x_t-x_1} \cdot \frac{x_3}{x_{num_t}-x_2} \cdot \quad ... \quad \cdot \frac{x_{num_t-1}}{x_{num_t}-x_{num_t-1}}) \\ & = \frac{P_t(0)}{r_i} \end{aligned}$$

Thus user having the sufficient attributes to satisfy the access policy can compute the universal point.

$$K = P_t'(0).r_i = \frac{P_t'(0)}{r_i}.r_i.P = P_t'(0).P$$

Thus $P_t(0) = P_t'(0)$ the above computation is elliptic curve scalar point multiplication possible only when the value $P_t'(0)$ is scalar.

6.2 Implicit Key Authentication

Theorem 1. An attribute-based authenticated group key agreement protocol provides implicit key authentication if each member is assured that no party outside the set can learn the group secret key means that the member not satisfying the access structure can not reveal the group key.

Proof: The proposed protocol satisfy the implicit key authentication as the participant not having sufficient attributes-values would not be able to construct the polynomials, so that the relation $P_t(0) = P_t'(0)$ can satisfy. Thus without getting $P_t(0)$ participant can not compute x_{in} (random number chosen by initiator) if intractability of ECDLP is hold in G.

6.3 Known Session Key Security

Theorem 2. A group key establishment protocol provides known session key security if each execution of the protocol should result in a unique secret session key. The compromise of one session key should not compromise other session keys.

Proof: In proposed protocol the random number r_i is chosen by trusted authority KGC and KGC is also computing S_j for all the verified attributes of the users. Hence the compromise of one session key should not effects on other session key.

6.4 Collusion Resistance

Theorem 3. An attribute-based authenticated group key establishment protocol provides collusion resistance if a set of members which is not satisfying the access policy, but having some partial attributes belonging to the access structure can not be collides in such a way they can be authenticated as a valid participant, even the union of their individual attribute are sufficient for an access structure.

Proof: The proposed protocol satisfies collusion resistance by dividing the attribute secrets by a random number chosen by KGC for each participants. The two participants may have same attribute-value, but KGC provide different secret value corresponding to their attribute. If the participants having partial attributes are trying to colludes they would not be able to construct the polynomials at user sides so that the relation $P_t(0) = P_t'(0)$ holds.

7 Conclusion

In this paper, we have presented an attribute based authenticated key agreement protocol without pairings. The proposed protocol introduces a new idea of attribute based authentication by using Lagrange Interpolation to authenticate the participants based on the access policy defined over the set of attributes. The protocol uses some elliptic curve point multiplications which reduces its complexity as compare to pairing based protocols. The proposed protocol avoid the collusion attacks, without defining the separate set of polynomials for each users. The security analysis of the proposed protocol has been done, under required security attributes, and it is found secure against different kind of attacks. The performance comparison shows that the proposed protocol requires significantly less computation and communication cost with attribute based authentication as compare to the other protocols found in the literature.

References

1. Smart, N.P.: Identity-based authenticated key agreement protocol based on Weil pairing. Electron. Lett. **38**(13), 630–632 (2002)
2. Boneh, D., Franklin, M.: Identity-based encryption from the weil pairing. In: Kilian, J. (eds.) Advances in Cryptology—CRYPTO 2001, pp. 213–229. Springer, Berlin, Heidelberg (2001)
3. Shim, K.: Efficient id-based authenticated key agreement protocol based on Weil pairing. Electron. Lett. **39**(8), 653–654 (2003)
4. Wang, S., Cao, Z., Choo, K.-K.R., Wang, L.: An improved identity-based key agreement protocol and its security proof. Inf. Sci. **179**(3), 307–318 (2009)
5. Ateniese, G., Kirsch, J., Blanton, M.: Secret handshakes with dynamic and fuzzy matching, January 2007
6. Sahai, A., Waters, B.: Fuzzy identity-based encryption. In: Advances in Cryptology – EUROCRYPT 2005. Lecture Notes in Computer Science, vol. 3494, pp. 457–473. Springer, Heidelberg (2005)
7. Vipul, G., Pandey, O., Sahai, A., Waters, B.: Attribute-based encryption for fine-grained access control of encrypted data. In: Proceedings of the 13th ACM Conference on Computer and Communications Security, CCS 2006, pp. 89–98. ACM, New York (2006)
8. Wang, H., Xu, Q., Fu, X.: Two-party attribute-based key agreement protocol in the standard model
9. Wang, H., Xu, Q., Ban, T.: A provably secure two-party attribute-based key agreement protocol. In: 2009 Fifth International Conference on Intelligent Information Hiding and Multimedia Signal Processing, pp. 1042–1045, September 2009
10. Gorantla, M.C., Boyd, C., González Nieto, J.M.: Attribute-based authenticated key exchange. In: Steinfeld, R., Hawkes, P. (eds.) Information Security and Privacy, pp. 300–317. Springer, Heidelberg (2010)
11. Steinwandt, R., Corona, A.: Attribute-based group key establishment. IACR Cryptology ePrint Archive 2010:235, August 2010
12. Li, J., Yao, W., Zhang, Y., Qian, H., Han, J.: Flexible and fine-grained attribute-based data storage in cloud computing. IEEE Trans. Serv. Comput. **10**(5), 785–796 (2017)

13. Han, J., Susilo, W., Mu, Y., Zhou, J., Au, M.H.A.: Improving privacy and security in decentralized ciphertext-policy attribute-based encryption. IEEE Trans. Inf. Forensics Secur. **10**(3), 665–678 (2015)
14. Qian, H., Li, J., Zhang, Y., Han, J.: Privacy-preserving personal health record using multi-authority attribute-based encryption with revocation. Int. J. Inf. Secur. **14**(6), 487–497 (2015)
15. Cheung, L., Cooley, J.A., Khazan, R.I., Newport, C.C.: Collusion-resistant group key management using attribute-based encryption. IACR Cryptology ePrint Archive 2007:161 (2007)
16. Bertino, E., Nabeel, M.: Poster: towards attribute based group key management, pp. 821–824 (2011)
17. Kumar, A.: Ternary tree based group key agreement protocol over elliptic curve for dynamic group (2014)
18. Tate, J.T.: The arithmetic of elliptic curves. Invent. Math. **23**(3), 179–206 (1974)
19. Bos, J.W., Halderman, J.A., Heninger, N., Moore, J., Naehrig, M., Wustrow, E.: Elliptic curve cryptography in practice. IACR Cryptology ePrint Archive, 2013:734 (2013)
20. Shamir, A.: How to share a secret. Commun. ACM **22**, 612–613 (1979)
21. Harn, L., Lin, C.: Authenticated group key transfer protocol based on secret sharing. IEEE Trans. Comput. **59**(6), 842–846 (2010)
22. Hu, S., Li, J., Zhang, Y.: Two-party attribute-based key agreement protocol with constant-size ciphertext and key. Secur. Commun. Netw., (2018)
23. Tian, Y., Yang, G., Mu, Y., Liang, K., Yu, Y.: One-round attribute-based key exchange in the multi-party setting. In: Chen, L., Han, J. (eds.) Provable Security, pp. 227–243. Springer, Cham (2016)

A Pairing Free Attribute-Based Authenticated Key Agreement Protocol Using ECC

Reshu Verma(✉) and Abhimanyu Kumar

National Institute of Technology, Uttarakhand, Srinagar 246174, India
{reshu.verma.cse17,abhi.kumar}@nituk.ac.in

Abstract. This paper presents a two party attribute based authenticated key agreement protocol without using bilinear pairing. Attribute based key agreement is required in several applications wherever a fine grained access control is required like cloud storage, military applications, broadcast encryption etc. In several applications a secure session between the participants is established without knowing their actual identities. In that case in place of identities a descriptive set of attributes (called as access policy) is used. The participants having sufficient set of attributes that satisfy the access policy are considered as authenticated and eligible for key establishment. The present paper uses Shamir Secret Sharing, Diffie Hellman Key exchange and elliptic curve arithmetic instead of bilinear pairing computations. The participant is authenticated based on the access policy defined over the set of attributes. The participant who satisfies the set of attributes perform Diffie Hellman key exchange with the other participant. Thus, a secure key is established between the participants. The authentication process of this protocol is information theoretically secure while the confidentiality rely on the intractability of Elliptic Curve Discrete Logarithm Problem (ECDLP). *abstract* environment.

Keywords: Attribute Based Encryption · ECC · Shamir secret sharing scheme · Diffie Hellman Key Exchange

1 Introduction

A key establishment scheme allows two (or more) number of participants to establish a secret common key (session key). Session key is required to establish a secure session via open public network in several applications. In some of the scenario one can often need to establish a secure session between the participants based on the possession of some specified set of attributes instead of their actual identity. In this scenario the participant's authentication is rely on the fulfilment of certain access control policy defined over a set of attributes. Satisfying the policy by a participant is equivalent to prove that they actually have the certain attribute-values over which the policy is defined.

N. Nain and S. K. Vipparthi (Eds.): ICIoTCT 2019, AISC 1122, pp. 105–114, 2020.
https://doi.org/10.1007/978-3-030-39875-0_11

The present paper proposes a key agreement protocol where both legitimate participants perform Diffie Hellman key exchange to jointly derive a common key. Here legitimacy of the participant is justified by possession of certain set of attributes defined by one of the participants. Thus the actual identity of the participants are need not be disclosed hence the protocol also provides the user anonymity features by default. The proposed protocol uses a novel approach to verify whether the participant satisfy the access policy or not and the key agreements is done between the participants only if they individually have sufficient attributes to satisfy the policy. As per the literature survey all existing attribute based protocols rely upon the computation of bilinear pairing in their operations. While the proposed protocol is free from bilinear pairing and uses elliptic curve arithmetic which is more faster and simpler than pairing.

In the proposed protocol one of the participants define the access policy based on necessary set of attributes and make it public. The protocol also needs an offline trusted authority called Key Generation Center (KGC) responsible for verifying the attributes of participants and computing attribute secret value for both the participants. The participant's authentication is done with the help of attribute secret value given by the KGC and lagrange interpolation of points.

The rest of the paper is organized as follows. In Sect. 2, related works on group key agreement protocol specially in attribute-based setting are discussed. The preliminaries related to proposed work have been addressed in Sect. 3. The Sect. 4 proposes the protocol, while Sects. 5 and 6 provides the performance comparison and security analysis of the proposed protocol respectively followed by the conclusion in Sect. 7.

2 Related Work

The key agreement protocol is used in establishing secure communication between two or more parties. Smart [1] presented the first two-party identity-based key agreement protocol (ID-KA) which uses the IBE scheme [2]. After that many identity based key establishment protocols are designed like [2–6] which has proved their security in different models. Later Sahai et al. [7] introduced the concept of Attribute Based Encryption (ABE) scheme as an special type of identity based cryptosystem, where the identity can be viewed as the set of descriptive attributes. Later et al. [8] extends the idea of ABE and introduces its two variants Key Policy Attribute-Based Encryption (KP-ABE) and Cipher Text Policy Attribute-Based Encryption (CP-ABE). From the literature, it is revealed that few key exchange protocols [9–12] under attribute-based authentication are available. Moreover, it is also noticed that two party attribute based key agreement protocols are also available, for e.g., [9,10]. In these protocols, two parties use attribute information of the other party instead of the traditional public key to generate the session key shared between them. These protocols uses Lagrange interpolation with bilinear pairing and generates private key corresponding to their descriptive attributes (functioned as his/her public key). However, these protocol can not be suitable to extend for multiple parties as these protocols are using bilinear pairing which increases the cost of computing common key. Recently, some protocols [13–16] are introduced based on CP-ABE scheme with efficient attribute revocation, which resists the user collusion attack and supports fine-grained access control. Some existing research proposals are available on attribute based group key management [17] for cloud computing mechanism, however, it does not focus on key establishment schemes.

Hence, a two party attribute based key agreement protocol has been proposed without bilinear pairing in order to decrease the complexity of computing common session key.

3 Preliminaries

In this section we define the preliminaries required for our proposed protocol.

3.1 Elliptic Curve Group

An elliptic curve E over a prime field F_p is denoted by E/F_p and defined by the equation $Y^2 = x^3 + ax + b$ where $a, b \in F_p$ An elliptic curve should be non singular and fulfil the following condition $\delta = (4a^3 + 27b^2)\ mod\ p \neq 0$ The points on E/F_p together with an extra point O called the point at infinity form a group based on the binary operation '+' called point addition operation defined between the two points on elliptic curve.

$$G = \{(x, y) \in E/F_p : x, y \in F_p\} \cup O$$

The detail description of Elliptic Curve Cryptography (ECC) can be found in various articles including [18, 19].

3.2 Diffie-Hellman Key Exchange

This protocol helps us to establish shared cryptographic key between the two users which helps in encryption and decryption of the message during communication over public channel [20].

3.3 Shamir Secret Sharing

It is a form of secret sharing, where a secret s is divided into n shares and shared among the shareholders in such a way that t or more that t shares are able to reconstruct the secret s but with less than t parts it cannot reconstruct the secret [21, 22]. Shamir secret sharing scheme is based on polynomial interpolation.

4 Proposed Protocol

Consider a set of n users $\{U_1, U_2, ..., U_n\}$ who wants to participant in the process of key establish based on an access policy defined over the set of attributes. All the user U_i $(1 \leq i \leq n)$ must register to an off-line trusted authority KGC to get the set of secrets corresponding to the attributes they have verified by the KGC. The proposed protocol can be defined in three phases given below:

4.1 Setup Phase

On taking a security parameter $k \in Z^+$ as input KGC choose a k bit prime p and determine the following tuple and publish it as public parameter

$$\{F_p, E/F_p, G, P, H, P_{pub}$$

where F_p is a prime field, E/F_p is an elliptic curve defined over F_p and P is a base point of E/F_p.

H is a hash cryptographic function defined as:

$$H : \{0,1\}^* \rightarrow Z_p^*$$

G represent an elliptic curve group as discussed in Sect. 3.

$P_{pub} = s.P$ where $s \in_R Z_p^*$ and keep s as master private key.

4.2 Attribute Secret Generation

- KGC calculates h_j for all the universal set of attributes $A = \{a_1, a_2, ..., a_m\}$ where m is the number of attributes in A as:

  $$h_j = H(a_j)$$

  for $1 \leq j \leq m$
- Every user U_i who wants to participates in the secure session must show their list of attributes to the KGC for validation.
- KGC first verify each of the attributes claimed by the user. The user needs to somehow prove that they actually holding the claimed attribute in front of KGC.
- After validating the attributes KGC chooses a random number r_i and calculate $R_i = r_i P$ for every user U_i.
- Now KGC computes the secret value for each verified attributes of every user as:
  - KGC computes $S_j = \frac{s.h_j \ mod \ p}{r_i}$ for each attribute belonging to user U_i where $1 \leq j \leq m_i$ and m_i is the number of attributes verified by KGC belonging to user U_i.
  - KGC delivers $\{S_j \forall 1 \leq j \leq m_i\}$ and R_i where $1 \leq i \leq n$ to the corresponding users U_i via a secure channel.
  - Every user can validate S_j received from KGC as $h_j.P_{pub} = S_j.R_i$ where $1 \leq i \leq n$

    The above relation is hold because

    $$h_j.sP = \frac{s.h_j \ mod \ p}{r_i}.r_iP$$

    $$sh_j.P = sh_j.P$$

4.3 Key Agreement Phase

Let there are two users Alice and Bob who wants to establish a common session key between them. The key agreement between them is done as follows:

Alice:

- Define the set of attributes needed for key establishment as: $\{A_1 = (a_1 \wedge a_2 \wedge a_3)\} \vee \{A_2 = (a_1 \wedge a_2 \wedge a_4 \wedge a_5)\} \vee \{...\} \vee \{A_w = (a_{w-2} \wedge a_{w-4} \wedge a_{w-1} \wedge a_w)\}$ where w is the number of set of attributes and make it publicly available.
- The index of attributes in the set $A_t; 1 \leq t \leq w$ are numbered from 1 to num where num is the number of attributes in the set A_t. So, points for constructing the polynomial is $\{(1, S_1), (1, S_2), (1, S_3)\}$ for set A_1 and $\{(1, S_1), (1, S_2), (1, S_4), (1, S_5)\}$ for set A_2 where $\{(S_1, S_2, ..., S_m)\}$ is the attribute secret value received from KGC. In the similar way points for constructing the polynomial is made for other sets of attributes.
- Construct polynomials P_t for each set of attributes A_t and compute

$$P_t(0) = \frac{a_t}{r_A}$$

  where r_A is received from KGC.
- Polynomial of degree $num - 1$ is to be constructed with each set of attributes where num is the number of attributes present in the set A_t.
- Multiply $P_t(0)$ with R_A received from KGC and get $P_t(0) = \frac{a_t}{r_A} r_A P = a_t P$.
- Choose a random number $x_1 \in Z_p^*$ and compute $x_1 P$.
- Now encrypt the packet C_t for each set of attributes A_t with the x coordinate of symmetric key $a_t P$ as: $C_t = Encrypt(a_t P + x_1 P); 1 \leq t \leq w$ and send this packet to Bob.

Bob:

- On receiving the encrypted packets from Alice, Bob chooses the set of attributes for which he is satisfying. Compute the polynomial P_t' with the attribute secret value S_j received from KGC for the set A_t he satisfies.
- Compute $P_t'(0) = \frac{a_t'}{r_B}$ where $P_t'(0)$ is the polynomial constructed by Bob and multiply it with R_B as

$$P_t'(0).R_B = \frac{a_t'}{r_B} r_B P = a_t' P$$

  where R_B is received from KGC.
- If $a_t' P == a_t P$, then decrypt the packet C_t belonging to the set of attributes A_t with the x coordinate of symmetric key $a_t P$.
- Compute inverse of the point $a_t P$ and add it with decrypted packet C_t as

$$(a_t P + x_1 P + (-a_t P)) = x_1 P$$

  and get $x_1 P$.

- Choose a random number $x_2 \in Z_p^*$ and multiply it with point x_1P and compute the common session key as:

$$K = x_1x_2P$$

- Now Bob sends the set of attribute A_t to which he satisfies and two packets $C_{b1} = (a_tP + x_2P)$, $C_{b2} = (a_tP + x_1P + x_2P)$ encrypted with a_tP, to Alice.

Alice:

- On receiving packets C_{b1} and C_{b2} from Bob, Alice see which set of attributes A_t Bob satisfies and choose a_tP for that set of attributes.
- Decrypt the received packets with the point a_tP, get (a_tP+x_2P) and $(a_tP+x_1P+x_2P)$.
- Compute inverse of a_tP and verify whether the packets came from the Bob or not as:
  - Add the inverse of a_tP in the packet C_{b1} as:

$$a_tP + x_2P + (-a_tP) = x_2P. \quad (1)$$

  - Add inverse of a_tP and x_2P in the packet C_{b2} as:

$$a_tP + x_1P + x_2P + (-a_tP) + (-x_2P) = x_1 \quad (2)$$

    If the received x_1P is same as point chosen by Alice then packets came from authenticated person i.e. Bob.
  - Now compute common session key as

$$K = x_1x_2P.$$

5 Performance Comparison

As per the literature survey made the proposed work is the first two party attribute based key agreement Protocol without bilinear pairing. However, this section also gives comparison of the proposed protocol with some other existing two party key establishment protocols in the Table 1. Based on the comparison given in Table 1 one can observe that proposed work have significant performance over the existing protocol and fulfill the necessary security features.

The overall computation cost of the proposed protocol is considered as low because as compare to other protocol it is free from bilinear pairing, more hashing requirement for the participants, modular exponentiations, etc. This protocol only require some scalar point multiplications over the elliptic curve which is a lighter operation as compared to the bilinear pairing, modular exponentiation, etc.

6 Security Analysis

This section explains that our proposed protocol is secure against different kind of attacks.

Table 1. Performance comparison with the existing protocols

Protocol	Security parameter	Operations required	Computation cost	Attribute based authentication
Wang et al. [9]	Bellare-Rogaway key agreement model	Bilinear pairing, lagrange interpolation	High	Yes
Wang et al. [10]	Bellare-Rogaway key agreement model	Bilinear pairing, lagrange interpolation	High	Yes
Li et al. [16]	aMSE-DDH	Bilinear maps	High	Yes
Proposed	ECDLP	Polynomial interpolation, ECPM, DHKE	Low	Yes

The Notations used in Table 1
ECPM: Elliptic Curve Scalar Point Multiplication
aMSE-DDH: Augmented multisequence of exponents decisional Diffie-Hellman
ECDLP: Elliptic Curve Discrete Logarithmic Problem
DHKE: Diffie Hellman Key Exchange

6.1 Correctness

An attribute-based key agreement protocol is correct if all the users are honest and they deliver all messages honestly. Every users having sufficient attributes to satisfy the access policy should be authenticated by KGC. Correctness of the protocol is rely on the authentication mechanism used. The entire process can be visualize as follows:

Alice construct the polynomial P_t for each set of attributes A_t of degree $num_i - 1$ where num_i is the number of attributes in the set of attributes A_t by interpolation of points. Let the following num_i points for the set A_t:

$$(x_1, \frac{y_1}{r_i}), (x_2, \frac{y_2}{r_i}), ..., (x_{num_i}, \frac{y_{num_i}}{r_i}) \tag{3}$$

U_i have to calculate a polynomial P_i' by interpolation of above points only as follows

$$\frac{y_1}{r_i}\left(\frac{x-x_2}{x_1-x_2} \cdot \frac{x-x_3}{x_1-x_3} \cdot \quad \cdots \quad \cdot \frac{x-x_{num_i}}{x_1-x_{num_i}}\right)$$
$$+\frac{y_2}{r_i}\left(\frac{x-x_1}{x_2-x_1} \cdot \frac{x-x_3}{x_2-x_3} \cdot \quad \cdots \quad \cdot \frac{x-x_{num_i}}{x_2-x_{num_i}}\right)$$
$$\vdots$$
$$+\frac{y_t}{r_i}\left(\frac{x-x_1}{x_t-x_1} \cdot \frac{x-x_3}{x_t-x_2} \cdot \quad \cdots \quad \cdot \frac{x-x_{num_i-1}}{x_{num_i}-x_{num_i-1}}\right)$$

Since the participant wants only the constant term of above expression i.e $P_i'(0)$ it can be calculated as:

$$P_i'(0) = \frac{y_1}{r_i}\left(\frac{x_2}{x_1-x_2} \cdot \frac{x_3}{x_1-x_3} \cdot \quad \dots \quad \cdot \frac{x_{num_i}}{x_1-x_{num_i}}\right)$$

$$+\frac{y_2}{r_i}\left(\frac{x_1}{x_2-x_1} \cdot \frac{x_3}{x_2-x_3} \cdot \quad \dots \quad \cdot \frac{x_{num_i}}{x_2-x_{num_i}}\right)$$

$$\vdots$$

$$+\frac{y_{num_i}}{r_i}\left(\frac{x_1}{x_t-x_1} \cdot \frac{x_3}{x_{num_i}-x_2} \cdot \quad \dots \quad \cdot \frac{x_{num_i-1}}{x_{num_i}-x_{num_i-1}}\right)$$

$$= \frac{P_i(0)}{r_i}$$

Thus user having the sufficient attributes to satisfy the access policy can compute the universal point.

$$K = P_i'(0).r_i = \frac{P_i'(0)}{r_i}.r_i.P = P_i'(0).P$$

Thus $P_i(0) = P_i'(0)$ The above computation is elliptic curve scalar point multiplication possible only when the value $P_i'(0)$ is scalar.

6.2 Implicit Key Authentication

Theorem 1. An attribute-based authenticated key agreement protocol provides implicit key authentication if each member is assured that no party outside the set can learn the secret key means that the member not satisfying the access structure can not reveal the secret key.

Proof: The proposed protocol satisfy the implicit key authentication as the participant not having sufficient attributes-values would not be able to construct the polynomials, so that the relation $P_i(0) = P_i'(0)$ can satisfy. Thus without getting $P_i(0)$ participant can not compute $K = x_1x_2P$ if intractability of ECDLP is hold in G.

6.3 Known Session Key Security

Theorem 2. A group key establishment protocol provides known session key security if each execution of the protocol should result in a unique secret session key. The compromise of one session key should not compromise other session keys.

Proof: In proposed protocol the random number r_i is chosen by trusted authority KGC and KGC is also computing S_j for all the verified attributes of the users. Hence the compromise of one session key should not effects on other session key.

6.4 Collusion Resistance

Theorem 3. An attribute-based authenticated group key establishment protocol provides collusion resistance if a set of members which is not satisfying the access policy, but having some partial attributes belonging to the access structure can not be collides in such a way they can be authenticated as a valid participant, even the union of their individual attribute are sufficient for an access structure.

Proof. The proposed protocol satisfies collusion resistance by dividing the attribute secrets by a different number r_i for each participants. The two participants may have same attribute-value, but KGC provide different secret value corresponding to their attribute. If the participants having partial attributes are trying to colludes they would not be able to construct the polynomials at user sides so that the relation $P_i(0) = P_i'(0)$ holds.

7 Conclusion

In this paper, we have presented a two party Attribute Based Authenticated key agreement protocol without pairings. The proposed protocol introduces a new idea of attribute based authentication by using Lagrange Interpolation to authenticate the participants based on the access policy defined over the set of attributes. The protocol uses some elliptic curve point multiplications which reduces its complexity as compare to pairing based protocols. The proposed protocol avoid the collusion attacks, without defining the separate set of polynomials for each users. The security analysis of the proposed protocol has been done, under required security attributes, and it is found secure against different kind of attacks. The performance comparison shows that the proposed protocol requires significantly less computation and communication cost with attribute based authentication as compare to the other protocols found in the literature.

References

1. Smart, N.P.: Identity-based authenticated key agreement protocol based on Weil pairing. Electron. Lett. **38**(13), 630–632 (2002)
2. Boneh, D., Franklin, M.: Identity-based encryption from the Weil pairing. In: Kilian, J. (ed.) Advances in Cryptology—CRYPTO 2001, pp. 213–229. Springer, Heidelberg (2001)
3. Shim, K.: Efficient id-based authenticated key agreement protocol based on Weil pairing. Electron. Lett. **39**(8), 653–654 (2003)
4. Wang, S., Cao, Z., Choo, K.-K.R., Wang, L.: An improved identity-based key agreement protocol and its security proof. Inf. Sci. **179**(3), 307–318 (2009)
5. Boyd, C., Cliff, Y., Gonzalez, Nieto J., Paterson, K.G.: Efficient one-round key exchange in the standard model. In: Mu, Y., Susilo, W., Seberry, J. (eds.) Information Security and Privacy, pp. 69–83. Springer, Heidelberg (2008)
6. Ateniese, G., Kirsch, J., Blanton, M.: Secret handshakes with dynamic and fuzzy matching, January 2007

7. Sahai, A., Waters, B.: Fuzzy identity-based encryption. Advances in Cryptology – EUROCRYPT 2005. Lecture Notes in Computer Science, vol. 3494, pp. 457–473. Springer, Heidelberg (2005)
8. Vipul, G., Pandey, O., Sahai, A., Waters, B.: Attribute-based encryption for fine-grained access control of encrypted data. In: Proceedings of the 13th ACM Conference on Computer and Communications Security, CCS 2006, pp. 89–98. ACM, New York (2006)
9. Wang, H., Xu, Q., Fu, X.: Two-party attribute-based key agreement protocol in the standard model
10. Wang, H., Xu, Q., Ban, T.: A provably secure two-party attribute-based key agreement protocol. In: 2009 Fifth International Conference on Intelligent Information Hiding and Multimedia Signal Processing, pp. 1042–1045, September 2009
11. Gorantla, M.C., Boyd, C., González Nieto, J.M.: Attribute-based authenticated key exchange. In: Steinfeld, R., Hawkes, P. (eds.) Information Security and Privacy, pp. 300–317. Springer, Heidelberg (2010)
12. Steinwandt, R., Corona, A.: Attribute-based group key establishment. IACR Cryptology ePrint Archive 2010:235, August 2010
13. Li, J., Yao, W., Zhang, Y., Qian, H., Han, J.: Flexible and fine-grained attribute-based data storage in cloud computing. IEEE Trans. Serv. Comput. **10**(5), 785–796 (2017)
14. Han, J., Susilo, W., Mu, Y., Zhou, J., Au, M.H.A.: Improving privacy and security in decentralized ciphertext-policy attribute-based encryption. IEEE Trans. Inf. Forensics Secur. **10**(3), 665–678 (2015)
15. Qian, H., Li, J., Zhang, Y., Han, J.: Privacy-preserving personal health record using multi-authority attribute-based encryption with revocation. Int. J. Inf. Secur. **14**(6), 487–497 (2015)
16. Hu, S., Li, J., Zhang, Y.: Two-party attribute-based key agreement protocol with constant-size ciphertext and key. Secur. Commun. Netw.
17. Bertino, E., Nabeel, M.: Poster: towards attribute based group key management, pp. 821–824 (2011)
18. Tate, J.T.: The arithmetic of elliptic curves. Inventiones Math. **23**(3), 179–206 (1974)
19. Bos, J.W., Halderman, J.A., Heninger, N., Moore, J., Naehrig, M., Wustrow, E.: Elliptic curve cryptography in practice. IACR Cryptology ePrint Archive, 2013:734 (2013)
20. Li, N.: Research on Diffie-Hellman key exchange protocol. In: 2010 2nd International Conference on Computer Engineering and Technology, vol. 4, pp. V4-634–V4-637, April 2010
21. Shamir, A.: How to share a secret. Commun. ACM **22**, 612–613 (1979)
22. Harn, L., Lin, C.: Authenticated group key transfer protocol based on secret sharing. IEEE Trans. Comput. **59**(6), 842–846 (2010)

Analysis of MEMS and Metamaterial Based Sensors and Its Involvement in Nanotechnology

Bhupendra Sharma[1](✉), Shraddha Gupta[1](✉), Ashwani Yadav[1](✉), and Rahul Runthala[2](✉)

[1] Amity School of Engineering and Technology, Amity University Rajasthan, Jaipur, India
bhupesh19@gmail.com, shraddhagupta10nov@gmail.com, ashwaniy2@gmail.com

[2] B.K. Birla Institute of Engineering and Technology, Pilani, Rajasthan, India
rahul.runthala@bkbiet.ac.in

Abstract. MEMS is very smart and highly demanding field of technology and wide range of applications are prevailing, it includes electrical and mechanical components with the micro range size. In this review paper different sensors are included with critical analysis of idea, Along with categorization of various sensors and their behavior enclosed with result analysis. The comparative study with various angles in terms of the sensing capability, application, and features with parameter Study in proper context. This paper is specifically focused on "metamaterial" based sensors it is newly generated technology and emerging field of research point of view that are the integral part of developing material, metamaterial as absorber, flexible metamaterial sensor domain technology with several applications. The major advantages, disadvantages and big challenges in the path of technology advancement have also been discussed in detail. Furthermore there is also expanded description of the applicability of MEMS and Metamaterial based sensors in versatile area like bionic sensors, structural health monitoring indicated in different fields with analytical manner.

Keywords: MEMS · Metamaterial · Accelerometer · Bionic sensor and structural health monitoring

1 Introduction

The objective of this work is to provide a short review on the recent trends technological sectors including different applications and MEMS is one of the most promising machinery with futuristic approach that makes life easy and safer because it is having wide range of application like air bag sensor, temperature analysis, pressure sensor, light sensor, security exploration, chemical detector, vacuum detector, door/window safety etc. It provides some mechanical respond after getting action in form of physical quantity from surrounding like heat, pressure, light, temperature etc. Moreover it is classified in various types according to their implementation. Accelerometer is a device that

N. Nain and S. K. Vipparthi (Eds.): ICIoTCT 2019, AISC 1122, pp. 115–125, 2020.
https://doi.org/10.1007/978-3-030-39875-0_12

identifies and measures the acceleration or rate of change of velocity. The gravitational acceleration plays a major role for determination of output.

One of the smartest technologies of Metamaterial is another newly explored area which has potential application in RF components for communication and sensing applications. These materials are artificial material with unusual electromagnetic properties that are not found naturally on earth. Every natural material have positive electrical permittivity & magnetic permeability, this is man-made structured materials termed as negative index materials (NIM) property which is very much important for high frequency electronics devices. With these uncommon material parameters, new kinds of miniaturized antennas and high frequency RF microwave circuit components or devices can be created for the wireless communications and the defense sectors, there are many more recent advancement in the field of metamaterial based sensing applications and property of metamaterial depends on the geometry of metamaterials elements rather than composition [1]. It has also extraordinary feature and conformal property, polyimide Kapton film used as a substrate, extensively used as a flexible Metamaterial absorber [6]. Today expansion of ultimate research on the MM methodology and recent approaches spreading in THz domain and Gold MM array fabricated on semiconductor substrate effectively controls the radiation [9]. High demanding properties of negative index artificial material is very significant for high frequency electronics devices because it provides high degree of sensitivity and resolution of sensors as well as explored new way of design features in sensor mechanism [1], it is defined as macroscopic composition having artificial, periodic elements are optimized arrangement in array structure. It is not available in nature, so embedded artificially in a specific host medium that will provide efficient desired results [34] and overcoming the intrinsic restrictions of conventional material as well as improving performance of existing devices with upgraded design by unique metamaterial properties [35]. The concentric loops are arranged in a certain way and each rings have cut, the capacitive loading creates resonating effects. When the dimensions are in the order of 1/10 of λ [34] this structure is called split ring resonator (SRR) (see in Fig. 1).

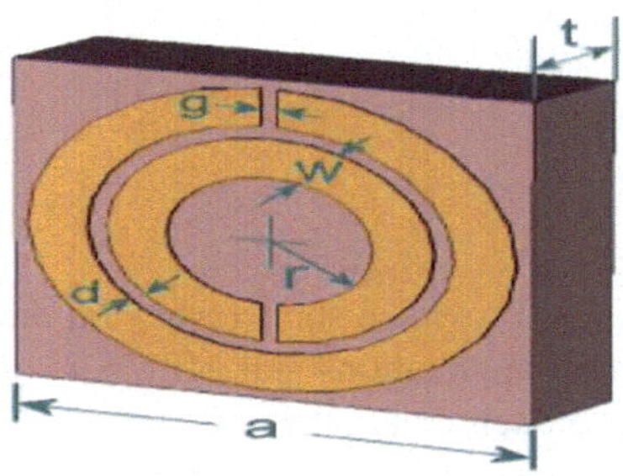

Fig. 1. Diagram of split ring resonator (SRR) [39]

2 Big Challenges

There are various big challenges in the field of MEMS technology and Metamaterial based sensors are to be discussed below-

- **MEMS sensors** - There are some challenges are facing along the attractive qualities of micro electro-mechanical system. The size reduction technique allows for higher process speed with less Responding time, reduction of switching activity. The whole batch process minimizes the manufacturing cost as well as improvement in reliability [14]. The micro mechanism operates at a size scale with typical mechanical devices that can generate some serious issue like friction and adhesion over the surface due to forces that leading sometime failure of the device [14] careful designing and testing is to reduce the unfavorable result during operation. Packaging of MEMS component done in well clean and dust free environment. The component should be uncontaminated, far from dusty domain. MEMS devices are very much sensitive, small amount of energy on the micro scale area of the sensor [12, 13]. It can also affected by temperature and vibration of molecules, thermal noise. Whenever the mechanical component starts motion during responding of electronics circuit must be clean and dust free.
- **Metamaterial Based Sensors** - The application of metamaterial in sensing area deliver better option than conventional technology, so the metamaterial field need advancement more in Mechanical, optical and Electromagnetic properties because it is based on mono molecular biosensor and high sensitivity and higher resolution features. It may face challenges such as fluctuations phenomena and it can be occurs by internal and external behavior of mechanism and surrounding, so the noise factor appears in sensor response [25]. External noise produced by shot noise [26], photo detector noise [27] and circuit operation. While intrinsic noise generated by Adsorption & Desorption (AD) effects it can be a part of major limiting factors in system performance, traditional photolithography technique is not suitable to apply [1], so advance fabrication steps should be involved with special material sensors creation.

3 Basic MEMS Sensor Mechanism

3.1 Accelerometer Function and Its Type

The common sensing mechanism of accelerometer depends on the capacitance according to the rate of Change in mass (see in Fig. 2). There is a string of fixed and movable Plates that vibration may be horizontal and vertical depends on the arrangement of structure.

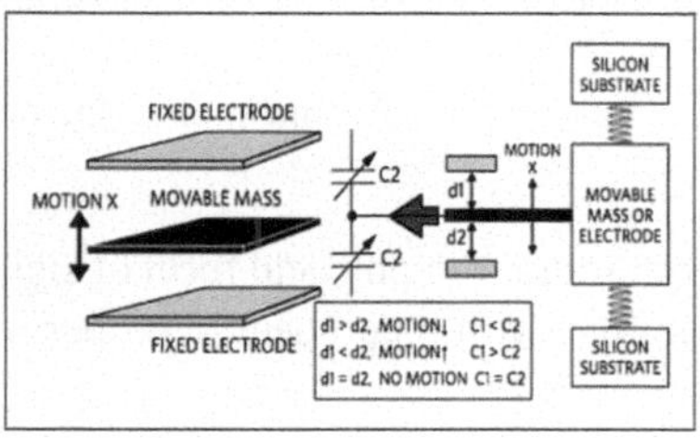

Fig. 2. Single moving mass accelerometer [7]

Accelerometer (Vertical Displacement)
Accelerometer vibration can cause this little mass to move around one it does shift the entire structure or when the count finger moves due to mass so it creates capacitance with respect distance changes as illustrate (see in Fig. 3). This electromechanical sensor that can sense the movement when the capacitance value varies and would be convert in the voltage form. A modern MEMS accelerometer having more fingures in arrangement then it increases the capacitance value as well as more sensitive and easy to detect. The suspended tiny weight vibrates the sensing system vertically and space changes in between plates it is very close but not touching to each other. The count finger diagram shown below there are fixed and movable fingers having gap in microns and the distance changes with the movement so respectively the capacitance changes and due to this the potential measured between the electrodes.

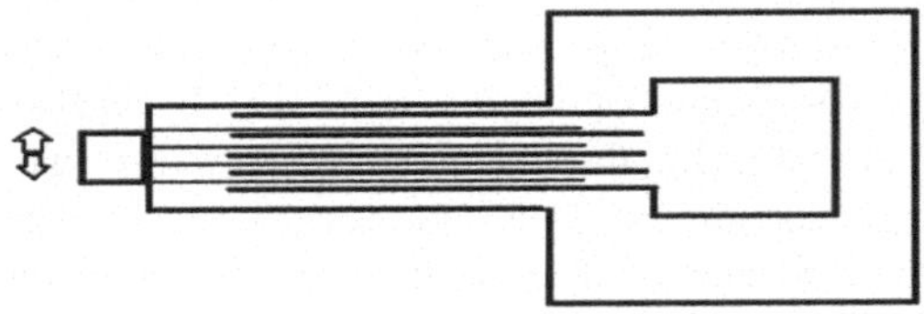

Fig. 3. Vertical displacement accelerometer

Accelerometer (Horizontal Displacement)
Accelerometer specifically measure the movement of horizontal axis. It basically works under various parameter like sensitivity bandwidth, dynamic range, frequency response [7].

Its movement is in horizontal direction the surface area changes between fingers, so due to displacement as represent (see Fig. 4). Capacitance changes that mechanism converted in readable voltage form by the help of signal conditioning process.

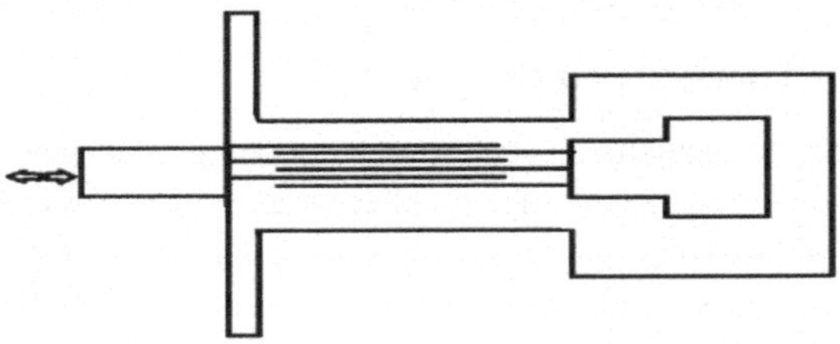

Fig. 4. Horizontal displacement of accelerometer

There are different energy sources in physical form of signals and various sensors to be used at particular applications with their specific properties are shown in following Table 1.

Table 1. Sensors analogy with different energy sources

Energy	Physical quantity	Application area
Mechanical [33]	Length, area, volume, acceleration	Inductive sensor, ultra sonic sensor
Thermal [33]	Temperature, entropy, heat flow, state of matter	RTD, Thermocouple, liquid flow sensor
Electrical [33]	Voltage, current, charge, resistance, inductance, capacitance, dielectric electric field	Voltage sensor, 0–25 V, capacitive touch sensor, DARE's frequency sensor
Electromagnetic [33]	Field intensity, flux density, magnetic moment, permeability	Wireless Door/window sensor, embedded fiber optical sensor, hall effect sensor
Chemical	Composition, concentration, reaction rate, pH	Soil moisture sensor, digital pH tester, clinical analyzer, IR gas sensor
Humidity [40]	Moisture and air temperature, environment humidity	Hygrometer, MEMS based humidity sensor, RFID humidity sensor
Acoustic	Sound detector, phonic	Microphone, piezoelectric resonator
Vibration [29]	Measure small shake, fluctuations	Accelerometer, mass air flow sensor

4 Methodology and Detailed Application of Different Sensors

4.1 Bionic Sensor

It is fundamentally used in biological application the system is much efficient in study and design of various physical analysis for detailed examine of body organs with highly efficient MEMS devices & sensors are wearable the detailed study of bionic sensors are shown in Table 2. Detection of human breath profile and find out for uncertain metabolites existence. It is biomarkers or chemical bio indicator that can diagnose disease, metabolic disturbances and provides health status of human being with faster response and effective treatment methodology of sensing technology [11, 15] as mention in Table 2. The advanced bionic technology enhance the capability of patient to regain the independence and confidence level increases the quality of life with happiness and one and only reason behind this is sensor technology and its feasibility. To collect information from the physical objects this analysis can be done by wearable accelerometer for determination or monitoring of physical activity in daily life [10]. Robust and adaptive healthcare systems are on the first priority to each and every human being so the physical health monitoring is one of the most prior and basic need. The new generation MEMS technology has various features like wireless connection with wearable sensor that can generate warning and communicate directly to the Doctor as well as patient during critical condition [16]. Along with here also specified the recently fashioned technology using distinct non-natural material based sensor with high efficiency and well performed behavior of metamaterial bio-sensor displayed from various sources are to be highlighted in following Table 2. It is an analytical idea of study with MEMS sensor as well as Metamaterial based applications.

4.2 Structural Health Monitoring

The particular wireless sensors are extensively used in buildings, dam, residential or heavy civil and mechanical projects. Structural health monitoring system provides the valuable knowledge of real time data and behavior of heavy construction stressed. It recognize and analyze real time structural health monitoring and collects the minute strain & damage signal from ageing or earthquake and this vibration response measures the primary data signals and sends to multiple post processing techniques [19]. MEMS technology also connected with wireless communication feature and auto warning signals is to be generated even during initial critical situation arises and inform to the central data center and it is make easy for to run all the system after the involvement of wireless technology with sensing applications along with metamaterial based sensors used for SHM are discussed as given in Table 3. It is an advanced technology that informs or detects little damages and categorized the approach for it with organized fashion. The metamaterial (MM) based sensor behaves as a unparalleled sensor or actuators with continue monitoring the input and output behavior with advanced monitoring of structure [28, 29].it is used as surface mounted or inside the structure [30] The following Table 3 provides a critical analysis about structural health monitoring in different domain.

5 Advantages of Modern Technology

5.1 Micro-Electro-Mechanical System (MEMS) Sensors

MEMS are made up of various components and techniques, the size of MEMS are lies 1 to 100 micrometers as the size consistently reducing using higher technical methodologies are using in micromachining process but the productive outcome by these applications are increasing day by day. Its size is just comparable to the wavelength of light.

5.2 Metamaterial Based Sensors

Metamaterial increase its demand and importance in various feasible applications such as medical field, industrial operations and in defense machinery due to higher sensitivity and resolution behavior as compared to traditional sensing structures. It have wider application range in near future by its unmatched and specialized applications like "super" lense [5], cloaking [22], negative refraction optics [23]. A "cloak" is basically a wrapper that encloses the object such that it (the cloak) and the target become "invisible" to an electromagnetic detecting device [24].the metamaterial based sensor is very popular candidate which elevated curiosity among various recent trends applications [31]. It is neither actuators nor sensors but is a designed pattern of array of patches shape on the surface of host assembly to be monitored [22].

Table 2. Observational details of bionic sensor

	Application area	Object	Sensing element	Features	Disease/Data detected by sensor	Sensor/Platform
MEMS based sensor	Optical [16]	Eye	Light	Wireless contact [16]	Diabetic retinopathy, dryness change in eye vision, blurriness [16]	IR sensor
	Acoustic [18]	Throat	Sound	Embedded system, high-fidelity throat microphone	Food intake recognition, chewing information	Proximity base ultrasonic sensor
	Olfactory [8]	Nose	Smell	Human breath gas analysis [15]	Electronic aroma detection [11]	Multi sensor of lactometer [8]
	Neural [17]	Skin	Erythemato-scuamous	Artificial neural network	Psoriasis, seborrheic dermatitis, pityriasis	Artificial neurons
Metamaterial based sensors	Cancer cell [34, 36]	Human body cells/Tissues	Tumor/Cells	Metamaterial element-OMEGA shape of ring resonator [36], Sense Fatal cells	Detect minor changes in water content of tissue can produce changes in permittivity and conductivity value [34], malignant tissue has higher water content than normal.	Biosensor resonator (Metamaterial base resonator operates in (GHz) [36]
	Hyperthermia [37]	Breast cancer	Control dose of heat to infected cell	Hyperthermia process-frequency uses according to stages of cancer (early stage with 6 mm tumor apply 6 GHz–41.6 V/cm source amplitude. with large focusing spot and 42 °C tumor temp. for 60 min), (late stage with 6 mm tumor apply 2.45 GHz with 53.5 V/cm source amplitude 42 °C temp. of tumor for 60 min [37]	Infected cells heated (above 42 °C) without heating surrounding normal or healthy cells [37], low loss metamaterial lenses, One flat LHM lenses different from conventional convex lenses [37], Four flat LHM lenses with four sources, special shape LHM lens applicator is more flexible to accommodate and easy to adjust focal point of source without varying LHM lenses [37]	Left handed metamaterial (LHM) lenses, microwave higher frequency range (6 GHz)
	Endoscopy [38]	Gastrointestinal monitoring/treatment [38]	Gastrointestinal monitoring	wireless endoscope with high-performance radio system for real-time data transmission, CSRR folded patch antenna, flexible substrate- RT/Duroid 5880	Measurement is performed using an Agilent E5071C VNA between 1 and 3 GHz, Return loss (S_{11}) = 11.7 dB, A 74% size reduction is attained after CSRR loading, f = 2.4 GHz [38]	Patch antenna loaded with CSRR fabricated on flexible substrate cylindrical shape

Table 3. Observational Detail of Structural Health Monitoring

	Application area	Sensing element/sensitivity	Features/Operational elements	Data observed	Sensor/Platform
MEMS Based sensor	Building oscillation	Displacement/acceleration	Wireless contact IEEE 802.15.4 and GPS [19]	Synchronized signal collection from different nodes	Colibris –MS9002 Single axis capacitive accelerometer [19]
	Infrastructure/Bridge	Inside the concrete structure of building	Working frequency is >30 kHz, too sensitive for vibration, self-generated and self-received signal [20]	Overcomes the defects of smart film, coaxial cable and camera lenses. respond faster and stable [20]	Piezoelectric element array [21]
	Wind Turbine [2]	Tested on Wind turbine	Sensitivity – 203.96 m $V.m^{-1}.s^2$, ADC-16bit, BW- 0.03 to 25 Hz	Amplitude detection, deflection shape identification	Accelerometer CXL01LF [2]
	Bridge health monitor [3]	Accelerometer sensor and ultra sonic sensor [3]	Wireless sensor-ADXL335, UART PIC16F877A	Pressure, level of water, acceleration information on android application system	Accelerometer sensor ADXL335 [3]
Metamaterial Based Sensors	Structural monitoring [4]	Contraction or elongation of building, probe is inserted in to structure	Comb like structure of Multiple Nested Split Ring Resonator (NSRR), measurement performed in CST software	This set is to detect Near field excitation response, high resolution (<1 μm), large dynamic range (over 7 mm), $S_{11} < 10$ dB, f = 400 MHz, high linearity ($R^2 > 0.99$ to 5 mm), sensitivity (>12.7 MHz/mm in 1-3 mm range) [4]	Displacement sensor, Rogers Duroid substrate ($\varepsilon_r = 3.2$),
	Focused on Structural monitoring – particularly near field experimental set up (leakage absent) [32] For far field it is essential to set up an anechoic chamber to control the reflections	Contact MM monitoring-circular MM unit Cell placed on the center of beam, Non-Contact MM monitoring set up-placed at certain distance below beam	Split ring resonator, VNA device for electric field measurement at output, customized software for frequency Vs transmission signal (S_{21}), GPIB cable, Spectrum analyzer, near field scanner, MM sample, Al-beam (32 cm. * 2.5 cm * 3 mm), medium – Air	Frequency in GHz or THz Range, S_{21} parameter, load monitoring – (50–80 N), wireless function, High sensitivity [32]	For both contact and non-contact experimental base setup MM unit acts as strain sensor and recognize as strain gauges or optical fiber strain sensor [32]

6 Conclusion

This review paper concept is fully inclined towards the new approaches of nanotechnology and its applications like MEMS and metamaterial based sensors. Different parameters have been covered along with analytical fashion. Both of the portions included intense study about different parameters with broad effective idea in diverse fields. Most of the important issues about nanotechnology sector have also been discussed with detail. This paper represents critical analysis of review of metamaterial base sensor applications and concentrated on analysis of new applications with smart materials.

References

1. Chen, T., Li, S., Sun, H.: Metamaterial application in sensing. Sensors **12**, 2742–2765 (2012). https://doi.org/10.3390/S120302742. ISSN 14248220
2. Swartz, R.A., Lynch, J.P., Zerbst, S., Sweetman, B., Rolfes, R.: Structural monitoring of wind turbines using wireless sensor networks. Smart Struct. Syst. **6**(3), 183–196 (2010)
3. Patil, P.K., Patil, S.R.: Structural health monitoring system using WSN for bridges. In: International Conference on Intelligent Computing and Control Systems, ICICCS. IEEE (2017). ISBN: 978-1-5386-2745-7/17
4. Ozbey, B., Unal, E.: Wireless displacement sensing enabled by metamaterial probes for remote structural health monitoring. Sensors **14**, 1691–1704 (2014). https://doi.org/10.3390/s14010169. ISSN 1424–8220
5. Liu, Z.W., Fang, N., Yen, T.J., Zhang, X.: Rapid growth of evanescent wave by a silver superlens. Appl. Phys. Lett. **83**, 5184–5186 (2003)
6. Griguer, H., Tentzeris, M.M., Nauroze, A., Drissi, M.: A novel ultra-thin flexible metamaterial absorber for human body protection from EMF hazards. In: 32nd URSI GASS, Montreal, 19–26 August 2017 (2017)
7. Dadafshar, M.: Accelerometer and Gyroscope sensors: operation, sensing and application, Maximum integrated (2014). http://pdfserv.maximintegrated.com/en/an/AN5830.pdf
8. del Cueto Belchi, A., Garcia Rodriguez, D., Rothpfeffer, N., Pelegri Sebastiá, J., Chilo, J.: Multi-sensor olfactory system. In: 2012 IEEE International Instrumentation and Measurement Technology Conference Proceedings, Graz, pp. 1139–1141 (2012)
9. Liu, R., Ji, C., Zhao, Z., Zhou, T.: Advance materials and material genome-review metamaterial: reshape and rethink (2015). https://doi.org/10.15315/j-eng-2015036
10. Gervais-Ducouret, S.: Next smart sensors generation. IEEE (2011). ISBN: 978-1-4244-8064-7/11/2011
11. Yuce, M.R.: Recent wireless body sensors: design and implementation. In: IEEE MTT-S International Microwave Workshop Series on RF and Wireless Technologies for Biomedical and Healthcare Applications (IMWS-BIO) (2013)
12. Wilson, A.D., Baietto, M.: Applications and advances in electronic-nose technologies. Sensors **9**, 5099–5148 (2009). ISSN 1424-8220
13. Pister, K.S.J.: Lecture notes "On the Limits and Applications of MEMS Sensor Networks", UC Berkeley. http://datasys.cs.iit.edu/reports/2002_WSU_csc8800_2002.pdf. Accessed 29 Apr 2018
14. Raicu, I.: MEMS a technology overview and limitations. In: CSC8800, p. 122, September 2004. https://compliantmechanisms.byu.edu/comtent/introduction-microelectromechanical-systemmemsComppiantMechanicalResearch. Brigham Young University, USA
15. Wilson, A.D.: Advances in electronic-nose technologies for the detection of volatile biomarker metabolites in the human breath. Metabolites **5**(140–163), 140–163 (2015)

16. Anitha, M.: Sensor grid based vision status monitoring in eye care system. Int. J. Future Comput. Commun. **1**(1), 57–61 (2012)
17. Filimon, D.-M.: Skin diseases diagnosis using artificial neural networks. In: 9th IEEE International Symposium on Applied Computational Intelligence and Informatics, Timisoara, Romania, 15–17 May 2014, pp. 189–194 (2014)
18. Bi, Y., Lv, M.: AutoDietary: a wearable acoustic sensor system for food intake recognition in daily life. IEEE Sens. J. **16**(3), 806–816 (2016)
19. Pérez, J.J., Saldarriaga, A.J., Bustamante, J.: A wireless body sensor network platform to measure vital signs in clinical monitoring. In: 2013 Pan American Health Care Exchanges (PAHCE), Medellin, Colombia, 29 April–4 May 2013 (2013). IEEE Catalog Number: CFP1318G-art, ISBN: 978-1-4673-6257
20. Giammarini, M., Isidori, D., Pieralisi, M.: Design of wireless sensor network for real-time structural health monitoring. In: IEEE 18th International Symposium on Design and Diagnostics of Electronic Circuits and Systems. IEEE (2015). https://doi.org/10.1109/ddecs.2015.45. ISBN: 978-1-4799-6780-3/2015
21. Li, X., Cui, H.: Experimental study of structural health monitoring methods based on piezoelectric element array. IEEE (2017). ISBN: 978-5090-5363-6/17
22. Schurig, D., Mock, J.J., Justice, B.J., Cummer, S.A., Pendry, J.B., Starr, A.F., Smith, D.R.: Metamaterial electromagnetic cloak at microwave frequencies. Science **314**, 977–980 (2006)
23. Pendry, J.B., Holden, A.J., Robbins, D.J., Stewart, W.J.: Low frequency plasmons in thin-wire structures. J. Phys.: Condens. Matter **10**, 4785–4809 (1998)
24. Marcus, S.W., Schwartz, C., Naor, M.: Metamaterials and their applications (2009). Manor division, Rafael, Haifa, Israel
25. Jakšic, Z., Jakšic, O., Matovic, J.: Performance limits to the operation of nanoplasmonic chemical sensors-noise equivalent refractive index and detectivity. J. Nanophoton **3**, 031770 (2009)
26. Homola, J., Piliarik, M.: Surface plasmon resonance (SPR) sensors. In: Homola, J., (ed.) Surface Plasmon Resonance Based Sensors, vol. 159, pp. 45–67. Springer, Berlin (2006)
27. Jakšić, Z., Djurić, Z.: Cavity enhancement of auger-suppressed detectors: a way to background-limited room-temperature operation in 3–14 μm range. IEEE J. Sel. Top. Quantum Electron. **10**, 771–776 (2004)
28. Annamdas, V.G.M., Soh, C.K.: Application of electromechanical impedance technique for engineering structures: review and future issues. J. Intell. Mater. Syst. Struct. **21**(1), 41–59 (2010)
29. Annamdas, V.G.M., Radhika, M.A.: Electromechanical impedance of piezoelectric transducers for monitoring metallic and non metallic structures: a review of wired, wireless and energy harvesting methods. J. Intell. Mater. Syst. Struct. **24**(9), 1019–1040 (2013)
30. Cho, S., Yun, C.B., Lynch, J.P., Zimmerman, A.T., Spencer Jr., B.F., Nagayama, T.: Smart wireless sensor technology for structural health monitoring of civil structures. Steel Struct. **8**, 267–275 (2008)
31. Smith, D.R.: A cloaking coating for murky media. Science **345**, 384 (2014)
32. Annamdas, V.G.M., Soh, C.K.: Influence of loading on the near field based passive metamaterial. Struct. Health Monit. (2015). https://doi.org/10.12783/SHM2015/81
33. Chidambaram, R.: Material Research and Technology Foresight, S.B. Krupanidhi, MRC, IISc, Bangalore Smart Sensors and Actuators page no. 543
34. Hosseinzadeh, H.R.S., Hosseinzadeh, H.: Metamaterials in medicine: a new era for future orthopedics. Ortho. Res. Online J. **2**(5) (2018). https://doi.org/10.31031/oproj.2018.02.000549
35. Bilotti, F., Sevgi, L.: Metamaterials definitions, properties, applications, and FDTD-based modeling and simulation. Int. J. RF Microw. Comput. Aided Eng. **22**(4) (2012). https://doi.org/10.1002/mmce

36. La Spada, L., Bilotti, F., Vegni, L.: Metamaterial biosensor for cancer detection (2011). ISBN: 978-1-4244-9289-3/11/$26.00 ©IEEE
37. Wang, G., Gong, Y.: Metamaterial lens applicator for microwave hyperthermia of breast cancer. Int. J. Hyperth. **25**(6), 434–445 (2009). https://doi.org/10.1080/02656730903061609
38. Cheng, X., David, E., Kim, C., Yoon, Y.-K.: A compact omnidirectional self-packaged patch antenna with complementary split-ring resonator loading for wireless endoscope applications. IEEE Antennas Wirel. Propag. Lett. **10**, 1532–1535 (2011)
39. Castro, P.J., Barroso, J.J., Leite Neto, J.P.: Experimental Study on split ring Resonator with different slit widths. J. Electromagn. Anal. Appl. **5**, 366 (2013). https://doi.org/10.4236/jemaa.59058
40. Karmakar, N.C., Amin, E.M., Saha, J.K.: Chipless RFID sensor for real time environment monitoring (2016). https://doi.org/10.1002/9781119078104.ch7

High Gain Patch Array Antenna for 5G Network Communication and IoT Applications

Kishana Ram Kashwan(✉)

Madanapalle Institute of Technology and Science, Angallu, Madanapalle 517325, AP, India
dean-ece@mits.ac.in

Abstract. A new array of patch antenna is proposed in this paper for upcoming 5G networks and IoT applications. The operating frequencies are optimized at 4 different bands ranging from 30 GHz to 50 GHz. Such design is suitable for 5G communication applications for Multiple Input Multiple Output (MIMO) systems. The proposed antenna has a simple structure and it consists of two patches separated by a distance which can be optimized for required different types of parametric combinations. A single point feed is divided by using the Wilkinson's power divider model to split or combine the power transmitted to or the power received from two patch elements respectively. The designed array has shown comparatively high performance and better directional radiation patterns. The four operating frequencies are observed at 31.0 GHz, 36.4 GHz, 43.4 GHz and 48.5 GHz with simulated return losses observed with approximate values at −44.9 dB, −33.4 dB, −29.2 dB and −23.9 dB respectively. The simulated results show that the efficiency varies from 18% to 23%. The VSWR is very close to unity indicating a good performance of the antenna array. The performance is encouraging and the array antenna is suitable for the applications in upcoming 5G network communication and other IoT applications.

Keywords: Patch antenna · 5G · IoT · High directivity · High gain

1 Introduction

1.1 Patch Antenna Preamble

The patch antennae are used for a number of applications for well-known and well documented advantages these have over other type of antennae [1]. Many authors have extensively explained about the features of patch antenna with benefits of the same. These includes ruggedness, compactness, low cost, easy to fabricate and replicate the same, high gain and directional properties, low maintenance and low failure rates, low power ratings etc. [2–22]. Patch antennas are also more suitable for modern communication technologies which need compact and low power consumption invariably in addition to all advantages of patch antenna in general. Patch antennas are specifically suitable for very high frequency operations such as millimeter wave communications in different generation of communications. The newer applications of patch antennas

N. Nain and S. K. Vipparthi (Eds.): ICIoTCT 2019, AISC 1122, pp. 126–132, 2020.
https://doi.org/10.1007/978-3-030-39875-0_13

are emerging quite frequently in recent times which includes the wearable electronics, medical diagnostics, body implantations, RF IDs, security and surveillance at a very high data transfer speeds [2–22]. As the technology for communication advances, more challenging designs and performance levels are desirable for meeting the expectations of industry [23].

1.2 Patch Antenna for 5G and IoT Applications

The 4G communication uses broadband with mobile communication at the center of interest. At that time, there was very little need felt by 4G communication developers to address issues of IoT and wearable electronics transmitting big data round the clock. The reason was that these requirements evolved initially slowly but suddenly rose exponentially after introduction of 4G communications protocols. These issues are now core of interest in 5G communications. The 5G communications is, therefore, required to handle from a very narrowband to a very broadband communication. The multiple input multiple output is the core of the 5G communications. It will require to handle extremely high speed data communications.

The 5G communication will require many times more capacity per area, more spectrum, more base stations and more spectral efficiency than Long Term Evolution (LTE) communication technology [23]. MIMO systems consists tens of hundreds of individually controlled antenna systems to handle massive data at very large speeds [23]. The antenna elements are used to multiplex data simultaneously to many users in same time frequency, called as spatial multiplexing [23]. This is specifically useful for intra and inter cell interferences. The theory applies for the IoT communications as well.

This research paper is an effort towards design of a two element array of patch antenna for multiple frequency bands. The new design is a step forward for single antenna MIMO System with multiple array element antenna on same substrate to realize the goad of 5G and IoT communications.

2 Theory of Proposed Patch Antenna Array

A two element antenna array is simulated on a common substrate for the better capacity and parametric gains. The simulation is done for a substrate of FR-4 with thickness of 0.8 mm. It has a simple design to receive quick results for desired frequency bands. The strip and the ground planes are simulated with copper as an ideal conductor with thickness of 17.5 μm. These dimensions are selected based on commercially available materials for fabrications with ease. The basic design with all dimensions mentioned is shown in Fig. 1.

The strip feed is designed as shown in the Fig. 1, by employing Wilkinson's power divider model [24]. The strip is divided into two symmetrical arms such that signal is divided over two arms equal in amplitude and in phase with each other leading to two antenna elements separately. Since both the feed strips to two antennas are at equal potential and therefore these are isolated from each other. The dimensions are selected such that these arms leading to the antenna elements are equal to the quarter-wave length transformers of the impedance. This leads to the impedance matching and also

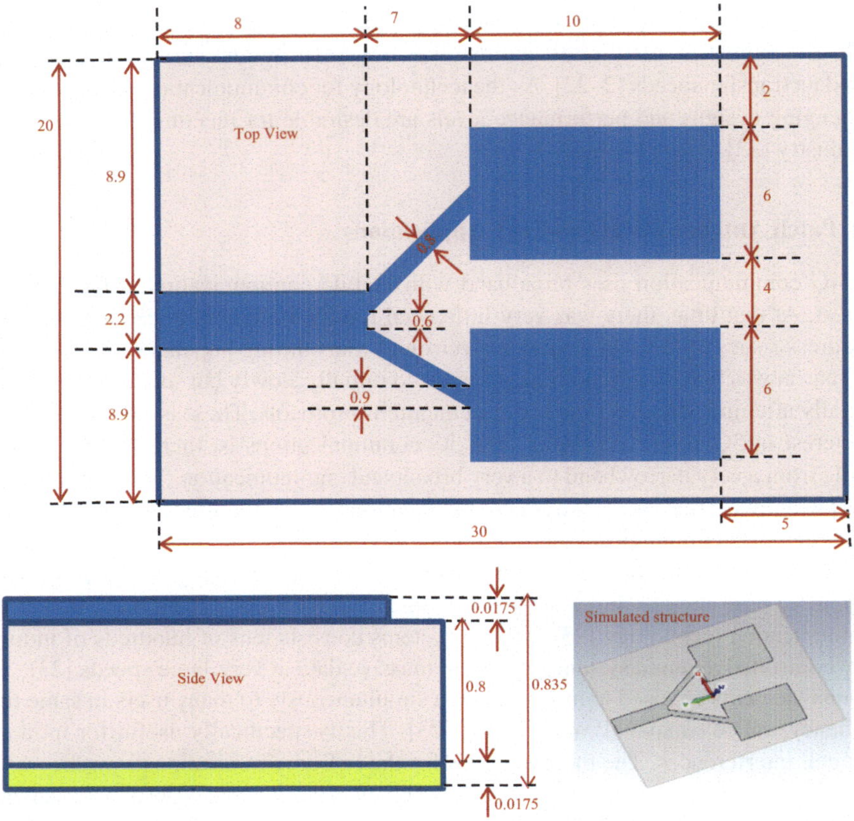

Fig. 1. A dimensional and structural view of design of two-element antenna array for MIMO and IoT communication systems. All dimensions indicated are in mm.

attenuations are almost insignificant, theoretically zero but in practice with very small values. It also leads to negligible resistive losses due to combination or split of the power at the junctions. This works as a three port network with two ports on antenna element side are at the same potential. The proposed antenna design is simulated on CST studio software using frequency and time domain analysis for the result reported.

The two elements are used to improve the performance in multiple ways. Firstly, these can be multidirectional and optimized individually for radiated transmissions for spatial multiplexing. Secondly these can be individually controlled for multiple transmissions for beam forming and coherently augmenting signal strengths at desired receivers.

This research work is to demonstrate the principal of spatial multiplexing signals for augmentation at receiver end which will support massive data quantity and data speeds for transmission in 5G and IoT applications. The same principle can be applied to increase the number of antenna elements in order to further achieve far superior results. Ideally, for MIMO applications in communications, hundreds of antenna elements may be required to make arrays for high end applications [23].

3 Simulated Results and Discussions

The results are encouraging with significant improvements observed in the performance of array for transmissions. Figure 2 shows the return loss (reflection coefficient s_{11}) for four frequency bands as indicated by markers on the graph.

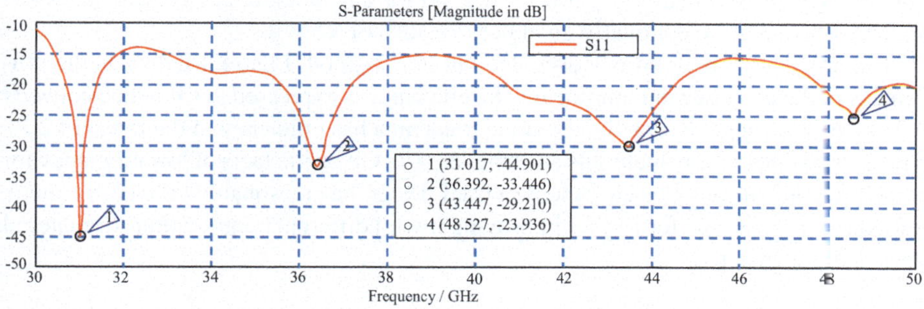

Fig. 2. The simulated results of two-element array antenna at four different frequency bands with return loss (s_{11}) as indicated in box.

The bandwidth for the resonance frequencies can be estimated easily by considering upper and lower half power frequencies. The wavelength for the resonance frequencies varies approximately from 0.9 mm to 0.6 mm which is normally called millimeter wave range. The 5G communication is destined to be of millimeter wavelength order.

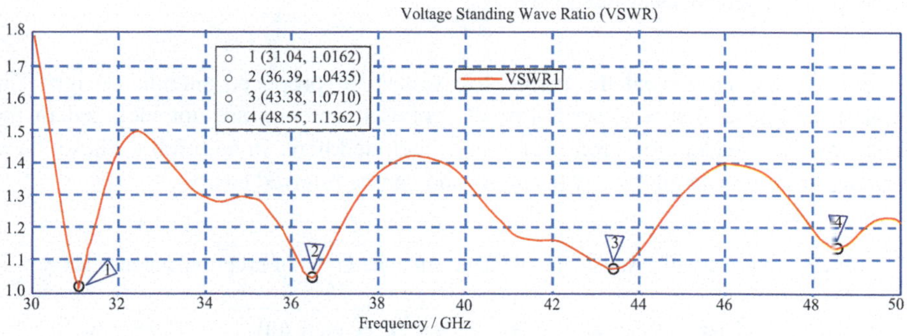

Fig. 3. The simulated results of two-element array antenna at four resonance frequency bands with voltage standing wave ratio (VSWR) as indicated in box.

Figure 3 shows the relation between resonance frequency and the voltage standing wave ratio (VSWR). The VSWR is expressed as the ratio of maximum to minimum voltage on a lossless line and is a number without any unit. It indicates the mismatch between load impedance and that of impedance of transmission lines or feedlines. The value of VSWR at perfect impedance match is 1 i.e. indicating no reflection wave and worst case scenario is at infinite value of VSWR at which complete reflection happens.

At infinite value of VSWR, the complete wave is reflected back to the source and the load is not able to absorb any power. Figure 3 shows simulated values of VSWR for the proposed antenna array for intended frequencies. The values are very close to unity for first three resonance frequencies and is also reasonable for the fourth resonance frequency. This validates the design for its superior performance of intended antenna array and makes strong candidate for MIMO suitability. Normally, as many literatures indicate, VSWR less than 2 is considered good for any antenna design. The proposed antenna has shown exceptionally good performance of VSWR.

The efficiency observed is reasonable but not so good. Figure 4 indicates the efficiencies of the proposed antenna array. The efficiency is expressed in dB which converts to 23% to 18% range. Normally, the smaller antenna has efficiency in the range of 20% to 60% or so due to lossy materials and smaller size. Given the facts of low efficiency for smaller antenna in mobiles etc., the proposed design has reasonable efficiency. There, however, exists a scope for further improving it with different combinations of material types and dimensions.

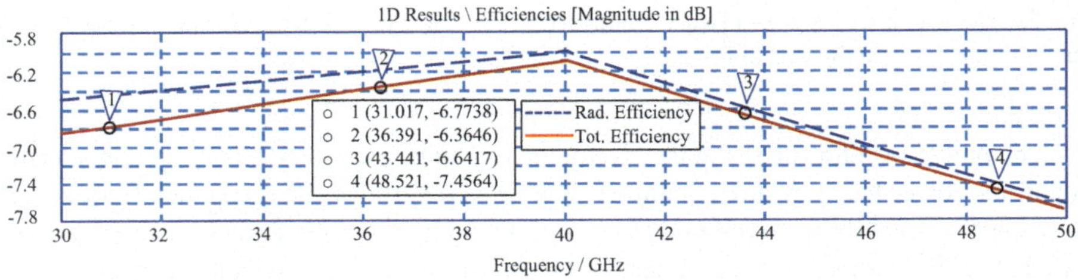

Fig. 4. The simulated results of two-element array antenna at four resonance frequency bands with radian and total efficiency indicated in box.

Table 1 summarized all the parametric values for proposed antenna design. The standard analytical theory descriptions of proposed antenna are not included in the paper, however, readers may refer references included here. In addition to these, there are many literatures available for the same and can be referred easily.

Table 1. Summary of results obtained from simulated tests of proposed antenna array

Frequencies (GHz)	Return loss (dB)	VSWR	Efficiency (dB)	Efficiency (%)
31.0	−44.9	1.01	−6.77	21.04
36.4	−33.4	1.04	−6.36	23.12
43.4	−29.2	1.07	−6.64	21.68
48.5	−23.9	1.13	−7.45	18.00

The proposed design is only simulated and has not been fabricated. It therefore has no validation scope available for the measurement. This forms the part of the future research work with extension of fabrication and validations.

4 Conclusion and Future Scope

This research paper has proposed a new two-element patch antenna array for 5G MIMO and IoT communication applications. A very simple and efficient design is tested by simulating tests. The simulated results show promising outcomes which are suitable for the intended applications of the antenna array. Results are tested at four observed resonance frequencies ranging from 30 GHz to 50 GHz. The antenna array operates in millimeter wave communications. Its return loss varies between −23 dB to −45 dB and VSWR is very close to unity. It has shown efficiency from 18% to more than 23%. The future work gives direction to improve upon efficiency and also test the fabricated antenna array for real time environment operations. This will validate the practical applications and usages of the antenna array.

References

1. Andriamiharivolamenaa, T., Lemaître-Augera, P., Tedjinia, S., Tirard, F.: Compact planar monopole antenna for wearable wireless applications. C. R. Physique **16**(9), 851–851 (2015). https://doi.org/10.1016/j.crhy.2015.07.008
2. Chenga, H., Ebadib, S., Renc, X., Gonga, X.: Wireless passive high-temperature sensor based on multifunctional reflective patch antenna up to 1050 degrees centigrade. Sens. Actuators A Physical **222**, 204–211 (2015). https://doi.org/10.1016/j.sna.2014.11.010
3. Emadeddin, A., Shad, S., Rahimian, Z., Hassani, H.R.: High mutual coupling reduction between microstrip patch antennas using novel structure. Int. J. Electron. Commun. (AEÜ) **71**, 152–156 (2017). https://doi.org/10.1016/j.aeue.2016.10.017
4. Kishore, N., Upadhyay, G., Tripathi, V.S., Prakash, A.: Dual band rectangular patch antenna array with defected ground structure for ITS application. Int. J. Electron. Commun. (AEÜ) **96**, 228–237 (2018). https://doi.org/10.1016/j.aeue.2018.09.039
5. Kumar, V., Gupta, B.: On-body measurements of SS-UWB patch antenna for WBAN applications. Int. J. Electron. Commun. **70**(5), 668–675 (2016). https://doi.org/10.1016/j.aeue.2016.02.003
6. Kushwaha, R.K., Karuppanana, P., Malviya, L.D.: Design and analysis of novel microstrip patch antenna on photonic crystal in THz. Physica B Condens. Matter **545**, 107–112 (2018). https://doi.org/10.1016/j.physb.2018.05.045
7. Matyas, J., Slobodian, P., Munster, L., Olejnik, R., Urbanek, P.: Microstrip antenna from silver nanoparticles printed on a flexible polymer substrate. In: International Conference on Nanosciences & Nanotechnologies and 9th International Symposium on Flexible Organic Electronics Materials Today: Proceedings, vol. 4, pp. 5030–5038 (2017)
8. Dellaouia, S., Kaabala, A., Halaouia, M.E., Asselman, A.: Patch array antenna with high gain using EBG superstrate for future 5G cellular networks. Procedia Manuf. **22**, 463–467 (2018). https://doi.org/10.1016/j.promfg.2018.03.071. 11th International Conference, Interdisciplinarity in Engineering, Tirgu-Mures, Romania
9. Raval, F., Kosta, Y.P., Joshi, H.: Reduced size patch antenna using complementary split ring resonator as defected ground plane. Int. J. Electron. Commun. **69**(8), 1126–1133 (2015). https://doi.org/10.1016/j.aeue.2015.04.013
10. Sawant, K.K., Suthikshn Kumar, C.R.: CPW fed hexagonal micro strip fractal antenna for UWB wireless communications. Int. J. Electron. Commun. **69**(1), 31–38 (2015). https://doi.org/10.1016/j.aeue.2014.07.022

11. Shang, F., Zhang, B.-J.: On study of the feeding method used for wide-banded micro-strip antenna. J. China Univ. Posts Telecommun. **18**(Suppl. 2), 158–160 (2011). https://doi.org/10.1016/s1005-8885(10)60154-9
12. Singh, A., Ansari, J.A., Kamakshi, Aneesh, M., Sayeed, S.S.: L-strip proximity fed gap coupled compact semi-circular disk patch antenna. Alexandria Eng. J. **53**, 61–67 (2014). https://doi.org/10.1016/j.aej.2013.12.001
13. Yu, L.C., Kamarudin, M.R.: Investigation of patch phase array antenna orientation at 28 GHz for 5G applications, vol. 86, pp. 47–50 (2016). http://doi.org/10.1016/j.procs.2016.05.012. 2016 International Electrical Engineering Congress, iEECON2016
14. Venkatesh, C.K., Naveenkumar, S.K.: A development of micro strip patch antenna from synthesized nanocomposites materials for defense applications. Mater. Proc. **4**(11), 12007–12011 (2007). https://doi.org/10.1016/j.matpr.2017.09.123. Part 3
15. Yun-xing, Y., Hui-chang, Z., Si, C., Shu-ning, Z.: Design of dual-frequency antenna for IoT applications. Cogn. Syst. Res. **52**, 891–895 (2018). https://doi.org/10.1016/j.cogsys.2018.09.015
16. Karatay, A., Orcan, D., Özkal, C., Yaman, F.: Implementation and experimental verifications of microstrip antennas for angular scanning of a Doppler radar. AEU Int. J. Electron. Commun. **101**, 76–84 (2019). https://doi.org/10.1016/j.aeue.2019.01.020
17. Abhilash, P., Roshni, S.B., Mohanan, P., Surendran, K.P.: A facile development of homemade substrate using 'quench free' glass-ceramic composite and printing microstrip patch antenna on it. Mater. Des. **137**, 38–46 (2018). https://doi.org/10.1016/j.matdes.2017.10.015
18. Anantha, B., Merugu, L., Rao, S.: A novel single feed frequency and polarization reconfigurable microstrip patch antenna. Int. J. Electron. Commun. **72**, 8–16 (2017). https://doi.org/10.1016/j.aeue.2016.11.012
19. Bhongale, S.R., Ingavale, H.R., Shinde, T.J., Vasambekar, P.N.: Microwave sintered Mg-Cd ferrite substrates for microstrip patch antennas in X-band. Int. J. Electron. Commun. **96**, 246–251 (2018). https://doi.org/10.1016/j.aeue.2018.09.040
20. Chung, K.L., Li, Y., Zhang, C.: Broadband artistic antenna array composed of circularly-polarized Wang-shaped patch elements. Int. J. Electron. Commun. **74**, 116–122 (2017). https://doi.org/10.1016/j.aeue.2017.02.006
21. Devesh, J.A.A., Siddiqui, M.G., Saroj, A.K.: Analysis of modified square sierpinski gasket fractal microstrip antenna for wireless communications. Int. J. Electron. Commun. **94**, 377–385 (2018). https://doi.org/10.1016/j.aeue.2018.07.027
22. George, J.N., Madhan, M.G.: Analysis of single band and dual band graphene based patch antenna for terahertz region. Physica E Low-dimensional Syst. Nanostruct. **94**, 126–131 (2017). https://doi.org/10.1016/j.physe.2017.08.001
23. Tölli, A., et al.: Massive multiple input multiple output (MIMO) Systems. In: Osseiran, A., Monserrat, J.F., Marsch, P. (eds.) 5G Mobile and Wireless Communications Technology, chap. 8, pp. 208–247. Cambridge University Press, Cambridge (2016). https://doi.org/10.1017/CBO9781316417744.009
24. Wilkinson, E.J.: An N-way power divider. IRE Trans. Microw. Theory Tech. **8**, 116–118 (1960)

Authentication and Privacy Preservation in IoT Based Forest Fire Detection by Using Blockchain – A Review

Sreemana Datta[1](✉), Ayan Kumar Das[2], Abhijeet Kumar[2], Khushboo[2], and Ditipriya Sinha[1]

[1] National Institute of Technology Patna, Patna, Bihar, India
sreemana.nitp@gmail.com, ditipriya.cse@nitp.ac.in

[2] Department of CSE, Birla Institute of Technology, Mesra, Patna campus, Mesra, Bihar, India
das.ayan@bitmesra.ac.in, abhijeet198771@gmail.com, khushi.krishna12345@gmail.com

Abstract. Internet of Things (IoT) has garnered a good chunk of limelight with the emerging advancements in technologies. It comprises of heterogeneous devices embedded with sensors, which are able to communicate directly with each other without any human intervention. In this paper, a survey has been done on authentication and privacy in IoT using Blockchain during Forest Fire. The purpose of Blockchain is to design a data structure to construct a distributed ledger of transactions that cannot be tampered by any third party. The blockchain forms by cryptographically linked blocks of transactions. Public-key cryptography is used to sign transactions among users in blockchain technology. It achieves privacy very strongly as it is very difficult to modify or delete blocks of data which are stored in the ledger of blockchain.

Keywords: IoT · Blockchain · Forest Fire · Validation · Privacy

1 Introduction

The Internet of Things (IoT) is simply a network of inter-connected devices that are able to collect and exchange data. The use of IoT is growing at a fast pace and in some of the researches it is predicted that it will scale up to billions till 2050 [1]. The concept of Internet of Things encompasses and unifies Internet and Things as the data is sensed from things and send through the internet. The increasing demand for IoT based systems and applications have raised the issue of security and privacy. In these applications the flow of user's personal data causes vulnerability of user's privacy. The authentication of all the end users and communicating devices in an IoT application is also importantly required.

N. Nain and S. K. Vipparthi (Eds.): ICIoTCT 2019, AISC 1122, pp. 133–143, 2020.
https://doi.org/10.1007/978-3-030-39875-0_14

Data transmission during Forest Fire is extremely difficult and a challenging task for the researchers. It is very much essential to transmit the sensed data regarding forest fire to the base station. During forest fires, sensitive data requires to be transmitted to the Base Station thereby enabling the Base Station to authenticate the sender and verify that no alternation is performed in transit. Here comes the requirement of Blockchain technology. In case of blockchain, no alteration is possible for any intruder. Thus, authentication and privacy of data is ensured in IoT using Blockchain during forest fires.

The benefits of Blockchain include:-

1. Greater Transparency and Enhanced Security.
2. Updated Traceability and Low costs compared to similar technologies.

The remainder of this paper is structured as follows: Sect. 2 presents Background Study; Sect. 3 focuses on the blockchain architecture, types and validation properties of the blockchain. Section 4 presents a review on the use of blockchain in forest fire scenario and its challenges. Section 5 evaluates the existing methodologies and concludes this work and providing an outlook on the future scope of work in this field.

2 Motivation

Motivations of the review work of this paper are as follows-

- **Secure data transmission during forest fire**-Most important challenge is to transmit the data securely during forest fire.
- **Preserving authenticity is also a concern**-When the data is transmitted to the base station, authentication is also necessary.
- **Privacy protection** - Most of the forest fires start because of human faults, carelessness or intentional fire-raising.so there should be a protocol that no attacker can tamper the sensitive data intentionally or unintentionally.

3 Forest Fire

Methodologies for secure data transmission in forest fires are depicted in Fig. 1.

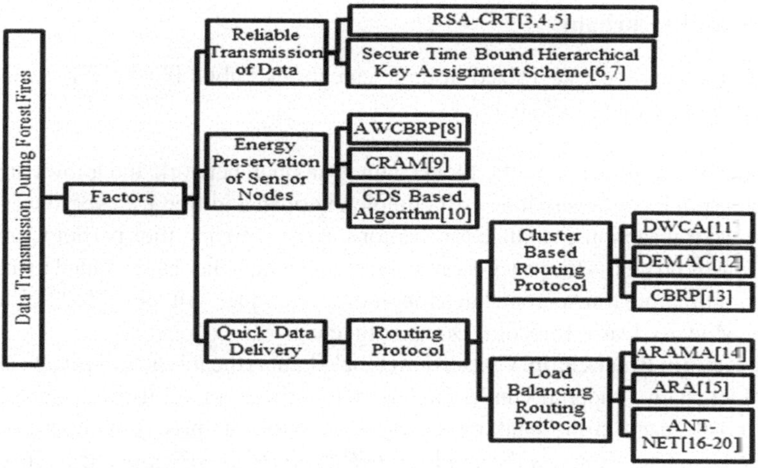

Fig. 1. Methodologies for secure data transmission during forest fires

4 Blockchain

It is a type of distributed and decentralized database which helps in cooperation between multiple authoritative domains. In typical blockchain architecture, every individual node maintains a local copy of the blockchain. Thus, there are multiple nodes in the internet that are connected with each other, and are maintaining a local copy of the global data sheet. The distributed ledger is made up of blocks of data that are chained together with Cryptography that makes it almost impossible to modify once something is recorded. By allowing digital information to be distributed but not copied, block chain technology created the backbone of a new type of internet.

4.1 Blockchain Architecture

A Block consists of two components: Block Header and List of Transactions. The header of a block connects the transactions; any change in any transaction will result in an alteration at the block header. The headers of subsequent blocks are connected in a Chain. Block header is the metadata about a block. It mainly consists of the following (Fig. 2):

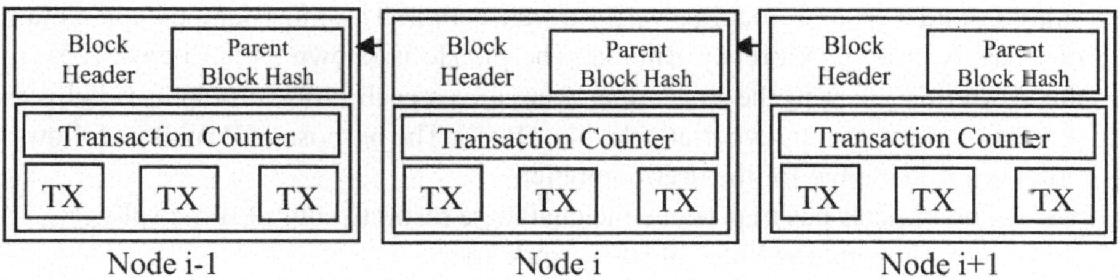

Fig. 2. Blockchain architecture

4.2 Types of Blockchain

The blockchain can be categorized in three categories-Public blockchain, Consortium blockchain and Private blockchain.

i. **Public Blockchain:** A public blockchain is an open network and allows anyone to take part in its core activities as it completely open and free for everyone. Anyone can read transactions, validate transactions in the network, thus participating in the consensus process – the process for determining which blocks get added to the chain. Transactions are transparent, but anonymous. Examples: Bitcoin [21, 22], Ethereum [23], Monero, Dash, Litecoin, Dodgecoin, etc.
ii. **Consortium Blockchain**: Consortium blockchain is the mixture of private and public blockchain. It operates under the leadership of a group. Consortium Blockchains are mostly used in the banking sector. The consensus process is controlled by a pre-selected set of nodes. Example: R3 (Banks), EWF (Energy), B3i (Insurance), Corda.
iii. **Private Blockchain:** Private Blockchain is intended for use of any single company or individual. In private blockchain, all the information are kept centralized. In private blockchain one has to seek permission to read or write data. This block chain is only accessible by those whose identities are known. It is lighter, faster and low energy consuming compared to public blockchain. Examples: MONAX, Multichain.

4.3 Validation Process

The validation processes for all the three types of blockchain is discussed in this section.

4.3.1 Public Block Chain

Four types of validation process for public blockchain is discussed in this subsection.

a. POW (Proof of Work): Proof of work is the process of solving a mathematical problem by using high computational power to validate the transaction in a block. The energy used in POW is much more and the chances of attack increases to 51% [24]. It gives more rewards to people with better and more equipment; it acts as a smart contract or a protocol to verify a transaction. The most famous application of POW is in cryptocurrency Bitcoin. The puzzle is known as hash cash [25]. In this reward is given to the first miner who solves each block problem. It helps in protection from various cyber-attacks like DDoS. The purpose of PoW is as follows-
1. Produces a challenge for the user/computer.
2. The user/computer has to produce the challenge result to submit some value.
3. Essentially showing the Proof of some work done

4. Eliminate entities which are slow or incapable to generate the proof of work.
5. The idea behind POW is that it is used to find or generate a value which is difficult to generate and yet easily verifiable.

b. POS (Proof of Stake): It is an alternate method for verifying the transaction in blockchain. It remains more concerned regarding validating the transaction rather than mining. It requires the user to show ownership of certain cryptocurrency units.
c. DPOS: Stands for Delegated Proof Of Stake. DPoS seeks to by speed up transactions and block creation, while not compromising the decentralized incentive structure at the heart of the blockchain.
d. Ripple: The Ripple network receives transactions from different nodes. The goal of consensus is to write the valid transactions to the ledger. After 50% of the nodes approve the transactions, the batch with approved transactions is pushed further for higher approval ratings (Table 1).

Table 1. Validation processes of public blockchain

Property	PoW	PoS	DPoS	Ripple
Energy Saving	×	Partial	Partial	√
Open Node Identity	√	√	√	√
Tolerated Power	<20%	<51%	<51%	<20%
Example	Bitcoin	Peercoin	Bitshares	Ripple

4.3.2 Consortium Blockchain

Hyperledger [26] is the example of validation process of consortium blockchain. It is the project under the Linux foundation for example node.js, dronecode etc. It is an open source platform which can be used by everyone.

4.3.3 Private Blockchain

Multichain [27] is the example of validation process of private blockchain. A simple API and a command-line interface is provided by Multichain. The Multichain defines miners to an identification set of entities. It introduces a criterion known as mining diversity that binds with $0 < \text{mining diversity} < 1$ (Table 2).

Table 2. Showing different works in block chain

Area	Issues	Technique	Result
Security, privacy [28–30]	Security and privacy threat like location tracking, hijacking of the vehicles	Light weight scalable block chain	To protect the privacy of user and to increase the security of the vehicular ecosystem
Majority is not Enough: Bitcoin Mining is Vulnerable [31, 32]	Selfish Mining	Backwards-compatible change to the Bitcoin protocol	A miner learns of competing branches of the same length, it should propagate all of them, and choose which one to mine on uniformly at random
Bubbles of Trust: A decentralized blockchain-based authentication system [33]	Authentication	Using Ethereum block chain	Create an efficient centralized authentication system
Block chains and smart contracts [34]	Deployment of block chain network in an IoT setting	Consensus algorithm	To show that the block chain IoT combination is powerful Significant and transformation
Blockchain with IoT [35, 36]	To give a brief history about block chain, consensus algorithm are also discussed	Consensus algorithm	To show that block chain and smart contracts could help to restrict misbehavior done by AI products

5 Blockchain in Forest Fire

The challenges of using blockchain in forest fire detection are discussed as follows.

i. **Security Issues:** It is one of the major challenges to transmit data securely. Because there can be various type of attacks like Sybil attack [37–39], DoS attack which can jeopardize the security of data transmission.
ii. **Privacy Leakage**: It can preserve a certain amount of privacy by using the key pair, but can't guarantee transactional privacy.
iii. **Scalability**: Blockchain can only process a few number of transactions per second which is a complex problem.

iv. **Storage Optimization**: Sometimes, the system has to redesign the blockchain in order to undergo storage optimization.

The data transmission methodologies are depicted in Fig. 3.

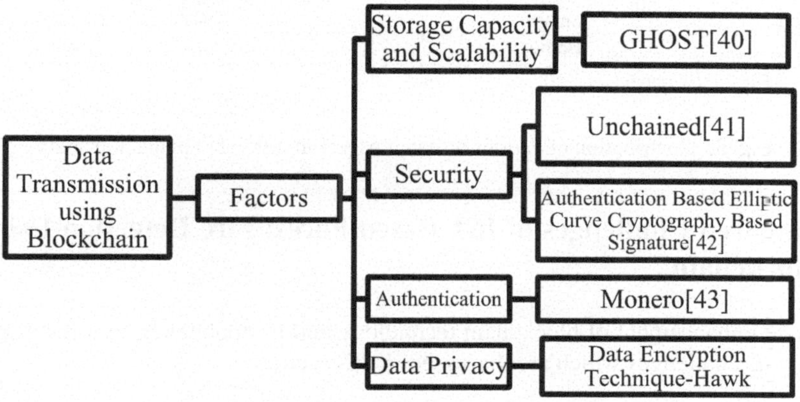

Fig. 3. Data transmission using blockchain

6 Results of the Literature Survey

The contribution of various research papers in the literature survey on various factors related to data transmission during forest fire is shown in Fig. 4. The factors are- Quick data delivery, Energy preservation of sensor nodes and reliable transmission of data. The occurrence of different research papers specific to different application areas of blockchain such as Smart Home, Crypto-currency (Bitcoin) etc. is shown in Fig. 5.

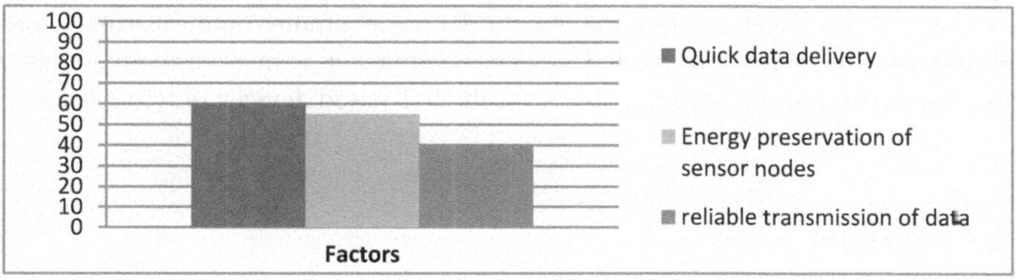

Fig. 4. Factor-wise contributions of various research paper

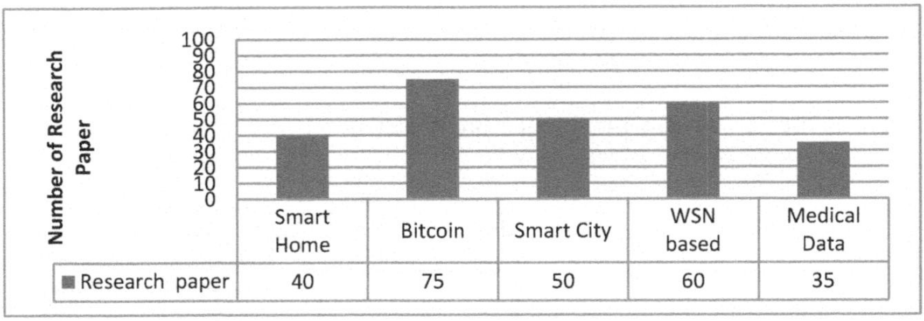

Fig. 5. Contribution of various research papers in different application areas

7 Research Challenges of IoT Based Forest Fire Detection Using Blockchain

The rapid enhancement of blockchain technology and its application to forest fires has faced some challenges which are discussed in this section.

a. **Scalability**: In the blocks of blockchain all the transactions should be stored for validation purpose. Performing several transactions within fraction of second will not be possible. Hence scalability of blocks needs to be maintained.
b. **Selfish mining:** Selfish mining is a strategy for mining bitcoin in which groups of miners collude to increase their revenue. Generally it is convinced that nodes with over 51% computing power can reverse the blockchain and reverse the happened transaction. However, recent research shows that the network is vulnerable even if only a small portion of the hashing power is used to cheat.
c. **Privacy leakage:** It can preserve a certain amount of privacy by using the key pair, but can't guarantee transactional privacy. All the transactions remain visible to all.
d. **Infrastructure Based Computing:** How to optimize performance and scarcity of resources for computing.
e. **Data extraction from complex environments**: How to sense information from complex environment, e.g. how to sense the data (temperature/humidity) in the forest.
f. **Power requirements:** All the IoT devices need power as most of them are battery operated. This issue is also a problem for the IoT based devices in forest fires.

8 Conclusions

This paper encompasses a broad overview of the different aspects of forest fire detection systems, as well as applications based on blockchain based approaches in the field. The security and privacy aspects have also been considered within the scope of study. This work presents a comprehensive overview on blockchain. It begins with an overview of blockchain technologies including blockchain architecture and key characteristics of blockchain. An analysis of the protocols used in data transmission during forest fire, its challenges has been performed. A discussion on the blockchain technology and related core technologies has been undertaken. Various current issues needs to be taken into account to use blockchain in forest fire environment. Even today, blockchain technology's growth is thwarted by several issues emanating, such as the security of data, scalability. The anonymity of user information needs to be ensured when using blockchain in IoT during forest fire and the user information needs to be completely deleted when removing the service.

Acknowledgement. This research is funded in parts by DST/SERB project ECR/2017/000983 grants. The authors are grateful to DST/SERB for the support.

References

1. Atzori, L., Iera, A., Morabito, G.: The internet of things: a survey. Comput. Netw. **54**(15), 787–2805 (2010)
2. Satendra, Kaushik, A.D.: Forest Fire Diaster Management. Published by National Institute of Disaster Management, Ministry of Home Affairs, New Delhi (2014)
3. Hefeeda, M.: Forest fire modeling and early detection using wireless sensor networks. ACM J. **5** (2009)
4. Rivest, R.L., Shamir, A., Adleman, L.: A method for obtaining digital signatures and public-keycryptosystems. Proc. CACM **21**, 120–126 (1978)
5. Sun, H.-M., Wu, M.-E.: An Approach Towards Rebalanced RSA-CRT with Short Public Exponent. Cryptology, ePrint Archive: Report 2005/053 (2005). http://eprint.iacr.org/2005/053
6. Wu, C.-H., Hong, J.-H., Wu, C.-W.: RSA cryptosystem design based on the Chinese remainder theorem. In: Proceedings of 2001 Asia and South Pacific Design Automation Conference ASP-DAC 2001. ACM, New York (2001)
7. Ateniese, G., De Santis, A., Ferrara, A.L., Masucci, B.: Provably-secure time-bound hierarchical key assignment schemes. In: Proceedings of CCS06, Alexandria, Virginia, USA. ACM 1-59593- 518-5/06/0010 (2006)
8. Karunakaran, S., Thangara, P.: An adaptive weighted cluster based routing (AWCBRP) protocol for mobile ad-hoc networks. J. WSEAS Trans. Commun. **7**, 248–257 (2008)
9. Zheng, Z., Wu, Z., Lin, H., Zheng, K.: CRAM: an energy efficient routing algorithm for wireless sensor networks. In: Proceedings of 19th International Symposium, Kemer-Antalya, Turkey (2004)
10. Liu, C., Wu, J.: Scalable routing in delay tolerant network. In: Proceedings of MobiHoc 2007 8th ACM International Symposium on Mobile Ad Hoc Networking and Computing, pp. 51–60. ACM Digital Library (2007). https://doi.org/10.1145/1288107.1288115. ISBN 978-1-59593-684-4
11. Chauhan, N., Awasthi, L.K., Chand, N., Chugh, A.: A distributed weighted cluster based routing protocol for MANETs. In: Proceedings of Second International Conference on Advances in Communication, Network and Computing, CNC 2011, Bangalore, India. CCIS, vol. 142, pp. 147–151. Springer, Heidelberg (2011). https://doi.org/10.1007/978-3-642-19542-6, ISBN 978-3-642-19541-9 (print), 978-3-642-19542-6 (online)
12. Majumder, A., Sarma, N.: DEMAC: a cluster-based topology control for ad hoc networks. IJCSI – Int. J. Comput. Sci. Issues **7**(5), 82–88 (2010). ISSN (Online) 1694-0814
13. Jiang, M., Li, J., Tay, Y.C.: Cluster based routing protocol (CBRP). In: INTERNET-DRAFT draft-ietf-manet-cbrp-spec01.tx, July 1999

14. Hossein, O., Saadawi, T.: Ant routing algorithm for mobile ad hoc networks (ARAMA). In: Proceedings of 22nd IEEE International Performance, Computing, and Communications Conference, April 2003, Phoenix, Arizona, USA, pp. 281–290 (2003)
15. Gunes, M., Sorges, U., Bouazizi, I.: ARA- the ant colony based routing algorithm for MANET. In: Proceedings of International Workshop on Ad-hoc Networking (IWAHN 2002), pp. 18–22, August 2002
16. Dorigo, M.: Ant colony optimization. http://iridia.ulb.ac.be/mdorigo/ACO/ACO.html. In Dynamic Routing in Traffic Networks Using AntNet
17. Ducatelle, F., De Caro, G., Gamardella, L.M.: Ant agents for hybrid multipath routing in MANETs. IDSIA, Galleria 2, CH-6928 Manno-Lugano, Switzerland (2005)
18. Camilo, T., Carreto, C., Silva, J.S., Boavida, F.: An energy-efficient ant-based routing algorithm for wireless sensor networks (2006)
19. Ghazy, A.M., Fatma, E.L.L., Hefny, H.A.: Threshold based AntNet algorithm for dynamic traffic routing of road networks Egypt. Inform. J. **13**, 111–121 (2012)
20. Lu, Y., Zhao, G., Su, F.: Adaptive ant-based dynamic routing algorithm (2004)
21. Nakamoto, S.: Bitcoin: A Peer-to-Peer Electronic Cash System
22. Conti, M., Kumar, E.S., Lal, C., Ruj, S.: A survey on security and privacy issues of bitcoin. Commun. Surv. Tutorials **20**, 3416–3452 (2018)
23. Wood, G.: Ethereum: A Secure Decentralised Generalised Transaction Ledger Eip-150 Revision
24. Li, X., Jiang, P., Chen, T., Luo, X., Wen, Q.: A survey on the security of blockchain systems (2017)
25. Adam Back: Hashcash - A Denial of Service Counter-Measure. http://www.hashcash.org/hashcash.pdf
26. Cachin, C.: Architecture of the Hyperledger Blockchain Fabric. https://www.zurich.ibm.com/dccl/papers/cachin_dccl.pdf
27. Gideon Greenspan, Founder and CEO, Coin Sciences Ltd, "MultiChain Private Blockchain—White Paper". https://www.multichain.com/download/MultiChain-White-Paper.pdf
28. Dorri, A., Steger, M., Kanhere, S.S., Jurdak, R.: Blockchain distributed solution to automated security and privacy. https://allquantor.at/blockchainbib/pdf/dorri2017blockchaindistributed.pdf
29. Feng, Q., He, D., Zeadally, S., Khan, M.K., Kumar, N.: A survey on privacy protection in blockchain system (2019)
30. Roulin, C., Dorri, A., Jurdak, R., Kanhere, S.S.: On the Activity Privacy of Blockchain for IoT. https://arxiv.org/pdf/1812.08970.pdf
31. Eyal, I., Sirer, E.G.: Majority is not Enough: Bitcoin Mining is Vulnerable. https://www.cs.cornell.edu/~ie53/publications/btcProcFC.pdf
32. Nakamoto, S.: Bitcoin: A Peer-to-Peer Electronic Cash System. https://bitcoin.org/bitcoin.pdf?
33. Conti, M., Kumar, E.S., Lal, C., Ruj, S.: A Survey on Security and Privacy Issues of Bitcoin. https://arxiv.org/pdf/1706.00916
34. Hammi, M.T., Hammi, B., Bellot, P., Serhrouchni, A.: Bubbles of trust: a decentralized blockchain-based authentication system for IoT. Comput. Secur. **78** (2018). https://doi.org/10.1016/j.cose.2018.06.004
35. Christidis, K.: Blockchains and Smart Contracts for the Internet of Things. http://ieeexplore.ieee.org/iel7/6287639/6514899/07467408.pdf
36. Dorri, A., Kanhere, S.S., Jurdak, R.: Towards an optimized blockchain for IoT. https://dl.acm.org/citation.cfm?id=3055003
37. Sharma, H.J., Dhawan, A.: An Enhanced and efficient mechanism to detect Sybil attack in Wireless Sensor Networks (2013)

38. Douceur, J.R.: The Sybil attack. Int. J. Adv. Res. Comput. Eng. Technol. (IJARCET) **2**(2), 792–796 (2013)
39. Mishra, A.K., Tripathy, A.K., Puthal, D., Yang, L.T.: Analytical model for sybil attack phases in internet of things. IEEE Internet Things J. **6**, 379–387 (2018)
40. Sompolinsky, Y., Zohar, A.: Secure high-rate transaction processing in bitcoin. In Financial Cryptography and Data Security, pp. 507–527. Springer, Heidelberg (2015)
41. Bochem, A., Leiding, B., Hogrefe, D.: Unchained identities: putting a price on sybil nodes in mobile ad hoc networks. SecureComm (2018)
42. Aydos, M., Sunar, B., Koc, C.K.: An elliptic curve cryptography based authentication and key agreement protocol for wireless communication. In: 2nd International Workshop on Discrete Algorithms and Methods for Mobile Computing and Communications Symposium on Information Theory (1998)
43. Sun, S.F., Au, M.H., Liu, J.K., Yuen, T.H.: Ring CT 2.0: a compact accumulator-based (linkable ring signature) protocol for blockchain cryptocurrency Monero. In: SCOPUS. LNCS (LNAI, LNB), vol. 10493, pp. 456–474. Springer, Heidelberg (2017)

An IoT Based Agri-Cloud Architectural Framework for Monitoring Presence of Fertilizer Under Multilayered Soil Farming

Nikhil V. Bhende(✉), Jayant K. Purohit(✉), and Anup A. Junankar(✉)

School of Engineering and Technology, Poornima University, Jaipur, India
nikhilbhende5@gmail.com

Abstract. Internet of Things in Precision Agriculture practices has made a long-lasting impact. Advanced agro-machineries helping farmers to increase their farm output and productivity. With addition of IoT enabled devices it is possible to integrate agro-machines and farmers to a network of connected things that shares valuable data with the help of which farmers can make sensible decisions on time. IoT devices boost up the smart farming through its efficient and dependent monitoring and controlling system. In this paper an IoT enable fertilizer applicator machine is proposed for the farmers of the countryside. The demand for the fertilizer is on high everywhere and so its associate problems like ineffective fertilizer placement methods, insufficient penetration of the fertilizer under the soil, unavailability of fertilizer for the crop roots during mid and late season growth stages of plant and many more. One of the possible solutions for these problems is to design and develop the machine which is able to place the fertilizers into different depths of soil so as to make it available to the crop during all growth stages of the crop. Further to monitor the amount and right placement of the fertilizer, an IoT enabled AgroCloud module architecture is proposed that will connect the person through mobile computing. This module will store the data and will communicate to the farmer for the information such as soil properties (pH value, EC), fertilizers presence (N, P, K), environmental condition (moisture, temperature) through smartphones and sensors. The complete model will benefit in increasing the yield through multi depth fertilizing concept along with incorporation of IoT assembly for better practices.

Keywords: Internet of Things · Precision Agriculture · Smart farming

1 Introduction

Implementation of Precision Agriculture is need of an hour to the agro-based country like India where many farmers are still untouched to the technology [1]. Looking at the current global population growth rate, it implicit that the annual requirement of the food will be growing exponentially and so as the requirement of the fertilizer. To gain higher productivity and farm output the placement and utilization of fertilizer is extremely important. It is observed that the practice of fertilizer placement in India is

N. Nain and S. K. Vipparthi (Eds.): ICIoTCT 2019, AISC 1122, pp. 144–150, 2020.
https://doi.org/10.1007/978-3-030-39875-0_15

performed manually although many agro based machines for placing seeds and fertilizers are available [2]. Even when these agro machineries are used, the problems such as insufficient uptake of nutrients by the seed or plant are observed during early growth stages of the plant. Also, uneven fertilizer distribution and insufficient fertilizer penetration under the soil makes them unavailable during mid season and late season growth stages of the plant. One probable solution to these problems is by designing and developing the fertilizer applicator equipment which can place right amount of fertilizer into the different depths of the farm soil simultaneously [14]. Placement of fertilizer (such as Potassium and Phosphorus) under multilayered of the soil makes them available at root zone. Briefly, the main aim of the fertilizer applicator will be to apply the fertilizers effectively, uniformly and cheaply in all growth stages of crop i.e. nursery, vegetative, mid season and late season stages. The best part is Precision agriculture methodology will assist the fertilizer applicator to monitor the fertilizer presence and its quantity under various depth of soil through integration of IoT based cloud architecture.

1.1 Precision Agriculture – A Gateway to Smart Farming

The first paragraphs that follows a table, figure, equation etc. does not have an indent, either. Precision agriculture is not new to the society as it was practised in 1980 already when equipment like fertilizer applicator was used with provision of checking chemical properties of the soil based on grid sampling [1]. It is now practiced as a new management concept for doing farming. Precision farming helps in efficient decision making processes such as when to apply the pesticides or at what amount of fertilizer to be placed. It can control the water system for farming; can reduce the amount of resources required for agriculture and thereby can escalate the overall output of the farm [4]. Precision agriculture do uses GPS or satellite navigation system but also incorporates the computing technologies like Internet-of-Things, mobile cloud computing and big data analysis. Farmers can now examine their farm and can have information and control over variable parameters like terrain type, moisture content in soil, organic carbon content, level of nitrogen, electrical conductivity (EC), soil fertility (pH value), amount of Potassium, Magnesium, Phosphorus and others. From few last year's technologies like Internet-of-Things are consistently upgraded and being assimilated in various parts of life [5].

1.2 Internet of Things – Technology for Farmers

Internet of things (IoT) is an integration of computing devices such as wireless network connectivity (Internet), RFID or advanced sensors with the other conventional systems [6]. These IoT enabled devices are developed for monitoring and controlling the various processes in the sectors like building smart cities, smart home appliances, health monitoring system, smart industries and many more. IoT have reached in agriculture domain too in which use of Wireless Sensor Network Technology are been practiced a lot. For monitoring soil properties Advanced WSN architecture are being developed, implemented and tested in smart farming management practices [7] (Figs. 1 and 2).

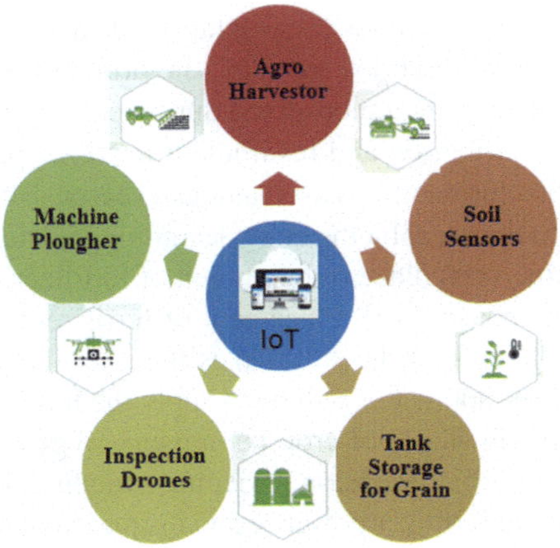

Fig. 1. IoT in smart farming

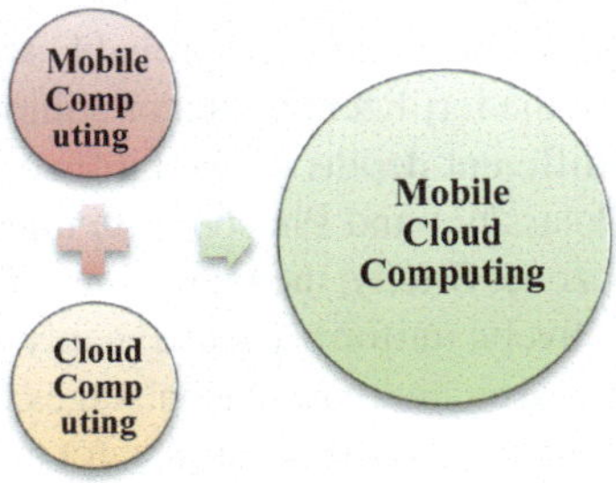

Fig. 2. Mobile cloud computing

1.3 Mobile Cloud Computing – Problem Solver in Farming

Mobile computing is a portable wireless network capable of transmitting information from source to user using mobile software, hardware and mobile communication network system. The related information can be communicated by sending messages to the farmers whenever required. On the other hand, cloud computing is an Internet enabled computing services to manage ones occupation [8]. Cloud computing is new business assistant which are now favored due to its low operating cost, ease of operation, speed delivery, global connectivity and security. Cloud services helps one to access the required resources through web and mobile like platform. Make use of cloud infrastructure which can protect, store, back up, analyze and recover data from anywhere. It aid with audio, video services and embed intelligence providing valuable insights in available information. Cloud computing have already made its way in Agricultural business [9]. On integrating the mobile and cloud computing systems with wireless sensor network a new upgraded and rich computational resource structure form called as Mobile Cloud Computing. The aim of the mobile cloud computing is to transport applications to mobile platform through cloud computing services. Agriculture requires huge amounts of data to be stored and its information needs to be shared through proper mechanism. Mobile cloud computing fulfils the supremacy of getting better the efficiency of agriculture in all facet. The two very useful services are Information as a Service (IaaS) and Software as a service (SaaS) can be provided to farmers for education and awareness purpose in well customized format for their ease of use.

1.4 Data Mining and WSN

Role of data mining is also important in agriculture which provides lot of information related to farming activities and in turn it uses this information to analyze other information for instance soil testing parameters [10]. Soil contains various ingredients that affect the crop yield and depending on present soil properties which acts like nutrients to the plant one can take better decision. Also, use of WSN (Wireless Sensor Network) for sensing, inspecting and recording the soil and environment parameters and collecting the gathered information to a cloud server at centre level are of great use now. One of the sensors used for measuring soil properties such as organic carbon and moisture is optical sensor [11]. It makes the use of light for its measurement and can be placed on equipments that perform work of seed sowing or fertilizing into the farm. Now a day's sensors are applied to drones and satellite also. Another sensor useful in precision agriculture practices is electrochemical sensor which provides important soil properties like soil nutrient stage and pH value. Soil contains specific type of ion which is detected by sensor electrodes helping in gathering, processing and mapping the chemical data of the soil [12]. For soil compaction a mechanical sensor is used through its probes inserted inside the soil clod and measures its resistive forces. Tractor pulling requirements for ground resistance can be used through such similar technology like tensiometer of Honeywell can detect the force applied by absorption of water by roots of plant. The amount of moisture present can be measured by estimating dielectric constant level in the soil through dielectric soil moisture sensor.

1.5 Multilayered Soil Farming

Multilayered soil farming process is the method to apply the fertilizers at various depths of farm soil during seed sowing process. The urge of such type method aroused due to insufficient penetration of fertilizers like nitrates, phosphate and potassium under the soil either by manual process or by fertilizer sowing equipments [15]. The multilayered soil farming technique will make fertilizer available for longer period of time under the entire growth rate of plant (Fig. 3).

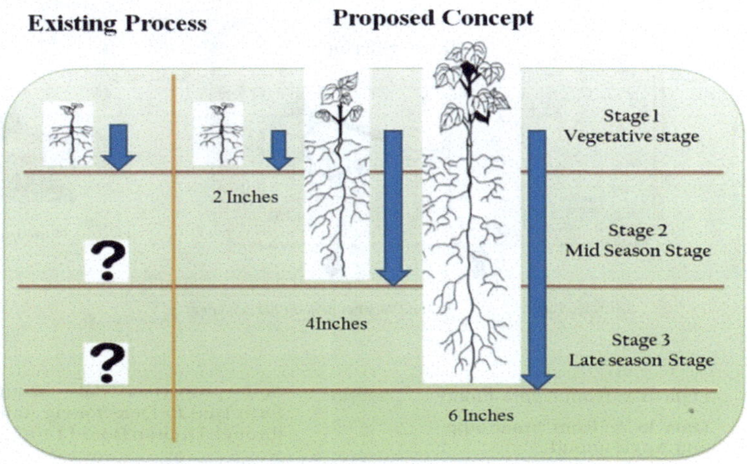

Fig. 3. The concept of multilayered soil farming (placing fertilizers at a depth of 2, 4 and 5 inches under the soil)

2 Proposed Framework of Integrated Model for Smart Farming

The proposed architecture of integrated model is given away in the following Fig. 4. The model consists of 4 units:

1. Detector-Rig Unit.
2. Smart-App Unit.
3. Aggie-Cloud Unit.
4. Big Data Mining and Info-structure Unit.

An IoT device with soil property measuring sensor comprises of Detector-Rig unit. For user interface the Smart-App unit is provide. For communicating the information

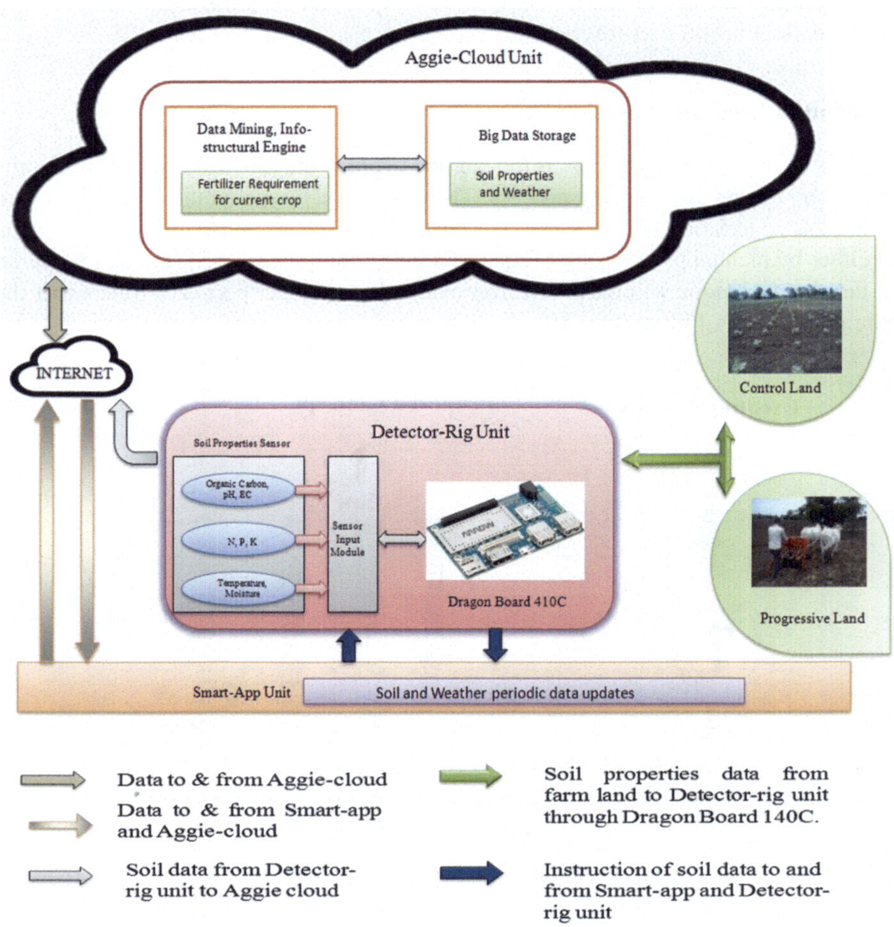

Fig. 4. Proposed architecture for an integrated agro-version for fertilizer amount prediction under precision agriculture.

with the user Aggie-Cloud unit is made available which will contain big data mining and storage.

2.1 Detector-Rig Unit

To get the accurate soil properties values periodically, this unit plays a very crucial role. It is responsible for the all the correct data from the soil that will be collected through the various soil properties detecting sensors. In Fig. 4 detector-rig kit is shown which incorporates the use of dragon board 140C which is an IoT based device with memory storage, GPS sensor for spotting positions and prócessing unit. Along with the single board computer the detector-rig unit consists of soil sensors connected to the kit. The sensing parameters considered are soil pH, organic carbon, moisture, temperature, nitrate, potassium, phosphorus channeling to the IoT platform.

2.2 Smart-App Unit

Smart-app unit is a mobile application which will be installed at end users cell phone consisting mainly user interface for farmer. The detector-rig unit will gather information related to soil farmland. The smart-app and detector-rig will interact to get information. All the gathered information will be received and sent big data storage of Aggie-Cloud unit. The data from the detector-rig unit is also stored in the aggie-cloud during fertilizer and seed sowing process as well any time during growth stages of the crop. The information received to farmers benefits in knowing the amount of fertilizer is placed in multi-depths of soil and the amount of nutrients uptake by the plant during life span of the crop till it get harvested.

2.3 Aggie-Cloud Unit

Aggie-Cloud unit comprises of big data storage which stores information like periodic soil properties. This helps in predicting requirement in fertilizer, soil quality and environment condition. The data will help in monitoring the soil parameters and in taking suitable decision on farming activity. Big data mining and info-structure engine plays vital role decision making and ultimately it helps in increase of crop yield based on the information available on soil properties, weather condition, water requirement. etc. By practicing such technology one gets enough data on soil properties and yield information details of the farm and a better inference with the help of data mining can be predicted for higher productivity and preserving fertility of soil for years.

3 Conclusion

To meet the worldwide demand of the food technologies under precision agriculture must be adopted. Such type of processes are inexpensive, flexible and user friendly in nature. This paper tries to integrate multidisciplinary approach towards smart farming through Internet of Things, mobile cloud computing, wireless sensing networks and big data analysis. The objective is to get real time data of the fertilizer placement by sowing beneath the soil at various levels. The proposed model helps in taking decision on controlling amount of fertilizer required at various level of crop growth and thereby increasing yield of the crop.

References

1. Khattab, A., Abdelgawad, A., Yelmarthi, K.: Design and implementation of a cloud-based IoT scheme for precision agriculture. In: 2016 28th International Conference on Microelectronics (ICM), pp. 201–204. IEEE, December 2016
2. Athani, S., Tejeshwar, C.H., Patil, M.M., Patil, P., Kulkarni, R.: Soil moisture monitoring using IoT enabled arduino sensors with neural networks for improving soil management for farmers and predict seasonal rainfall for planning future harvest in North Karnataka—India. In: 2017 International Conference on I-SMAC (IoT in Social, Mobile, Analytics and Cloud) (I-SMAC), pp. 43–48. IEEE, February 2017
3. Shaikh, F.K., Zeadally, S., Exposito, E.: Enabling technologies for green internet of things. IEEE Syst. J. **11**(2), 983–994 (2017)
4. Channe, H., Kothari, S., Kadam, D.: Multidisciplinary model for smart agriculture using internet-of-things (IoT), sensors, cloud-computing, mobile-computing & big-data analysis. Int. J. Comput. Technol. Appl. **6**(3), 374–382 (2015)
5. Mekala, M.S., Viswanathan, P.: CLAY-MIST: IoT-cloud enabled CMM index for smart agriculture monitoring system. Measurement **134**, 236–244 (2019)
6. Rao, R.N., Sridhar, B.: IoT based smart crop-field monitoring and automation irrigation system. In: 2018 2nd International Conference on Inventive Systems and Control (ICISC), pp. 478–483. IEEE, January 2018
7. Raut, R., Varma, H., Mulla, C., Pawar, V.R.: Soil monitoring, fertigation, and irrigation system using IoT for agricultural application. In: Hu, Y.C., Tiwari, S., Mishra, K., Trivedi, M. (eds.) Intelligent Communication and Computational Technologies, pp. 67–73. Springer, Singapore (2018)
8. Jayashankar, P., Nilakanta, S., Johnston, W.J., Gill, P., Burres, R.: IoT adoption in agriculture: the role of trust, perceived value and risk. J. Bus. Ind. Mark. **33**(6), 804–821 (2018)
9. Ahmed, O.M.E., Osman, A.A., Awadalkarim, S.D.: A design of an automated fertigation system using IoT. In: 2018 International Conference on Computer, Control, Electrical, and Electronics Engineering (ICCCEEE), pp. 1–5. IEEE, August 2018
10. Liu, P., Li, B.: Water and fertilizer integration intelligent control system of tomato based on internet of things. In: International Conference on Cloud Computing and Security, pp. 209–220. Springer, Cham, June 2018
11. Arshad, J., Saleem, S., Badar, M.S.U., Khalid, S., Mumtaz, Z., Ullah, S., Illyas, Z., Madni, H.A.: An Intelligent Monitoring and Controlling of Greenhouse: Deployment of Wireless Sensor Networks and Internet-of-Things (2018)
12. Babu, S.M., Lakshmi, A.J., Rao, B.T.: A study on cloud based internet of things: CloudIoT. In: 2015 Global Conference on Communication Technologies (GCCT), pp. 60–65. IEEE, April 2015
13. Andrew, R.C., Malekian, R., Bogatinoska, D.C.: IoT solutions for precision agriculture. In: 2018 41st International Convention on Information and Communication Technology, Electronics and Microelectronics (MIPRO), pp. 0345–0349. IEEE, May 2018
14. Schneider, F., Don, A., Hennings, I., Schmittmann, O., Seidel, S.J.: The effect of deep tillage on crop yield–what do we really know? Soil Tillage Res. **174**, 193–204 (2017)
15. Shukla, S.K., Yadav, R.L., Gupta, R., Singh, A.K., Awasthi, S.K., Gaur, A.: Deep tillage, soil moisture regime, and optimizing N nutrition for sustaining soil health and sugarcane yield in subtropical India. Commun. Soil Sci. Plant Anal. **49**(4), 444–462 (2018)

Analysis of Web Usage Patterns to Identify Most Frequently Accessed Web Page by Multiple Users

Priyanka Verma[1(✉)] and Nishtha Kesswani[2]

[1] IIS (deemed to be) University, Mansarovar, Jaipur, India
priyanka.gianchandani@iisuniv.ac.in
[2] Central University, Bandar Sindri, Ajmer, India

Abstract. All Data related to web sites that we access is stored in web logs. Increase in browsing these days has led to increase in size of these web log files. Web Mining is one technique that can be applied to these log files to mine navigational patterns. There are various types of web mining depending upon data mined Content, Usage or Structure. In this paper we focus on Mining of usage patterns: Web Usage Mining to discover most frequently accessed web page by multiple users after preprocessing of log file.

Keywords: Web · Usage · Patterns · Navigation · Log file · Web Usage Mining

1 Introduction

Web Usage mining is identifying users navigational patterns by implementing various mining techniques on log files stored on client, server or proxy. In this paper we focus on server log files. These log files have to preprocessed before pattern discovery and analysis. There are four stages of web usage mining (Refer Fig. 1). Amongst all four phases preprocessing is considered to be the most time consuming stage. During preprocessing various columns/data is cleaned, users and sessions are identified. At end of preprocessing we get session file which can be further used to discover patterns and analyze them. In this paper we focus on what is web usage mining, what are its stages, methods of pattern discovery and analysis. We have designed software using Net beans and SQL Server that will perform analysis of web usage patterns to discover most frequently accessed web page.

2 Web Usage Mining

2.1 Web Usage Mining

Web usage mining aims to study user clicks. The objective is to capture and model behavioral patterns and profiles of users who interact with a Web site. Such patterns can be used to better understand the behaviors of different user segments

N. Nain and S. K. Vipparthi (Eds.): ICIoTCT 2019, AISC 1122, pp. 151–159, 2020.
https://doi.org/10.1007/978-3-030-39875-0_16

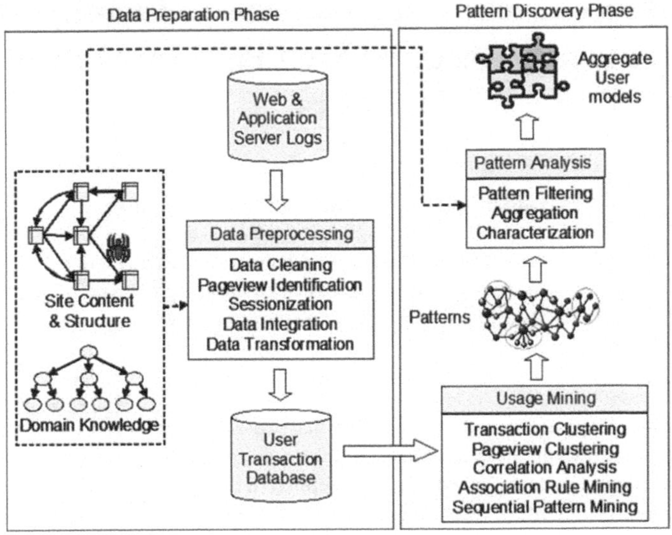

Fig. 1. Stages of WUM [2]

- To improve the organization and structure of the site,
- To create personalized experiences for users by providing dynamic recommendations of products and services.

This type of analysis involves the automatic discovery of meaningful patterns and relationships from a large collection of primarily semi- structured data, often stored in Web and applications server access logs, as well as in related operational data sources [2, 5].

2.2 Stages of Web Usage Mining [WUM]

The overall Web usage mining process can be divided into three inter- dependent stages [Refer Fig. 1].

In the **pre-processing** stage, the click stream data is cleaned and partitioned into a set of user transactions representing the activities of each user during different visits to the site.

In the **pattern discovery** stage, statistical, database, and machine learning operations are performed to obtain hidden patterns reflecting the typical behavior of users, as well as summary statistics on Web resources, sessions, and users.

In the **pattern analysis** stage the discovered patterns and statistics are further processed, filtered, possibly resulting in aggregate user models.

2.3 Log File

Log file that we have used is server log file in National Center for Supercomputing Applications (NCSA) Format. (NCSA) is a fixed (cannot be customized), ASCII (American standard code for information interchange) format, available for Web sites but not for FTP sites. Access log files contains information as given by [6]

- Client IP Address
- User Identifier
- Date - The format is DD-MM-YYYY
- Time - The time format is HH:MM:SS
- Protocol
- Method
- HTTP request field (URL accessed)
- Protocol version
- Status Code - Success (200 series), Redirect (300 series), Failure (400 series) and Server Error (500 series).

  - Transfer Volume (Bytes) Field
  - Referrer Field

When you open an NCSA Common format file in notepad, it appears as shown in Fig. 2 below.

File Edit Format View Help
10.40.1.243 - 14956heena [31/Jul/2015:00:01:00 +0530] "CONNECT clients1.google.com:443 HTTP/1.1" 200 0 "HTTPS" "outgoing via 10.20.20.8"
10.40.1.249 - guest [31/Jul/2015:00:01:01 +0530] "CONNECT igcdn-photos-h-a.akamaihd.net:443 HTTP/1.1" 200 0 "HTTPS" "outgoing via 10.20.20.8"

Fig. 2. Log file

3 Preprocessing

When loading a particular web page, the browser also requests for all the objects embedded in that page, such as.gif or.jpg graphics files. Each such request is logged separately including requests for.gif or.jpeg files. Thus Clickstream data requires substantial preprocessing before the user's behavior can be analyzed. Preprocessing is opening and most essential step. Input to the preprocessing phase of Web usage mining is Web Log file. It is large in size and contains number of raw and irrelevant entries and therefore cannot be directly used in Web usage mining process. Preprocessing of log fie is a very complex and painstaking job and it takes 80% of the total time of web usage mining process as whole. Various stages of preprocessing are as mentioned below:

3.1 Data Cleaning

Data cleaning is second stage after the extraction and storage of log file in table. This stage is performed to remove

- Unnecessary content which includes requests for images, styles and scripts or other files [6]
- Requests made by the crawlers (Robots.txt)
- Records with status code over 299 or under 200
- Records with POST or HEAD method
- Removing blank lines [6]
- Removing lines starting with #-comments [6]

This stage is the most simplest of all stages as it consists of only filtration of the data. Final file after cleaning contains accesses made by standard visitors only [1, 3]. After data cleaning of collected log file as mentioned in the experimental setup there is reduction of almost 63% in size of log file as shown in Fig. 3 below.

Fig. 3. Records before and after cleaning

3.2 User Identification

Since user visits a site more than once, the server logs record multiple sessions of each user. Sequence of activities logged of same user is called as user activity record. Unique way of identifying user is through client side cookies which are sometimes disabled by the user. IP address alone is not enough to identify user due to proliferation of ISP proxy servers which assigns rotating IP address to clients when they browse web.

For identifying user we need to check [4]

- Users IP address
- Web site topology
- Browsers edition
- Operating system

Figure 4 below shows total number of users identified in log file after user identification using Users IP Address.

3.3 Session Identification

Session is a sequence of requests made by a single user with a unique IP address on a Web site during a specified period of time. Sessions are identified either through time

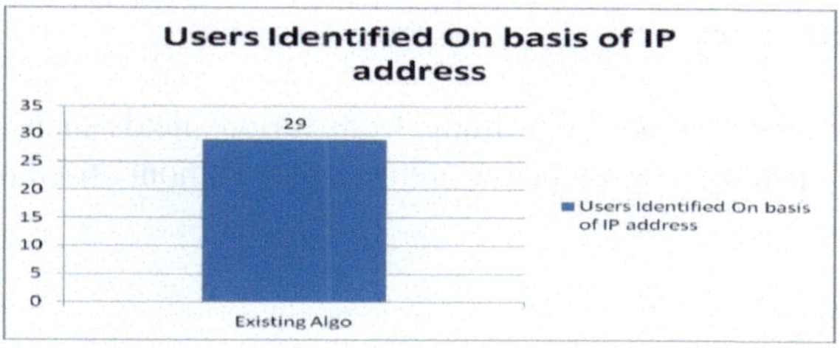

Fig. 4. Users identified after user identification

based heuristics or navigation based heuristics. "Time oriented heuristics is based on total session time or page-stay time. If user 1 has accessed two pages p and q then both belong to same session if difference in timestamp of p and q does not exceed threshold value (30 min). Threshold value can change according to needs (25.5 min, 10 min). Navigation oriented approach makes use of uses the web graph constructed by using hyperlinks among web pages" [10, 11]. New user means new session. If referrer page is null then also we can say that it is new session [8]. Figure 5 below shows sessions identified per date on a particular date. For session identification method used is time based which uses 30 min as timeout value. Figure 6 below shows URL's accessed by various users on a particular date which can be used to track traffic on a particular date. Figure 6 makes it quite clear that highest traffic is on 30/7/2015 by user 8.

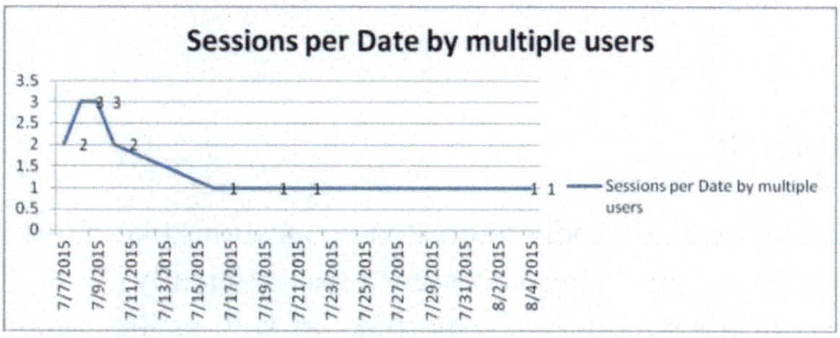

Fig. 5. Sessions identified per date

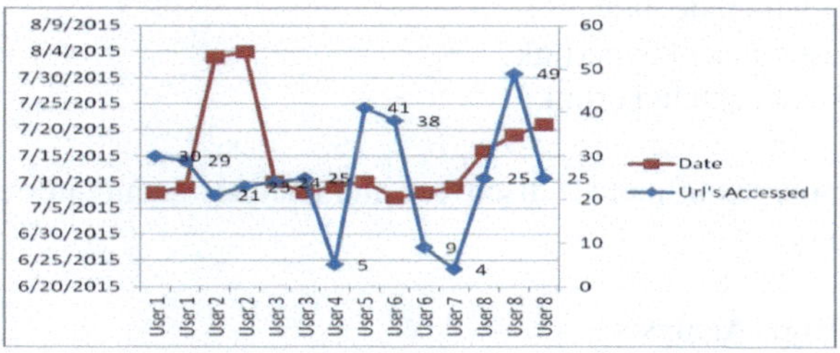

Fig. 6. URL access per session

4 Pattern Discovery

Pattern Discovery uses methods, algorithms from various fields such as statistics, data mining to discover patterns. Given below are various commonly used pattern discovery methods

- Statistical Analysis
  It is the most commonly used technique to extract knowledge about visitors on website. By using session file we can perform various kinds of statistical analysis such as mean, median, frequency etc. on variables such as viewing time, Session id, and Page number.
- Association Rules
  It can be used to relate pages that are most often referenced together in session. For example using Apiori algorithm we can reveal a correlation between users who visited page containing electronic accessories. This data after discovery is useful for various business and marketing applications.
- Classification
  Developing category of users belonging to particular class. It is done by using various algorithms such as decision tree, k nearest neighbor classifiers etc.
- Clustering
  It is technique to group together a set of items having same characteristics. There are two kinds of clusters that can be discovered. Usage clusters or page clusters. Clustering of users helps to cluster users having same characteristics. Page clustering helps to cluster pages having same content.

5 Pattern Analysis

Web site administrators are extremely interested in questions like "How are people using the site?", "Which pages are being accessed most frequently?" etc. These questions require the analysis of the structure of hyperlinks as well as the contents of the pages. The end products of such analysis might include:

- The frequency of visits per document?
- Most recent visit per document?
- Who is visiting which documents?
- Frequency of use of each hyperlink?
- Most recent use of each hyperlink?

Most common types of pattern discovery and analysis techniques employed in the Web usage mining are:

- Session and Visitor Analysis
  It is statistical analysis of pre-processed session data used to gain knowledge about visitor behavior. Reports based on this type of analysis may include information about most frequently accessed pages, average view time of a page, average length of a path through a site, common entry and exit points, and other aggregate measures.

- Cluster Analysis and Visitor Segmentation
  Clustering is a data mining technique that groups together a set of items having similar characteristics. In the usage domain, there are two kinds of interesting clusters that can be discovered:

  - Usage clusters: Establish groups of users exhibiting similar browsing patterns
  - Page clusters: Clustering of pages having similar content

- Association and Correlation Analysis
  Association rule discovery and statistical correlation analysis can find groups of items or pages that are commonly accessed or purchased together. Most common approaches to association discovery are based on the Apriori algorithm. This algorithm finds groups of items (page views appearing in the preprocessed log) occurring frequently together in many transactions (i.e., satisfying a user specified minimum support threshold). Such rules can also be used to optimize the structure of the site.
- Analysis of Sequential and Navigational Patterns
  The technique of sequential pattern mining attempts to find inter-session patterns such that the presence of a set of items is followed by another item in a time-ordered set of sessions or episodes. By using this approach, Web marketers can predict future visit patterns which will be helpful in placing advertisements aimed at certain user groups.
- Classification and Prediction based on Web User Transactions
  Classification is the task of mapping a data item into one of several predefined classes. Classification can be done by using supervised learning algorithms such as decision trees, naive Bayesian classifiers, k-nearest neighbor classifiers, and Support Vector Machines. For example, given a set of user transactions, the sum of purchases made by each user within a specified period of time can be computed.

6 Experimental Setup

- Log file is collected for proposed tenure i.e. one month from server (7th July–6th August)
- Total No of Records: 30,296
- Algorithms selected for Field Extraction [12], Data Cleaning [7], User identification [7] and Session Identification [13] are implemented using Net Beans and SQL Server 2005

Table 1 below shows Testing Environment used to perform preprocessing on the collected log file.

6.1 Input and Output of Preprocessing Algorithm

- Input-Log file
- Output-Preprocessed session File

Table 1. Testing environment

Main memory	2.00 GB
CPU	Intel Core i3
Operating system	Windows 7(32 bit OS)

After implementing algorithms selected for various preprocessing stages as given in Sect. 6 above output retuned in terms of records cleaned, users and sessions identified is given in Table 2 below.

Table 2. Testing environment

Dataset	Duration	Original records	Records after preprocessing	Users identified	Distinct URL's identified
University log file	7th July 2015-6th August 2015	30295	18962	29	196

6.2 Pattern Discovery Through Statistical Analysis

Using Statistical analysis technique on session file created after running preprocessing algorithm that performed Data Cleaning, User Identification, Session Identification it was found out that there are

Distinct URL's: 196
Distinct Users: 29
Most frequently accessed URL: http://gmail.com.

6.3 Pattern Discovered Through Session and Visitor Analysis

We have performed Session and Visitor Analysis doing statistical analysis of pre-processed session data to gain knowledge about visitor's behavior. This analysis gave us information about most frequently accessed pages which is gmail.com.

7 Discussion

Data preprocessing is significant and prerequisite phase in Web Usage Mining. Various heuristics are used in various phases so as to remove irrelevant items and to identify users and sessions along with other browsing information. This would result in creation of user session file which can then be used for various mining tasks in the subsequent pattern discovery phases. In paper above

- Algorithms for various preprocessing stages were identified
- Data cleaning, user identification and session identification was performed
- Session file created after preprocessing is then used to identify most frequently accessed web page through session and visitor analysis

8 Scope of Future Research

Currently only one form of pattern analysis: session and visitor analysis has been done. There are other forms of analysis which can be done as discussed in Sect. 5. For example clustering of similar types of sites accessed by users can be done.

References

1. Reddy, K.S., Partha Saradhi Varma, G., Babu, I.R.: Preprocessing the web server logs an illustrative approach for effective usage mining. ACM SIGSOFT Softw. Eng. Notes **37**, 1–5 (2012)
2. Cooley, R., Mobasher, B., Srivastava, J.: Web mining: information and pattern discovery on the World Wide Web. In: Proceedings of Ninth IEEE International Conference, pp. 558–567 (1997)
3. Frias-Martinez, E., Karamcheti, V.: A customizable behavior model for temporal prediction of web user sequences. In: WEBKDD, pp. 66–85 (2002)
4. Huiying, Z., Wei, L.: An intelligent algorithm of data pre-processing in web usage mining. In: Proceedings of the 5th World Congress on Intelligent Control and Automation, pp. 3119–3223. IEEE (2004)
5. Fayyad, U.M., Piatetsky-Shapiro, G., Smyth, P.: From data mining to knowledge discovery: an overview. In: Advances in Knowledge Discovery and Data Mining, pp. 1–34. AAAI/MIT Press (1996)
6. Langhnoja, S., Barot, M., Mehta, D.: Pre-processing: procedure on web log file for web usage mining. Int. J. Emerg. Technol. Adv. Eng. **2**(12), 419–423 (2012)
7. Suguna, R., et al.: User interest level based preprocessing algorithms using web usage mining. Int. J. Comput. Sci. Eng. (IJCSE) **5**(09), 815–822 (2013)
8. Vijayashri, L., Madhuri, J.: Data preprocessing in web usage mining. In: Proceedings of International Conference on Artificial Intelligence and Embedded Systems (ICAIES 2012), Singapore, pp. 1–5 (2012)
9. Mobasher, B.: Web usage mining in web data mining exploring hyperlinks, contents, and usage data, Chicago, USA, Ch. 12, pp. 449–482. Springer, Heidelberg (2012)
10. Chitraa, V., Davamani, D., Selvdoss, A.: A survey on preprocessing methods for web usage data. Int. J. Comput. Sci. Inf. Secur. (IJCSIS) **7**, 78–83 (2010)
11. Cooley, R., Mobasher, B., Srivastava, J.: Data preparation for mining world wide web browsing patterns. Knowl. Inf. Syst. **1**, 1–27 (1999)
12. Patil, P., Patil, U.: Preprocessing of web server log file for web mining. World J. Sci. Technol. **2**, 14–18 (2012). Proceedings of National Conference on Emerging Trends in Computer Technology (NCETCT 2012)

A Composite Technique to Fortify Security for DaaS Services in Cloud Environment

Samarjeet Yadav(✉), Pratishtha Saxena, Neelam Dayal, and Shiv Prakash

Department of Computer Science and Engineering, Center for Advanced Studies, AKTU Lucknow, Lucknow, India
17mcs11@gmail.com

Abstract. Cloud computing gives an extensive amount of virtual storage to users. The demands, the importance and the usage of cloud computing are increasing day by day. Therefore, security problems and threats are also increasing. An individual and any business organization or a company may have a huge amount of data; they cannot store it physically because the cost of that huge data storage will be more. Cloud computing solves this problem by providing the platform to store huge amount of data. The data storage on the cloud can be blend with the other users' data. This can lead to data protection issues in cloud storage. As confidentiality is the main parameter of security, if confidentiality is broken then it will result in loss of data for individuals and organizations. This data needs to be protect with confidentiality measures before being store on cloud. To provide security of this data we are using composite technique i.e. obfuscation with encryption to fortify the security level. In this paper, we propose a composite technique to increase the security level while reducing the data size, and the time of execution.

Keywords: Internet of things · Cloud computing · Encryption · Obfuscation · Integrity · Data security · Data storage

1 Introduction

IOT devices generates huge amount of data, for this data cloud computing provides a pathway, so that data travel to its destination. This data needs to be secured in order to ensure privacy and security to the network users.

Cloud computing is a big shift from the long-established way businesses used to think about IT resources. Cloud computing having important features, such as providing a huge amount of storage for storing data, is attracting organizations towards it. Along with storage, it facilitates many services such as minimizing cost, escalating speed, productivity, performance, global accessibility, and security. Physical storage of large amount of data would not be feasible for an organization, hence the cloud computing comes as an aid to them. In future, cloud computing is going to be internet based and computing system will provide reliable and custom services for the users to access. Cloud computing works with different types of cloud applications, it also gives a way

N. Nain and S. K. Vipparthi (Eds.): ICIoTCT 2019, AISC 1122, pp. 160–169, 2020.
https://doi.org/10.1007/978-3-030-39875-0_17

to access and store the data of cloud from anywhere in the world only by connecting the cloud applications via the internet.

Cloud storage provides virtual space to store the major portion of the data; however, the owner of the data has not much control over their data to secure it on cloud. Cloud provider have all the control over the user's data security.

The security is one of the major challenges in cloud computing, due to its nature of outsource computing. If a vigorous security scheme is not implemented on the cloud system then it may be exposed to several vulnerabilities. A number of security parameters should be used to secure the cloud data in order to provide confidentiality, integrity and all the security levels of data shield security.

The main pillar of security is confidentiality, which ensures that data is accessible by an authenticated cloud user only. It also protects the data from insider and outsider attackers.

Either insiders or outsiders can do the attacks on cloud data. Outsiders can be hackers and can attack data from the outside cloud service provider, but insider act as an administrator. It is hard to identify both type of attackers but insiders are much difficult to identify than the outsider. Therefore, the user should be very careful while storing any data on cloud.

If any of the attack occurs then entire data is exposed to third party Therefore, it is necessary to encrypt all data before storing it on cloud.

In Fig. 1 Presents, risk categories of cloud security, which includes data security, physical security, technological and compliance & audit; which are further sub-categories as per their requirement.

In this paper, we focused on data security, which is one of the big challenge for cloud security. CIA (confidentiality, integrity, and availability) is the main pillar of security, along with that data loss & leakage, user access, data segregation, access management are also the challenges for data security.

The overall paper is organized in five sections. In Sect. 2 we are briefly discussing various security proposals until date. In Sect. 3 our proposed technique is discussed. In Sect. 4, we are providing a detailed discussion on the results with security analysis, which is followed by Sect. 5 conclusion.

2 Literature Review

Arockiam et al. [1] proposed a technique to achieve security with the help of confidentiality. The main aim of this research was to provide confidential data by using the combination of obfuscation and encryption techniques. The author also used private keys while encryption so that only permitted user get that key for the decryption. Therefore, for the security purpose, the keys were generated and used to secure the data with the help of the obfuscation mechanism.

Suthar et al. [2] proposed a new framework which was the combination of encryption and obfuscation technique. Confidentiality for the data in network was provided by cloud encryption. The author also proposed tracking of a user key, along with the hash of the file with the help of secured storage server. For cloud providers, they proposed an efficient

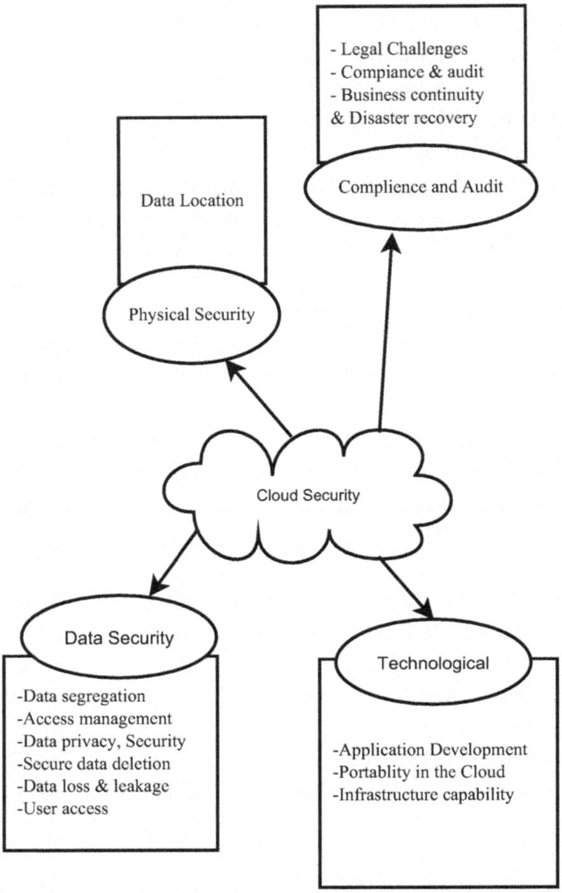

Fig. 1. Cloud security risk categories and subcategories.

technique by which the users secret information such as passwords, contact details and other personal information etc. not modified by the third party

Velumadhava Rao et al. [3] in the year 2017 proposed a technique which was used for securing bot numerical and non-numerical data with the help of obfuscation and encryption in public cloud storage. Due to that technique, the size of the data reduce, it took less cost service. There were two private keys, which were generated from the cloud server to ensure the confidentiality for numerical data. The obfuscation technique was executed at users side before uploading that data and by using mathematical function and programming logic such that mul (), pow () etc. were proposed. The main advantage is to extract the data size by this technique. Basu *et al.* [4] proposed an algorithm named as AROcrypt to secure the data storage in the cloud. They described (SEaaS) security as a service in the cloud environment. The author also conducted the simulation by using analysis tools and its existing techniques. The technique proposed by the author provided by SEaaS model. Data stored on the cloud after encryption while the private key or secret key retained with the owner and the user was granted corresponding decryption keys.

The proposed technique by author (AROcrypt) was based on the technique of symmetric encryption. Hence, the data got encrypted before getting uploading or sending on the cloud.

Different authors from [6–9] discussed, they showed in their paper about the security on cloud computing. They used encryption and obfuscation technique to secure it but these techniques were taking time to encrypt and decrypt the data separately whereas [10] showed the full survey on privacy and security in cloud computing, they showed many methods by which one can secure their data. Partheeban et al. [11] showed in their research paper about the security of deliver models such as PaaS, SaaS and IaaS. [20–27] discussed the algorithm based on cloud storage security. They used different types of techniques and methods to execute it and to provide security for data in a cloud environment.

3 Proposed Technique

In this paper, we mainly focused on three domains of cloud computing that are uploading, integrity management and secure access control with proper rights management.

The proposed composite technique is used to secure the numerical data in the cloud environment. If the user encrypts the numerical data with the help of obfuscation technique then the data will be obfuscated, hence it is suitable and convenient to secure the data. There are two private keys K_1 and K_2, which are used in this proposed composite algorithm for encryption, decryption respectively. The proposed composite technique uses five different types of mathematical operations such as mul (), pow (), rotate (), random value (1024), mod (), and ASCII () on numerical data. The secret keys K_1 and K_2, are generated from the cloud side and are forwarded to the users. Key Management as a Service (KMaaS) is used for generating the secret keys, which will be maintained at service provider.

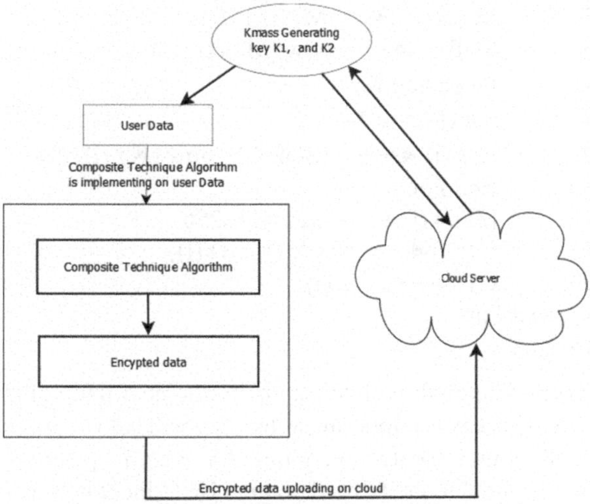

Fig. 2. Flow chart of composite technique.

Figure 2 shows the overall process of composite technique, Kmass and cloud servers are linked to each other and Kmass will generate secret key K1 and K2. Composite technique will take the data of user and it will implement the process and give the output data, which will be in encrypted form. Further, this encrypted data will be uploaded on cloud.

Algorithm 1 describes the process of data encryption mechanism proposed in this paper. PT which is plaintext, NS, is the size of plaintext and K_1, K_2 are the secret keys. Cipher text CT is generated as the output of the algorithm.

Line 1 in algorithm begins the process, following the line 2 which is fetching the key K_1, line 3 starts the for loop for variable i having the range from 0 to (NS − 1), further the multiplication of key K_1 and PT(i) is stored in the variable $M_x(i)$ for every i^{th} position which is shown in line 4. Now, we are taking random value for creating confusion while decrypting the file by third party. Here we take 1024 as random value, as shown in line 5 we are adding this value with $M_x(i)$ and storing it in variable M_{add}. In line 6, we are doing square of M_{add} so that the number become large of every i^{th} position and storing it in variable SQR.

Algorithm 1: Composite Technique for Data Security

Input: PT ⟵ Plaintext
NS ⟵ the size of (PT)
K_1, K_2 ⟵ secret keys

Output: Cipher Text, CT

1. Begin
2. Fetch key K_1
3. For (i=0 to (NS-1))
4. M_x (i) ⟵ K_1 * PT (i)
5. M_{add} (i) ⟵ M_x (i) +1024
6. SQR ⟵ pow (M_{add} (i), 2)
7. Fetch key K_2
8. For (j= 1 to NS)
9. Res (i) ⟵ (SQR (i) +pow (K_2 (j), 2))
10. End For
11. MOD (i) ⟵ Res (i) %256
12. CT (i) ⟵ ASCII (MOD (i))
13. CT ⟵ CT (i)
13. End For
14. End

In line 7, the second key will be fetched and in the next line 8, the range for K_2 is defined from 1 to NS. This key is important, as this key will help to provide more security for data. Key K_2 will be incremented every time for every i^{th} position and it will keep incrementing until it is equal to the NS or less than NS. Line 9 is showing the square of value K_2 for every incremented value and adding these i^{th} value to the corresponding

i^{th} value of SQR(i), the addition of these values are stored in variable Res(i). In line 10, the MOD of Res(i) for every i^{th} position is taking out and further this MOD value is converting into ASCII character as shown in line 11, which gives our output in the form of cipher text CT. The ASCII character will be obtained for each mod value.

4 Result Analysis and Discussion

To have the better understanding of the proposed method a sample run, with sample data is explained as follows:

INPUT:

Let us consider the plaintext that is the salary of the faculty.

PT-33 56 98 66 74 56 99 11 23

Now, find the size of values that is in the PT and put it as NS.

NS-9

K_1 and K_2 are secret keys.

DEFINE:

I is the variable and having range from 0 to (NS − 1).

J is the variable and having range from 0 to NS.

Step 1: Fetch the value key K_1.

Step 2: Now get the generated key K_1 value, which is multiplying by the plain text (PT), and put it as M_x. For example, the value of K_1 is 11. Now, multiply the K_1 and PT as shown in Table 1.

Table 1. Multiplication of PT and K_1.

Mx(i)	PT(i) * K1
33	363
56	616
98	1078
66	726
74	814
56	616
99	1089
11	121
23	253

$K_1 = 11$

Step 3: Add random value 1024 into $M_x(i)$ and store it in $M_{add}(i)$

Step 4: Now, a square of value for M_{add} values are calculated: Now, Find the square SQR(i) for Madd(i) as shown in Table 2.

Table 2. Squaring value of M_{add}.

SQR	Pow (Madd(i), 2)
1387	1923769
1640	2689600
2102	4418404
1750	3062500
1838	3378244
1640	2689600
2113	4464769
1145	1311025
1277	1462165

Step 5: K_2 the secondary is generated and here the sample $K_2 = 2$.

$K_2 = 2$

Step 6: Square the value of K_2 with each increment and add that value in SQR(i) as shown in Table 3.

Table 3. Multiplication of SQR and K_2

Res(i)	SQR(i) * pow(K_2++, 2)
1923769	7695076
2689600	24206400
4418404	70694464
3062500	76562500
3378244	121616784
2689600	131790400
4464769	285745216
1311025	106193025
1462165	163072900

Step 7: By dividing Res(i) by 256 we can find the Modulus value. The Mod values can be calculated by dividing the Res(i) values by 256, and the ASCII character will be produced for each and every mod values. These ASCII characters are the cipher text of the (PT) plaintext as shown in Table 4.

Table 4. MOD value for Res

Res%256	Mod value
7695076	114
24206400	192
70694464	144
76562500	116
121616784	152
131790400	192
285745216	8
106193025	185
163072900	90

Step 8: The Converted MOD(i) into ASCII Code to produce the cipher text CT is **CT= rÀ t˜À¹Z**

CT= rÀ t˜À¹Z

The proposed composite technique executed successfully and produced the cipher text (CT) with the 256 types of ASCII character codes. The following solutions are drawn from the above execute results and sample data inputs. Plain text to Cipher text:

The Plain text is: 33 56 98 66 74 56 99 11-23
Cipher Text: rÀ t˜À¹Z

With the help of obfuscation technique, the size of data is also reduced.

The data size for the above same plaintext (PT) is 27 bytes. (The plaintexts are 33 56 98 66 74 56 99 11 23). However, the size of the cypher text data (The Cipher text are **rÀ t˜À¹Z**) for the same plaintext is 12 bytes. It decreased by one-third (1/3).

From the findings, it is very clear that confidentiality is maintained and therefore the security is enhanced.

Table 5 shows, it shows that the execution time of real time technique result is more than the proposed composite technique result. It also shows data size is being reduce in proposed composite technique.

Table 5. Real time and proposed technique comparison.

Parameters	Real time technique result	Proposed composite technique result
Execution time	8 s	3 s
Data size	18 byte (27 bytes)	12 byte (27 bytes)

5 Conclusion

As discussed in the paper, cloud computing is one of the areas in which the problems regarding data security is increasing day by day. Large data cannot be stored physically because of the storage problem. Encryption and obfuscation are two techniques, which reduce cost and size of the data. It also enhance the security level of data. In this paper, we have proposed a technique by combining both techniques to minimize the cost and size of the data by encrypting it before storing it on the cloud. The proposed method enhances the security level of cloud data and plays a vital role in DaaS.

References

1. Arockiam, L., Monikandan, S.: Arocrypt: a confidentiality technique for securing enterprise's data in the cloud. Int. J. Eng. Technol. **7**, 245–253 (2015)
2. Suthar, K., Patel, J.: Proceedings of International Conference on Communication and Networks, vol. 508 (2017)
3. Velumadhava Rao, R., Selvamani, K.: Data security challenges and its solutions in cloud computing. Procedia Comput. Sci. **48**, 204–209 (2015)
4. Basu, S., et al: Cloud computing security challenges & solutions-a survey. In: 2018 IEEE 8th Annual Computing and Communication Workshop and Conference, pp. 347–356 (2018)
5. Suthar, K., Patel, J.: ObfuCloud : An Enhanced Framework for Securing DaaS services using Data Obfuscation Mechanism in Cloud Environment, pp. 1–10 (2017)
6. Saravana Kumar, N., Rajya Lakshmi, G.V., Balamurugan, B.: Enhanced attribute-based encryption for cloud computing. Procedia Comput. Sci. **46**, 689–696 (2015). ICICT 2014
7. Oli, S.A., Arockiam, L., IEEE: Confidentiality Technique to Obfuscate the Numerical Data to Enhance Security in Public Cloud Storage (2017)
8. Celesti, A., Fazio, M., Villari, M., Puliafito, A.: Adding long-term availability, obfuscation, and encryption to multi-cloud storage systems. J. Netw. Comput. Appl. **59**, 208–218 (2016)
9. Hosseinzadeh, S., Hyrynsalmi, S., Conti, M., Leppänen, V.: Security and privacy in cloud computing via obfuscation and diversification: a survey. In: Proceedings of the IEEE 7th International Conference on Cloud Computing Technology and Science, CloudCom 2015 (2016)
10. Shankarwar, M.U., Pawar, A.V.: Security and privacy in cloud computing: a survey. In: Advances in Intelligent Systems and Computing (2015)
11. Partheeban, P., Kavitha, V.: A study with security concerns in service delivery models of cloud computing. Int. J. Appl. Eng. Res. **10**, 42219–42230 (2015)
12. Deshpande, P., Sharma, S.C., Kumar, P.S.: Security threats in cloud computing. In: International Conference on Computing, Communication, and Automation, ICCCA 2015 (2015)

13. Chauhan, K.K., Sanger, A.K.S., Verma, A.: Homomorphic encryption for data security in cloud computing. In: Proceedings of the 2015 14th International Conference on Information Technology, ICIT 2015 (2016)
14. Mell, P., Grance, T.: The NIST Definition of Cloud Computing (Draft) Recommendations of the National Institute of Standards and Technology, NIST Special Publication (2011)
15. Collberg, C., Thomborson, C., Low, D.: A taxonomy of obfuscating transformations. Technical report (1997)
16. George Amalarethinam, D.I., Fathima Mary, B.: Confidentiality technique for enhancing data security in public cloud storage using data obfuscation. Int. J. Control Theory Appl. **9**, 107–113 (2016)
17. Maheshwari, V., Nourian, A., Maheswaran, M.: Character-based search with data confidentiality in the clouds. In: CloudCom 2012 - Proceedings of the 2012 4th IEEE International Conference on Cloud Computing Technology and Science, pp. 895–899 (2012)
18. Duggins, S., Tsui, F., Karam, O., Kubanyi, Z.: Semantic Obfuscation and Software Intention. Cse.Spsu.Edu
19. Kaur, M., Singh, R.: Implementing encryption algorithms to enhance data security of cloud in cloud computing. Int. J. Comput. Appl. **70**, 16–21 (2013)
20. Yang, K., Jia, X.: An efficient and secure dynamic auditing protocol for data storage in cloud computing. IEEE Trans. Parallel Distrib. Syst. **I**(9), 2–5 (2014)
21. Younis, Y.A., Kifayat, K., Merabti, M.: An access control model for cloud computing. J. Inf. Secure Appl. **19**(1), 45–60 (2014)
22. Velumadhava Rao, R., Selvamani, K.: Data security challenges and its solutions in cloud computing. Procedia Comput. Sci. **48**(C), 204–209 (2015)
23. Subashini, S., Kavitha, V.: A survey on security issues in service delivery models of cloud computing. J. Netw. Comput. Appl. **34**(1), 1–11 (2011)
24. Khan, K.M., Shaheen, M.: Data obfuscation for privacy and confidentiality in cloud computing. In: Proceedings of the 2015 IEEE International Conference on Software Quality, Reliability and Security-Companion, QRS-C 2015 (2015)
25. Zhou, J., Cao, Z., Dong, X., Vasilakos, A.V.: Security and privacy for cloud-based IoT: challenges. IEEE Commun. Mag. **55**, 26–33 (2017)
26. Fang, J., He, J., Pathan, M.S., Gao, M.: A novel storage and search scheme in cloud computing. In: Proceedings of the 12th International Conference on Computational Intelligence and Security, CIS 2016 (2017)
27. Arockiam, L., Monikandan, S.: Efficient cloud storage confidentiality to ensure data security (2014). ieeexplore.ieee.org

Performance of Dual-Input Storage Based Induction Motor Drive for Electric and Fuel Cell Hybrid Electrical Vehicle Applications

Narayan Yadav[1](✉), Sushma Gupta[2](✉), and Tripta Thakur[2](✉)

[1] MANIT, Electrical Engineering, Bhopal, India
narayanyadav777@gmail.com
[2] MANIT, Department of Electrical Engineering, Bhopal, India
sush_gupta@yahoo.com, tripta_thakur@yahoo.co.in

Abstract. This paper presents the design and control of a two-input bi-directional DC-DC converter (BDC) and implements it for driving a Fuel-cell based three-phase induction motor drive. The Fuel-cell hybrid electric vehicle (FC-HEV) powertrain mainly consists of dc-source, DC-DC converter, inverter, and a driving motor. Usually the single source is used for powering electric vehicles (EVs). However, due to a single source the range of EVs is limited. In this study, two separate dc sources (battery and fuel-cell) are hybridized using a multiport BDC. This BDC is utilized to power FC-HEV drivetrain. The FC-HEVs have to operate in powering mode and braking mode, the performance of induction motor drive in forward powering mode and regenerative braking mode has been discussed. Moreover, for controlling purpose the v/f control is implemented to enhance the performance of EV drive and the simulation results have been obtained in MATLAB environment.

Keywords: EVs · FC-HEVs · Multi-input bi-directional DC-DC converter · Constant v/f control

1 Introduction

Automobiles are growing rapidly in developing countries due to improving human development index. With the rising urbanization, automobiles are becoming pervasive. On the one hand, increasing numbers of vehicles show development while on the other hand it also increases the number of hazardous gases in the atmosphere. Conventional fossil fuel based vehicles are one of the major sources of greenhouse gas emission, so there is a rising demand for shifting vehicular technology to the alternate one, to cope up with climate change. Hybrid vehicles are one of the alternatives to reduce carbon emission in the atmosphere. Electric Vehicles (EVs) can solve the problem of air pollution [1] as well as minimize declination of fossil fuel, a battery can be charged by renewable energy source. Furthermore, EVs overall efficiency is better than gasoline based vehicles.

N. Nain and S. K. Vipparthi (Eds.): ICIoTCT 2019, AISC 1122, pp. 170–176, 2020.
https://doi.org/10.1007/978-3-030-39875-0_18

Basic construction and classification of Hybrid electrical vehicle (HEV) are described in [2]. Electrical vehicle (EVs) and fuel cell hybrid electric vehicle (FC HEVS) are becoming more popular due to zero emission and simple propulsion. The basic block diagram of FC-HEV drive is consists of energy sources (primary and secondary), bidirectional DC-DC converter (BDC), PWM controlled inverter, motor and controlling unit which is shown in Figs. 1 and 2 respectively.

The recent development of EVs and their key components highlights in [3, 4].

In battery driven EVs, the voltage level of batteries are low, thus DC-DC converter is used to boost the voltage level. DC-DC converters with bidirectional capability are essential for working the EVs in a regenerative braking mode. Bidirectional converters (BDC) with multiple inputs are gaining popularity nowadays. BDC can be of two types namely isolated BDC and non-isolated BDC. Isolated BDC is mostly used in renewable power supply due to their high voltage gain whereas Non-isolated BDC [5, 6] used in EVs/HEVs applications because of reduced size.

Earlier only batteries were used to power electric motors, but with the recent advancement fuel cell, ultra-flywheel and ultra-capacitors are also used as alternative energy sources [7–9]. A fuel cell has specification similar to a battery. However, a fuel cell produces electricity by using hydrogen and oxygen, whereas electrolyte presents in battery produce electricity.

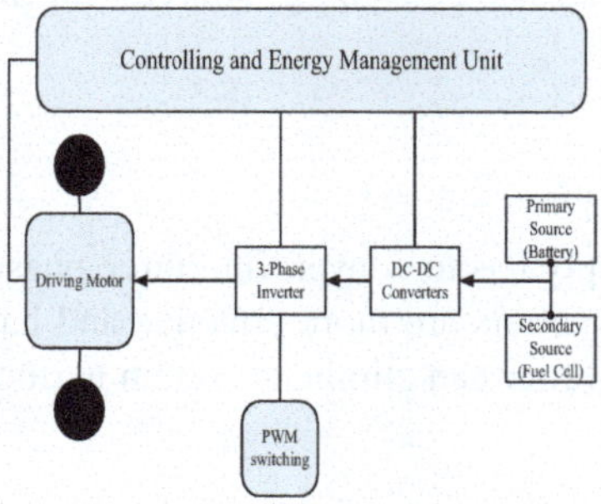

Fig. 1. Forward motoring mode of FC EV.

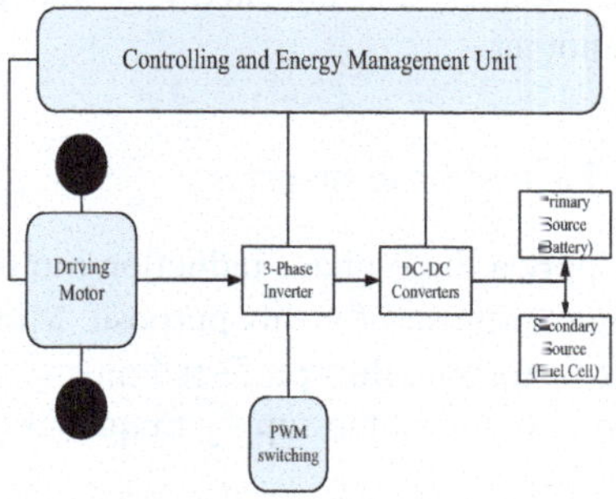

Fig. 2. Regenerating braking mode of FC HEV

For running an electric vehicle, a driving motor is required i.e. induction motor, permanent magnet synchronous motor (PMSM), brushless dc motor (BLDC) and switched reluctance motor (SRM) [10]. To control this motors various controlling scheme for speed control of induction machine and PMSM are available now [11, 12]. Controlling schemes of machines can be classified into two types, Scalar control, and vector control. Scalar control deals with the magnitude and generally used for controlling the speed of induction motor drives. Vector control is a more complex controlling scheme, used for speed control of PMSM drives. In this study scalar control [13] of induction motor is used in which the ratio of voltage and frequency is kept constant to maintain the flux constant.

2 Key Components of EVs and FC HEVs

2.1 Energy Storage Unit

The batteries are generally used as primary storage energy source. A good battery should have characteristics like high charging and discharging efficiency, high power density, less internal resistance and better life. In earlier days lead-acid batteries were used to power the electric vehicles, but recently with advancement in battery technology, various batteries with the better power density and lesser volume are available i.e. lithium-ion (Li-ion), nickel-cadmium (Ni-Cd) and Nickel-Metal Hydride (Ni-MH) [7–9]. Moreover newer technology i.e. Fuel cell, ultra-capacitors are also available to replace the batteries.

2.2 Bidirectional DC-DC Converter

To match with the voltage level of driving motor, BDC is required to increase or decrease the voltage level according to the need of a vehicle. A DC-DC converter transforms unregulated voltage to regulated dc voltage. The DC-DC converter can be classified into the unidirectional and bidirectional converter. Unidirectional DC-DC converters are used where power fed from source to load is required, whereas bidirectional can support power flow from both directions. BDC design in [5] can interface two input supply (main source and auxiliary source) at a time, also both energy sources can charge one another.

2.3 Three-Phase Inverter

To supply a three-phase induction motor, dc supply has to convert into three phase ac, inverters are used for this purpose. Multilevel inverters are more effective and having lesser distortion, thus preferred mostly. PWM inverter can eliminate certain harmonics as well as control the supply frequency.

2.4 Driving Motor

According to the type of supply, driving motor can be classified as ac motor and dc motor. Earlier only dc motors were used to drive electrical vehicles because of their easy controlling. Recently with the development of better control technique, ac motors are preferred over dc motors. Among ac motors, permanent magnet synchronous motor (PMSM) and three-phase induction motors are used in large numbers for vehicles manufacturing.

3 MATLAB Sımulınk Model of Electrıc Vehicle Drive

The electric vehicle mainly consists of energy storage, DC-DC converters for maintaining voltage level, inverter, and driving motor. A Simulink model of EVs drive is shown in Fig. 3. The voltage obtained from battery and fuel cell is low voltage un-regulated DC, but the required voltage is regulated high-level voltage. Therefore, the DC-DC converter implemented in the system has become a coupling medium to provide a high level

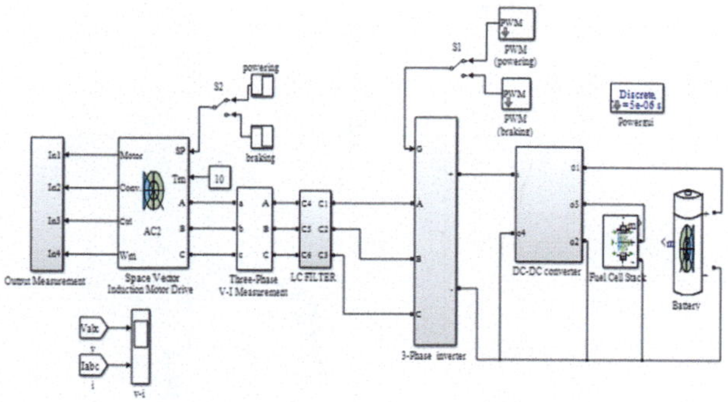

Fig. 3. A Simulink model of FC-HEV drive.

regulated DC supply at the output port. The output voltage obtained from BDC in the form of regulated DC which is converted through the DC-AC converter in the form of three-phase AC. Moreover, to obtain the precise control a space vector control technique has been implemented.

A bi-directional DC-DC converter with dual-input energy storages is shown in Fig. 4. For energy supply, a Lithium-ion battery of 98 V and fuel cell stack of 48 V are used. A dc voltage of 410 V is obtained at the output terminal.

A current control based scheme for controlling of all the five IGBTs switches is shown in Fig. 5.

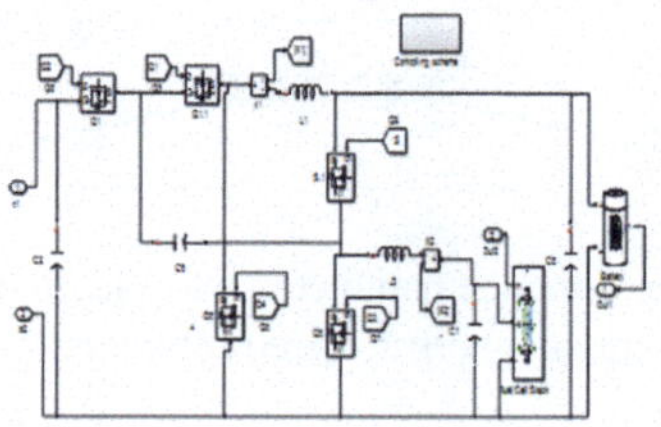

Fig. 4. A dual-input DC-DC bidirection converter

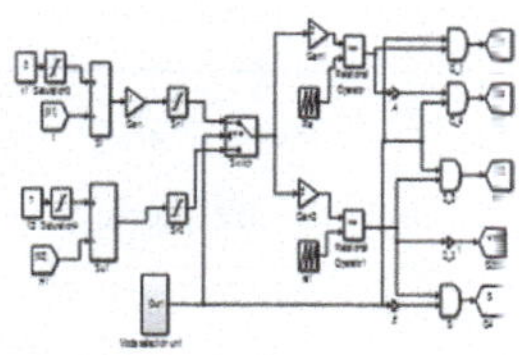

Fig. 5. Controling scheme of dual-input DC-DC converter of FC HEV

4 Result and Discussion

Figure 6 shows the speed characteristics of Induction. The simulation time is t = 2 s. During the forward powering mode, speed is increased from 0 to 1500 rpm, the motor attains the set value 1500 rpm at t = 0.6 s. The induction motor runs as a constant speed drive till t = 1.5 s, after t = 1.5 s, the motor now operate in regenerative braking mode thus speed decreases from 1500 rpm to 0 rpm when time reach t = 2.5 s (Fig. 7).

The v/f control is one of the most popular method for speed control of induction motor. A complete v/f control speed response of induction motor is shown in Fig. 8. In forward powering mode the value of voltage and frequency is increased proportionally

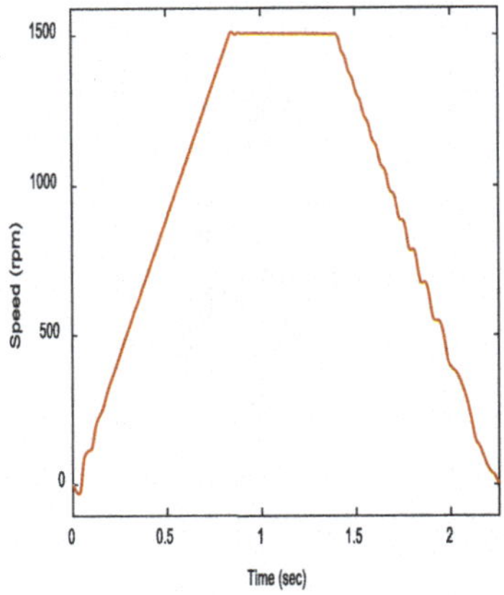

Fig. 6. Characteristics of induction motor

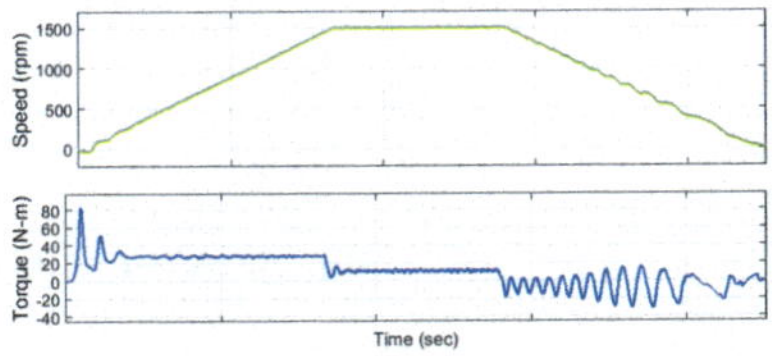

Fig. 7. Speed and torque characteristic of induction motor

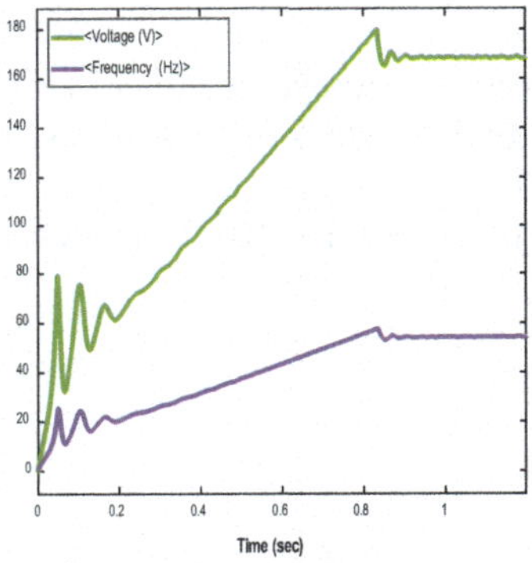

Fig. 8. Voltage and frequency response

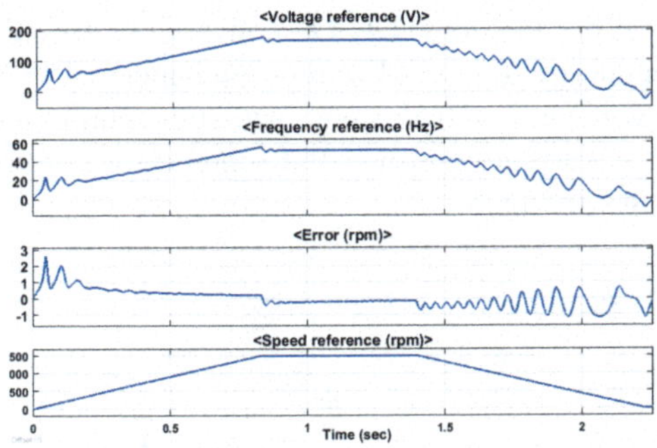

Fig. 9. Performance parameters of induction motor Voltage; frequency; error; speed reference

in such a way that ratio of v/f remains constant. Similarly, in regenerative braking mode, frequency and voltage are decreased, as a result, induction motor started to operate in regenerative braking mode.

To operate the induction motor in forward powering mode, the supply voltage is raised to 180 V and frequency up to 50 Hz which is shown in Fig. 9. The speed of induction is increased to 1500 rpm in forward powering mode, the reference speed has been set to 1500 rpm and the actual speed of induction motor follows the reference speed, the error generated is the difference between both, error between actual operating speed and reference speed is shown in Fig. 9. Also, the shape of voltage, as well as frequency, is observed the same which justifies the realization of constant v/f control.

4.1 Forward Powering Mode

In forward powering mode lithium-ion battery and fuel cell stack supply energy to induction motor. In this mode, induction machine working as a motor with positive slip. To maintain the flux constant ratio of V/f is maintain constant which is shown in Fig. 9.

4.2 Regenerative Braking Mode

In this mode by using PWM-inverter supply frequency is reduced below rated frequency so that slip becomes negative and induction machine works as an induction generator. To maintain flux constant, voltage and frequency are reduced proportionally.

5 Conclusion

A bi-directional DC-DC converter for FC-HEVs with multiple energy sources is developed in this study. The BDC has been utilized to convert the lower voltage level of multiple sources to a higher voltage level with high voltage gain. The converter has six switches which can be controlled independently to dive the induction motor in powering and braking mode. From the result obtained in MATLAB/Simulink tool, it has been observed that when induction motor is operated in forward powering mode, the voltage and frequency linearly increase and linearly decreases when operating in braking mode. Moreover, while switching from powering mode to braking mode, the smooth speed control has been achieved. The performance of this scheme makes it suitable for FC-HEVs.

References

1. Habib Ullah, M., Gunawan, T.S., Sharif, M.R., Muhida, R.: Design of environmental friendly hybrid electric vehicle 2012. In: International Conference on Computer and Communication Engineering (ICCCE), Kuala Lumpur, pp. 544–548 (2012)
2. Kebriaei, M., Niasar, A.H., Asaei, B.: Hybrid electric vehicles: An overview. In: 2015 International Conference on Connected Vehicles and Expo (ICCVE), Shenzhen, pp. 299–305 (2015)
3. Butler, K.L., Ehsani, M., Kamath, P.: A Matlab-based modeling and simulation package for electric and hybrid electric vehicle design. IEEE Trans. Veh. Technol. **48**(6), 1770–1778 (1999)
4. Chan, C.C.: The state of the art of electric and hybrid vehicles. Proc. IEEE **90**, 247–275 (2002)
5. Lai, C., Lin, Y.-J., Hsieh, M.-H., Li, J.-T.: A newly-designed multiport bidirectional power converter with battery/supercapacitor for hybrid electric/fuel-cell vehicle system. In: 2016 IEEE Transportation Electrification Conference and Expo, Asia-Pacific (ITEC Asia-Pacific), Busan, pp. 163–166 (2016)
6. Nahavandi, A., Hagh, M.T., Sharifian, M.B.B., Danyali, S.: A nonisolated multiinput multioutput DC–DC boost converter for electric vehicle applications. IEEE Trans. Power Electron. **30**(4), 1818–1835 (2015)
7. Khaligh, A., Li, Z.: Battery, Ultracapacitor, Fuel Cell, and Hybrid Energy Storage Systems for Electric, Hybrid Electric, Fuel Cell, and Plug-In Hybrid Electric Vehicles State of the Art. IEEE Trans. Veh. Technol. **59**(6), 2806–2814 (2010)

8. Bauman, J., Kazerani, M.: A Comparative Study of Fuel-Cell–Battery, Fuel-Cell–Ultracapacitor, and Fuel-Cell–Battery–Ultracapacitor Vehicles. IEEE Trans. Veh. Technol. **57**(2), 760–769 (2008)
9. Cook, B.: Introduction to fuel cells and hydrogen technology. Eng. Sci. Educ. J. **11**(6), 205–216 (2002)
10. Yang, Z., Shang, F., Brown, I.P., Krishnamurthy, M.: Comparative study of interior permanent magnet, induction, and switched reluctance motor drives for EV and HEV applications. IEEE Trans. Transp. Electrification **1**(3), 245–254 (2015)
11. Pindoriya, R.M., Mishra, A.K., Rajpurohit, B.S., Kumar, R.: Performance analysis of control strategies of permanent magnet synchronous motor. In: 2016 IEEE Region 10 Conference (TENCON), Singapore, pp. 3224–3227 (2016)
12. Wang, C.-C., Fang, C.-H.: Sensorless scalar-controlled induction motor drives with modified flux observer. IEEE Trans. Energy Convers. 18(2), 181–186 (2003)
13. Griva, C., Profumo, F., Ilas, C., Magureanu, R., Vranka, P.: A unitary approach to speed sensorless induction motor field oriented drives based on various model reference schemes. In: IAS 1996. Conference Record of the 1996 IEEE Industry Applications Conference Thirty-First IAS Annual Meeting, San Diego, CA, USA, 1996, USA, vol.3, pp. 1594–1599 (1996)

A Lightweight Authentication Scheme for RFID Using ECC

Atul Kumar(✉) and Ankit Kumar Jain

National Institute of Technology, Kurukshetra, Haryana, India
atul.kumar1995@gmail.com

Abstract. Internet of things (IoT) contains a massive number of "things" which are communicated without human interaction. The "things" refers to a physical device such as bar code, tag, smart card, etc. These devices contain a limited amount of memory and a small processing unit. Hence, these devices require a lightweight scheme to provide authentication. However, there are various security issues with these tags such as tracking, cloning, replay attacks, disclosure attack, etc. To remove these security issues, many researchers proposed various authentication schemes. However, there are various security issues, large storage and communication requirement in these schemes. A new lightweight authentication scheme for RFID tag has been proposed which uses hash function and elliptic curve cryptography to authenticate tag and server. The hash function provides irreversibly to the scheme. The performance analysis of the scheme is analyzed in terms of communication, storage, and memory required to store the data. The scheme requires 3 hash function and 1 multiplication to mutual authentication between tag and reader. The scheme provides mutual authentication, forward security and also resistance from tracking attack, cloning attack.

Keywords: IoT · RFID · Mutual authentication · Untraceable privacy

1 Introduction

Radio frequency identification (RFID) is used to identify the object in the network. Nowadays, there is a huge amount of RFID tag available in the market. These tags can be used in various places such as purchasing, logistics & supply chain, industry, service industries, health care, commercial, transportation, etc. RFID contains various operating bands such as a low - frequency band, medium frequency band and ultra-high frequency band. The low-frequency band operates in a frequency range from 30 kHz to 300 kHz. The medium-frequency band can operate from frequency range 3 MHz to 30 MHz. High-frequency band can operate from 300 MHz to 3 GHz.

RFID contains two components: tag and a reader. The reader is a device that consists of a processing unit, transmitter, and receiver. The tag (transponder)

N. Nain and S. K. Vipparthi (Eds.): ICIoTCT 2019, AISC 1122, pp. 177–183, 2020.
https://doi.org/10.1007/978-3-030-39875-0_19

is an electronic device which communicates with the reader. There are three types of transponders such as passive transponder, semi-passive transponder, and active transponder. The passive transponder contains no power battery. It receives power from an electric or magnetic field of the reader. Hence, the passive transponder moved out of the range of the reader, it contains no power supply. The passive transponder usually operates lightweight operation such as XOR, ROT, AND, OR. The semi-passive transponder contains a small power battery, this battery is used when the transponder is not in the range of the reader. When the transponder comes in the range of the reader, then it receives power from the reader through an electric or magnetic field of the reader similar to passive tag. The semi-passive tag contains longer life span than active tags. The active tag contains battery power and used to transfer and receive data from the reader. Active tag can compute heavily operation rather than passive tags [3].

In authentication of the RFID contains four classes such as full-fledged class, simple class, lightweight class, and ultra-lightweight class. In full-fledged class contains heavily operation such as asymmetric operation, hash function. This class requires large battery power and huge bandwidth. The simple class contains hash function and pseudo-random number. This class contains lessor battery power than full-fledged class. The lightweight class consists of pseudo-random number, checksum, and elliptic curve. The ultra-lightweight class contains simple operation such as AND, OR, Rot, XOR, etc. The ultra-lightweight class requires less battery power and memory rather than other classes. This paper proposed a lightweight mutual authentication scheme for RFID. The scheme consists of hash function and elliptic curve to authenticate tag and reader, provides lessor communication and lesser memory capacity for RFID tags. Also, this scheme provides mutual authentication, forward security and resistance against tracking attack, cloning attack.

This paper consists of 6 sections. The second section describes existing schemes for authenticating RFID tag and reader. The third section describes proposed lightweight authentication scheme for RFID. The fourth section consists of the security analysis of the scheme. The fifth section consists of performance analysis of the scheme. Sixth section describes conclusion of overall paper.

2 Related Work

In 2008, Lee et al. [5] proposed a lightweight scheme in which he uses elliptic curve to authenticate tag by the server. In this scheme, the server transmits random number to tag. After receiving random number, the tag computes authenticate message using random number and shared key. Then, the tag transmits authentication message to the server. The server validates tag by computing inverse of message. Later on, Bringer et al. [1] found various security issues in Lee et al. scheme such as tag impersonation attack and tracking attack. Liao et al. [6] proposed a lightweight mutual authentication scheme and also implementing elliptic curve to compute T_s. Then, it transmits T_s to the tag. This scheme provides mutual authentication between tag and server.

Zhao et al. [9] found key management security issues in Liao et al. [6] scheme. Then, Zhao et al. [9] proposed scheme consists of x, y coordinate of shared key K is used to generate authenticate message. This scheme provides mutual authentication and resistance from various security issues such as tracking attack, cloning attack, forward security and replay attack. Then, Chien et al. [2] proposed a scheme consisting of an elliptic curve and a hash function. This scheme can't provide security against various attack such as replay attack and spoofing attack [8].

3 Proposed Scheme

The proposed scheme consists of two phases: initialization phase and mutual authentication phase. The Fig. 1 shows exchange of messages between tag and reader during mutual authentication.

3.1 Initialization Phase

The server chooses X_T as random number ($X_T \in Z_n$) and computes Z = X_TP. The server chooses ID_i for each tag and computes H(ID_i). Server transmits messages Z, $H(ID_i)$ through a secure channel to the tag. After receiving messages, the tag stores Z, $H(ID_i)$ in its memory. Then, the server computes Y = X_sP. The server stores $H(ID_i)$, X_s, Y, Z in its database. The Table 1 shows notation which is used in the scheme.

Table 1. Notation

Symbol	Meaning
X_T	Private key of the tag
ID	Unique identity of the tag
Xs	Private key of the server
$\oplus$	Exclusive or operation
Z	Shared key between tag and server
Y	Public key of the server
H(ID)	Hash value of the ID
n, m	96 bit random number generated during mutual authentication
p	It is generator of group G over an elliptic curve
T_H	Time require computing single hash function
T_M	Time require computing single elliptic multiplication
$(n.Z)_x, (n.Z)_y$	X and y coordinates of n.Z

3.2 Mutual Authentication Phase

1. The server chooses random number "n" and computes A, $Auth_s$ using shared key Z = (Z_X, Z_y)

$$A = n \oplus (Z)_x \tag{1}$$

$$Auth_s = H((Z)_y \oplus (n.Z)_x) \tag{2}$$

Then, the server transmits A, $Auth_s$ messages to the tag.

2. After receiving A, $Auth_s$, the tag computes n from A such that:

$$\mathrm{n} = A \oplus (Z)_x$$

$$Auth_s' = \mathrm{H}((Z)_y \oplus (n.Z)_x)$$

and verify $Auth_s == Auth_s'$ and after successful validation, then tag chooses a random number "m" and computes B, C, $Auth_T$:

$$B = m \oplus (Z)_y \tag{3}$$

$$C = H(ID_i) \oplus H(m) \oplus (n.Z)_y \tag{4}$$

$$Auth_T = H(H(ID_i) \oplus m \oplus (n.Z)_x) \tag{5}$$

Then, tag transmits B, C, $Auth_T$ to the server.

3. After receiving messages from the tag, the server computes m from B and computes $H(ID_i^*)$

$$\mathrm{m} = B \oplus (Z)_y$$

$$H(ID_i^*) = C \oplus H(m) \oplus (n.Z)_y$$

Then, the server finds $H(ID_i^*)$ in its database, if the tag is not found in its database. Then, the server terminates the session. Otherwise, the server computes $Auth_T'$

$$Auth_T' = \mathrm{H}(H(ID_i) \oplus m \oplus (n.Z)_x)$$

After computing $Auth_T'$, the server validates $Auth_s == Auth_s'$ and successful validation, mutual authentication takes place between tag and reader.

4 Security Analysis

4.1 Mutual Authentication

In mutual authentication, genuine server and genuine tag authenticate each other. The tag validates server from the $Auth_s$ using shared key Z which is generated by genuine server. Also, the server validates the tag from $Auth_T$ which is generated by the valid tag. So, genuine tag and genuine server validate each other. Hence, the scheme provides mutual authentication between genuine tag and reader.

4.2 Integrity

The adversary eavesdrops A, $Auth_s$, B, C, $Auth_T$ packets between the networks. Then, the adversary modifies A and transmit A, $Auth_s$ to the tag. Then, the tag computes n from X and computes $Auth_s^{'}$, which is different from the $Auth_s$. Hence, this protocol provides integrity of the message.

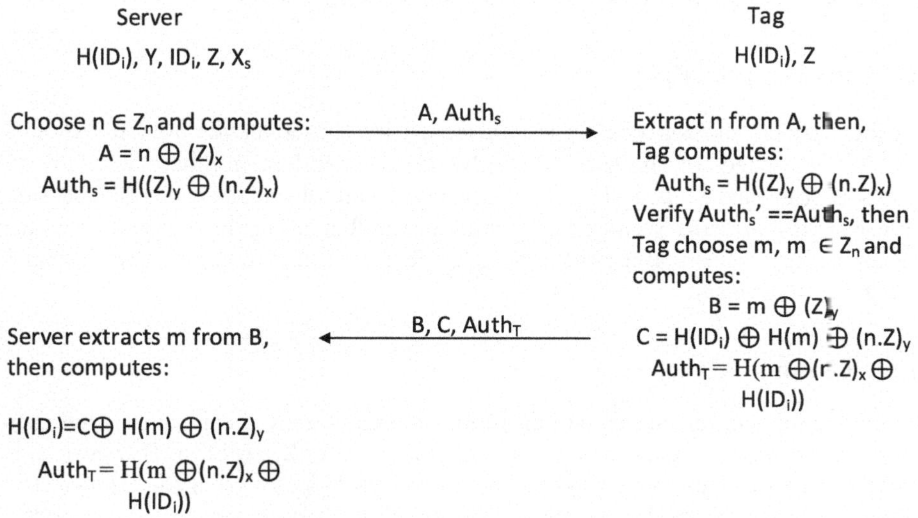

Fig. 1. Proposed scheme for RFID

4.3 Tracking Attack

In this attack, the adversary computes identity (ID) of the tag. To compute identity of the tag, Jules and Weis [4] proposed model contained three phases: learning phase, challenges phase and guessing phase.

1. Learning phase consists of execute query in which adversary eavesdrops packet between communication network.
2. In challenge phase contains adversary is given two different tag ID such as ID_1 and ID_2 where $ID_1 = 0 \bmod 2$ and $ID_2 = 1 \bmod 2$.
3. The adversary guesses one ID i.e ID_b from these two different ID_1 and ID_2 and adversary wins the game if it identifies correct ID of the tag with $\in$ advantage.

$$Adv_A(K) = |Pr[Adv_{wins}] - 1/2| \tag{6}$$

The adversary guesses the identity of the tag using $Auth_T$, C. In these messages, hash function is used to protect identity of the tag. The hash value is irreversible.

Hence, the probability of the adversary to compute identity of tag using $Auth_T$ message is 1/2. Then,

$$Adv_A(K) = |Pr[Auth_T^{'} == Auth_T] - 1/2| = |1/2 - 1/2| = 0 < \in (K) \quad (7)$$

Where $\in$(K) is negligible function $\in$(.) and K is security parameter, i.e. bit length of the secret key. The adversary can't able to compute the identity of the tag. Hence, the scheme provides resistance from tracking attack.

4.4 Forward Security

In this attack, the tag can be compromised by the adversary and it extracts previous messages of the tag. This scheme uses random number "n", "m" to compute the shared key (K). The adversary can physical access to the tag. However, the adversary can't know random number n and m to guess previous messages between tag and reader. Hence, this protocol provides security against the forward security.

4.5 Cloning Attack

In cloning attack, the adversary compromises some tags and computes other tags secret information. In the scheme, the tag stores only Z, $H(ID_i)$. However, this scheme uses random value to compute Z and $H(ID_i)$ for each tag. Hence, the adversary can't guess the other tag shared key by compromising some tag. So, this scheme provides security against the cloning attack.

5 Result and Comparison

In the scheme, total memory required by the tag to compute authentication is equal to the memory required by hash function and key = H(ID_i) + z, where H(ID_i) = 160 bits, Z = (Z_x, Z_y) = 160 + 160 = 320 bits. So, Overall cost for memory required by the tag = 160 + 320 = 480 bits. The communication cost of the scheme from server to tag is A, $Auth_s$ where A = 160 bits, $Auth_s$ = 160 bits. So, overall communication cost from server to tag is 320 bits. The communication cost from tag to server is from B, C and $Auth_T$, where B = 160 bits, C = 160 bits and $Auth_T$ = 160 bits. Therefore, the communication cost from tag to server is 160 + 160 + 160 bits = 480 bits. Hence, the total communication during mutual authentication is 320 + 480 = 800 bits. The total number of hash functions used by the tag is 3, and the total number of multiplications is 1. So, the Computational time requires by the tag is equal to $3{*}T_H + 1{*}T_M$, where T_H defines time to compute hash function and T_M defines time to compute elliptic curve. Table 2 shows various analyses of the scheme with existing schemes.

Table 2. Analysis of various authentication schemes for RFID

Analysis	Liao et al. [6]	Ryu et al. [7]	Chen et al. [2]	Ours
Communication cost	1280	960	1440	800
Tag Computational time	$5 * T_H$	$2 * T_H + 2 * T_M$	$3 * T_H + 4 * T_M$	$3 * T_H + 1 * T_M$
Mutual authentication	Yes	No	Yes	Yes
Untraceability	No	Yes	Yes	Yes
Forward security	Yes	Yes	Yes	Yes
Server spoofing	Yes	Yes	No	Yes
cloning	No	Yes	Yes	Yes

6 Conclusion

RFID contains various tags and a reader. These tags contain a limited amount of memory, low power battery, low computational power, etc. Hence, these tags require a lightweight scheme to provide authentication between the tag and reader. In this paper, lightweight authentication scheme for RFID has been proposed which consists of hash function and elliptic curve to authenticate between tag and reader. This scheme provides lessor communication and lessor memory capacity for RFID tags which are discussed in this paper. The scheme consists of three hash function, and one elliptic curve to implement authentication. The scheme also provides security against tracking attack, cloning, and forward security.

References

1. Bringer, J., Chabanne, H., Icart, T.: Cryptanalysis of EC-RAC a RFID identification protocol. In: International Conference on Cryptology and Network Security, pp. 149–161. Springer (2008)
2. Chen, Y., Chou, J.S.: ECC-based untraceable authentication for large-scale active-tag RFID systems. Electron. Commer. Res. **15**(1), 97–120 (2015)
3. Finkenzeller, K.: RFID Handbook: Fundamentals and Applications in Contactless Smart Cards Radio Frequency Identification and Near-Field Communication. Wiley, Hoboken (2010)
4. Juels, A., Weis, S.A.: Defining strong privacy for RFID. ACM Trans. Inf. Syst. Secur. (TISSEC) **13**(1), 7 (2009)
5. Lee, Y.K., Batina, L., Verbauwhede, I.: EC-RAC (ECDLP based randomized access control): provably secure RFID authentication protocol. In: 2008 IEEE International Conference on RFID, pp. 97–104. IEEE (2008)
6. Liao, Y.P., Hsiao, C.M.: A secure ECC-based RFID authentication scheme integrated with ID-verifier transfer protocol. Ad Hoc Netw. **18**, 133–146 (2014)
7. Ryu, E.K., Kim, D.S., Yoo, K.Y.: On elliptic curve based untraceable RFID authentication protocols. In: Proceedings of the 3rd ACM Workshop on Information Hiding and Multimedia Security, pp. 147–153. ACM (2015)
8. Shen, H., Shen, J., Khan, M.K., Lee, J.H.: Efficient RFID authentication using elliptic curve cryptography for the Internet of Things. Wireless Pers. Commun. **96**(4), 5253–5266 (2017)
9. Zhao, Z.: A secure RFID authentication protocol for healthcare environments using elliptic curve cryptosystem. J. Med. Syst. **38**(5), 46 (2014)

Lean and Industry 4.0 Strive to Create Smart Factory Through Integration of Systems: An Exploratory Review

Hardik Majiwala[1(✉)], Suresh Sharma[2], and Pankaj Gandhi[3]

[1] School of Engineering, P P Savani University, Surat, Gujarat, India
hardikmajiwala@gmail.com
[2] Quality Circle Forum of India, Surat Chapter, Surat, Gujarat, India
suresh_sharmas@rediffmail.com
[3] Research and Development, STPL, Surat, Gujarat, India
pnpj71@gmail.com

Abstract. Lean manufacturing concept is being widely used for several decades by industries and service organizations to ease operational complexity and enhance productivity. It is so despite the debated criticism about drifting definition and priorities of lean with time. In 1980s, focus of lean was on reduction of waste. Then in 1990s, priority shifted to quality, cost and delivery. Again, in 2000 the focus shifted to customer value. On the other hand, lean experts and authors advocated such changes to be evolutionary and natural with the passing time, keeping the intent of lean intact. On analysis, it is found that lean operates with multi-model advantage using a mix of tools and practices best suited for the company's position & the type of industry. The lean approach raised excellence of organization to higher level through process standardization, continuous improvement culture development and workers empowerment on production line. Once again, with the changing market expectations many companies require now new approaches to deal with multiple process complexity and operational challenges. Consequently, a usual set of innovative digital technology known as Industry 4.0 and has appeared as new approach in the manufacturing sector. It claims to create a smart system of machines, components, products, properties, process and ICT systems in the whole value chain to partake a smart factory. Lean tools have been using simpler devices to detect abnormalities and stop assembly line or machine to improve human reliability. Systems and devices like 'andon', Jidoka, contact sensors, thermocouples, and photoelectric devices are some examples. Industry 4.0 differs in technologies to achieve the common goals as lean but with automation.

Keywords: Industry 4.0 · Lean production · System integration

1 Introduction

The Industry 4.0 term was coined by the National Academy of Science and Engineering (acatech) its working group was founded in the year 2011on the Hanover exhibition [1].

N. Nain and S. K. Vipparthi (Eds.): ICIoTCT 2019, AISC 1122, pp. 184–195, 2020.
https://doi.org/10.1007/978-3-030-39875-0_20

The topic is hot and relevant across the world as can be seen from the acceptance of such need all over. Many countries have launched similar strategies with different names. United States has 'National Strategic Plan for advance manufacturing. For the U.K. it is, 'The coming of manufacturing: a novel era of opportunities and challenges'. Japan addresses it as overview of the white paper on engineering Industry. "*Made in China 2025*" makes obvious mention of the similar strategies. 'Make in India' also is India's vision towards Industry 4.0. Though the concept is widely researched, a workable road map to implement industry 4.0 is not available yet. Some are too common to put in the training; others are too elaborate to modulate for special industry [2]. Industry 4.0 is structured to make increasing use of information and communication technology (ICT) for the autonomy within manufacturing environment and even end-to-end digital combination of the engineering across whole value chain. The purpose of this paper is to discuss many questions underpinning the present study: (1) What is Industry 4.0? (2) Will it be really called in future as 4th Industrial Revolution? (3) Can Industry 4.0 and lean production work together? In addition, considering the fact that Industry 4.0 is significantly ICT oriented, the author through this study takes his earlier published paper title, 'Scope of optimizing ICT objectives applying lean principles' to next level [3].

2 Methodology

Methodology as a system helped in designing the study process for the selected field of lean management and Industry 4.0 technologies integration. Methodology also supported in determining the appropriate research methods. Qualitative method was adopted for stating the research problem and questions and for selecting the appropriate methods to collect facts and Figureures. Quantitative approach was used in analyzing the statistical data to answer the framed question on the topic of lean and industry 4.0.

The present exploratory review has been conducted to determine the connection between Industry 4.0 and Lean and to be conceptualize possible model for integration of these two systems together. As regards course of actions and their sequence, four steps module has been developed:

- Extensive study of lean manufacturing for determining its appropriate structure in terms of definition and dimensions of the concept.
- Challenges of lean implementation in the perspective of system integration plan.
- Extensive study of Industry 4.0 vision from theoretical framework together with various components and technologies involved.
- Implementation of the lean through Industry 4.0 and vice versa i.e.; supporting implementation of Industry 4.0 through lean principles including value addition and waste reduction.

The study captured the opinion of professionals and practitioners of lean from all over the world, those looking for Industry 4.0. During the current study, conducted secondary research to find what most popular experience and findings are on the selected topic of lean and Industry 4.0. The study was presented through conference [4].

3 Literature Review

The major part is divided the theoretical framework towards smart factory and the conceptual framework to integrate lean tools and techniques with industry 4.0.

3.1 Theoretical Framework

Theoretical framework contains the basics of lean principles in manufacturing sector and the factors to be considered in context of the industry 4.0. & other part includes the concept of industry 4.0 in the manufacturing unit.

3.1.1 Lean Manufacturing

Evolution of manufacturing system according to Womack [5] took place gradually passing through three distinct phases; craft production (one off), mass production (Henry Ford started), and then Lean production (TPS - Toyota Production System). Therefore lean was industrialized by Japanese industry. It was done under the challenge to re-erect Japanese economy structure after world war – II. A concept was not known outside Japan until 1970 and people came to know about the technology through working experience in U.K. based Japanese companies like Toyota and publication of books started after 1990. Liker, J. in his book 'The Toyota Way, [6] at page 11 addressed Toyota Production System (TPS) from where lean manufacturing evolved. It is "a system designed to provide the tools for people to continually improve their work." TPS is based on 14 principles in four sections;

Philosophy - Each organization must take a longstanding philosophy.
Process - The true process will generate a correct results.
People and Partners - Add worth to a business by emergent people.
Problem Solving - Incessantly resolving root cause ambitions directorial knowledge.

Based on a survey, many challenges and obstacles in total transition of lean production were experienced by industries in the initial stages; *Giant mass-producer* – like General Motors, Renault, Fiat were not able to reform their way totally. However, G.M opened its new complex with ambitious effort to implement the full set of lean technique for product development, supplier co-ordination and factory operation removing old ways of thinking. *Every mass*-producer did not have a lean competitor located nearby so they could afford to continue on old pattern. Inward focus of the Japanese lean producer, was also the initial obstacle. Financers giving preference to mass producers posed obstacle to medium and small companies. Out dated thinking about the world economy was also the real obstacle. Imparting know-how to all the levels of the organization and remove fear of change have always been big challenge in adapting lean manufacturing system. Shortage of resources to re-engineer the process effectively has been a challenge in Indian industries. Lean bears fruit only if the flow of product is maintained smoothly throughout the chain, from a supplier's contractor to the customer's client and therefore long supply chain management is still a challenge. Natural calamities also affected the efficiency of the logistics and lean system. Lean production has been described in several ways by different authors and researchers as the developments taking place with the passing time. Therefore lean manufacturing is a manufacturing approach including of many engineering practices engaged towards recognizing value addition processes after the view of the customers.

In this paper 'ten factors' approach has been selected for a specific purpose of integrating lean manufacturing principles and an Industry 4.0 systems together. Discussions on integration is addressed at Sect. 3. B. The approach propagated by Shah and Ward [7] is in reference who identified the dimensional arrangement of lean manufacturing techniques and advanced consistent values to determine the significance. They enumerated the conceptual classification and measurement of lean using ten influences. These factors emerge from four different groups as shown below;

Supplier Factors

Provider Input: Critics and exhibitions of items and administrations got from clients to be occasionally imparted back to providers, for viable exchange of data.

In the Just in time (JIT) conveyance by providers: Only required amount of items to be conveyed by providers at the predefined time when clients require them.

Provider Advancement: Suppliers to be created alongside the producer, to stay away from irregularity or crisscross in ability levels.

Client Association: Customers are the prime drivers of a business, their requirements and desires ought to be given high need.

Process Factors

Force Creation: A commencement of need from the successor through kanban should empower the progression of generation from the ancestor, implied as JIT creation.

Ceaseless Stream: A streamlined progression of items without huge ends ought to be built up over the production line.

Setup Time Decrease: Time required to adjust assets for varieties in items ought to be kept up as least as could reasonably be expected.

Control and Human Factors

Complete profitable/preventive support: Failure of machines and hardware ought to be stayed away from by viable periodical upkeep methods. If there should arise an occurrence of disappointment low correction time is to be kept up.

Measurable procedure control: Quality of items is of prime significance, no deformity ought to get permeated from a procedure to a consequent one.

Representative inclusion: With sufficient inspiration and qualification, workers are to be engaged for a general commitment towards the firm.

3.1.2 Industry 4.0

Industry 4.0 is an internationally accepted area of research in the field of Internet of Things (IoT). The central object of Industry 4.0 is smart factory. This concept is designed to use advanced information and communication technology (ICT). In turn it enables machines, human beings, products and funds to interconnect information with individually. Industries 4.0 and its smart factories are part of a worldwide Internet of Things and facilities according to commendation for executing the premeditated initiative of

Industry 4.0 by Kagermann [1]. Vision of the Industry 4.0 is to move the industrial production from a physical to an integrated cyber-physical system of production with ICT support. The cyber-physical systems combine the physical world with the cyber world by embedded computer controlled feedback loops. Technically this amounts to a total integration and connection of data acquiring sensors and data based controlling actuators [8]. A big data analysis in industrial informatics by Lee is appended below to indicate the requirement to promote independence at all levels by the companies striving to reach level of smart factory [9].

Comparison of a Factory Today and an Industry 4.0 Factory

As regards Component (e.g., sensor), today's factory key attribute is precision for which key technologies in use are smart sensors and fault detection. Whereas, Industry 4.0-smart factory's hallmark is self-governing action created by own calculations and the corresponding main technologies must provide squalor observing and residual valuable life prediction. Similarly, for Machine the important attribute today can be producibility and performance (quality and throughput) and technologies used are condition-based monitoring and diagnostics. These should be achieved using key technologies for functioning time record with predictive fitness monitoring. With regard to Manufacturing System (e.g., manufacturing execution systems), today's key attributes are productivity and overall equipment effectiveness for which key technologies used are lean operations aiming work and waste reduction. Whereas, smart factory must independently configure and organize deploying key technologies for low-maintenance, self-adapting production systems Cyber-Physical Systems (CPS). Digital Physical Production System alludes to "physical and virtual, neighborhood and worldwide, on a level plane and vertically arranged frameworks, dynamic frameworks limits, halfway or complete self-governance, dynamic continuous control, and thorough collaboration among human and system" [11]. It involves keen machines, warehousing frameworks and creation offices that have been grown carefully and highlight start to finish Information Communication Technology – based joining, from inbound coordinations to generation, advertising, outbound coordinations and administration. The reference depends on the last report of the Industry 4.0 Working Group [1]. Digital Physical System is the changed substance for the utilization of Internet of Things (IoT). General concentration for the use of Cyber-Physical System lies not just on the social occasion of ongoing information from the physical condition to the advanced framework, yet more on the structure examination of information hotspots for self-ruling and self-sorted out procedures. With the foundation of Cyber-Physical Systems, Smart Factory is empowered to be conceivable in organizations. Keen Factory "establishes – key highlights of Industry 4.0 [1]. Under the Smart Factories, "notwithstanding condition checking and flaw determination, parts and frameworks can increase mindfulness and self-productiveness, which will give the executives more knowledge on the status of the processing plant. Likewise, distributed correlation and combination of wellbeing data from different segments give an exact wellbeing forecast in part and framework levels and power processing plant the board to trigger required upkeep at the most ideal time to achieve in the nick of time support and addition close to zero personal time" [10]. Taken together, it is of extraordinary significance for the acknowledgment of Industry 4.0 in organizations [12].

As per Naoaki Fujino and Yasushi Konno of Nomura Research Institute [13] CPS goes for structure a component that empowers not just the making of the 3D plan illustrations of items and parts, yet in addition the representation of every single corporate movement as models just as the reproduction of such exercises including corporate structures. CPS is one of those models in the internet that are utilized to comprehend different work difficulties happening in physical space by using a huge swath of PC assets in advanced space. Here work alludes to a progression of assembling items in the assembling business and incorporates item arranging, item structure, physical investigation (heat conduction, swaying, stress, and so forth.), fabricating process arranging, plan of creation offices, plan of generation lines and reenactment of generation exercises. CPS intends to construct an instrument whereby all such work should be possible in computerized space, and an assortment of cutting edge sensor innovations are utilized to keep up connections with physical space on an ongoing premise. In this way the idea of CSP is that recreation, for example, a well known pretending diversion in a type of computer game ends up conceivable in generally speaking corporate work. To accomplish the objectives of the CPS system, the acatech (German Academy of Science and Engineering) report proposes the execution of decentralization with an abnormal state of reconciliation in the accompanying three key perspectives: Horizontal incorporation through esteem arranges i.e., organize empowering the worldwide division of work that are firmly connected with one another past organization and nation visitors. Vertical coordination and arranged assembling frameworks through "savvy mother processing plant". Start to finish advanced combination of designing over the whole esteem chain to accomplish the brilliant industrial facility function [13].

3.2 Conceptual Framework

Under conceptual frame work of this study, the plan of action related to Lean manufacturing and Industry 4.0 together with the integration modules of these systems have been reviewed and discussed.

Integration of Lean Manufacturing and Industry 4.0

Advocating the importance, Daniel Kupper suggests the integrated application of Lean manufacturing and Industry 4.0 as the greatest operative way to reach the high level of operative superiority [14]. Many companies are indeterminate about how to syndicate the dual for thoroughgoing benefit. Therefore, this is considered as an imperative research area to be expansively reconnoitered. The ten dimensions of lean manufacturing under four groups as recommended by Shah and Ward [7] have been reviewed in Sect. 3. In this section those lean dimensions have been included in the integration module with corresponding components of Industry 4.0. The objective is to discussed and evaluate in what way the machineries and models of Industry 4.0 act to these dimensions of lean.

Supplier Factors: Effect of Industry 4.0 parts that help and support provider factors are communitarian assembling and improvement condition for ideal asset use. Other three measurements under this gathering are additionally looked into and assessed. Figure 1 demonstrates effect of Industry 4.0 on provider factors.

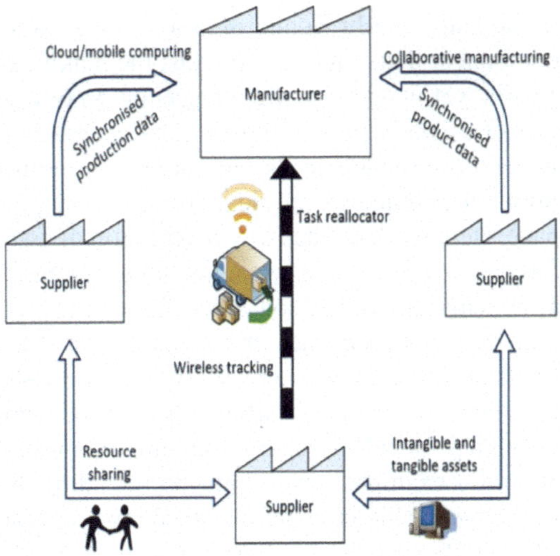

Fig. 1. Impact of Industry 4.0 on supplier factors [15]

Provider Feedback: The prior conventional correspondence systems between all gatherings in the business are currently revamped through distributed computing and versatile processing administrations. Cell phones and tablets are associated with the web and normal cloud. Just-in-time Delivery: Right travel routes and reliability in logistics have been developed. Supplier has access to wireless tracking and smart reallocation system in the event of emergency.

Supplier Development: Technological network is established for sharing imperceptible properties such as investigation and acquaintance in terms of records and information. Also tangible resources such as equipment, machines, and human experts from other organization for common goal are made available. Through virtual organization and standardized interface lean empowered suppliers get synchronized with manufacturers. Challenges of hardware - software compatibility issues are then eliminated.

Control Factor: Industry 4.0 is focused on catering to the needs of customer to support lean manufacturing.

Customer Involvement: Manufacturing implementation system B2C solicitations keep the client informed and involved from specification freeze to end use. Traditional quality function deployment (QFD) having limitations of product quantity requirements and product design requirements are overcome by Big Data that facilitates complicated calculations.

Process Factors: Figure 2 indicates effect of Industry 4.0 on three procedure factors.

Force Production: e-Kanban framework perceives absent and void canisters consequently through sensors and triggers for prompt recharging. Remote data and correspondence framework through radio recurrence ID (RFID) labels to screen the status, number and area of materials with e-Kanban offices of Industry 4.0.

Constant Flow: Industry 4.0 arrangements utilizing RFID innovation help by continuous precise following of stock. Holonic producing framework, drawing in multi operator

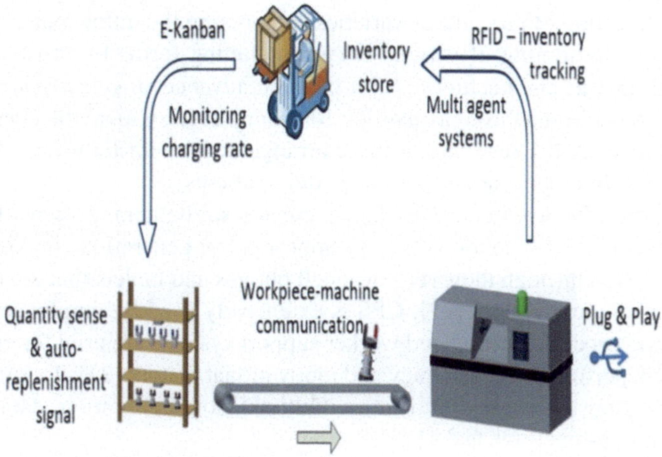

Fig. 2. Impact of Industry 4.0 on process factors Adopted from [15]

framework for material taking care of, arranging and control, the stream framework gets more modularized. Basic leadership gets moved from concentrated structure to decentralized through specialists [16]. A material appropriation strategy dependent on Internet of Things (IoT) in a JIT creation condition for a blended model get together workshop was proposed by [17]. An intelligent optimization algorithm was also developed to provide solution.

Setup-Time Lessening: “Single Minute Exchange of Die (SMED)” of lean manufacturing technology will be now automated with Industry 4.0 expertise, play and plug, where disseminated systems are fortified with self-optimizing and machine learning behavior. Moving parts with RFID tags enable the machine with receiver that organizes reduced setup time through self optimization of machines and work piece-machine communication [18].

Control and Human Factors

Absolute Productive/Preventive Maintenance: With further developed investigation and enormous information condition, machines are prepared to act naturally mindful and self-kept up. Such machines evaluate their own wellbeing and debasement and use information from different machines to maintain a strategic distance from genuine upkeep issues [19]. Undertaking asset arranging framework (ERP) have included systems for prescient support. It coordinates between machine information, ERP information, tangible information and prescient calculations [20].

Slip-up Proofing and Statistical Process Control: For the situation of Industry 4.0, shrewd items accompany insights regarding the tasks to be done on them. The succession of activities to be performed on an item is as of now stacked onto the bearer of that item. This data is promptly passed on to the machine for computerized tasks and appeared better perception interfaces for manual activities. Improved man-machine interfaces likewise present data in an increasingly fitting way, and maintain a strategic distance from the likelihood of committing errors in the generation procedure [21]. RFID empowers programmed detecting of procedures varieties by perusing the inflow put away in RFID labels. IOT aids the joining of various esteem including forms by brushing data and information from various machines. A lot of three advances together add to the large scale periods of Six Sigma through qualities, for example, recognizability, perceivability, memory, and area [22]. Every one of these arrangements with Industry 4.0 guarantee that imperfection free items are created and sold to clients.

Representative Involvement: This factor creates strengthening of workers. In the workplace of Industry 4.0, laborers give prompt input of generation conditions as continuous information through their very own cell phones and tablets that are coordinated with the organization's system [21]. CPS is extensively used by employees to help in decision making process. Specialized worker support systems are used for evaluation of workers' speed, performance, accuracy and motivational factors [23]. To mitigate workers' monotony, they are trained to acquire multi-skill on; calibration, data processing and other such non-repetitive tasks [21].

4 Data Analysis and Discussion

To conduct an exploratory review of the topic, the databases selected include; Science Direct, Business Foundation Complete and Hypothetical Search Ample. Search provided the abstract, title and keywords in most of the cases. Objective was to find; to what level Lean Production principles are comprised and addressed in the reference model. Three out of 41 articles presented or discussed on Lean tree and Industry 4.0 integration.

Concept of an Industry 4.0 impact matrix on lean production system was rarely found. However, one cluster of Industry 4.0 provided a matrix with 11 lean technologies on 'Y' axis and 8 components of Industry 4.0 on 'X' axis. Highest possible impact from the component of Industry 4.0 on the corresponding lean production technologies emerge as; Big Data and Analytics of industry 4.0 impact Kaizen. J,I,T. Jidoka, standardization and Takt time. The people and teamwork were found having low positive impact of most components of Industry 4.0 except in Human Machine Interaction (HMI), Virtual Reality and Augmented reality [8]. Table 1 shows the matrix discussed. This data and finding will support the development decision for Industry 4.0 solution in the early phase. This data and finding will support the development decision for Industry 4.0 solution in the early phase.

Integration Solutions to Achieve Significant Benefits: An ongoing worldwide review by the Boston counseling bunch found that driving ventures perceive the significance of both lean administration and digitalization in their long haul arranging. In a review of in excess of 750 creation administrators (for the most part car respondents), 70% said that plant digitalization would be profoundly applicable in 2030, contrasted with 13% who said that today is significant.

Evaluating the improvement potential, the above study additionally finds that the improvement capability of a coordinated methodology is huge. Likewise discovered that when either approach-lean activities or Industry 4.0 – is connected alone, it can diminish transformation costs by roughly 15%. Nonetheless, as assessed, organizations that utilization incorporated Lean and Industry 4.0 methodology can lessen transformation costs by

Table 1. Industry 4.0 influence matrix on Lean production system

	Data Acquisition and Data Processing				Machine to Machine Human-Machine Communication (MZM)		Human-Machine	
	Sensors and Actuators	Cloud Computing	Big Data	Analytics	Vertical Integration	Horizontal Integration	Virtual Reality	Augmented Reality
5S	+	+	+	+	+	+	++	+++
Kaizen	+	++	+++	+++	+++	+++	+++	+++
Just-in-Time	++	++	+++	+++	+++	++	+	++
Jidoka	+	+++	+++	+++	++	++	+	+
Heijunka	++	++	+++	+++	+++	++	+	+
Standardisation	++	+++	+++	+++	++	++	+++	+−+
Takt Time	+	+	+++	+++	+++	+++	+	+
Pull Flow	++	+	+	+	+++	+++	+	+
Man Machine Seprtion	+	+	+	+	+	+	+++	+−+
People and Teamwork	+	+	+	+	+	+	+++	+−+
Waste Reduction	+	+	++	+++	+++	+++	+	+

40%. Organizations have additionally utilized an incorporated way to deal with decrease costs related with low quality by 20% and work-in-process stock by 30%.

The coordinated methodology permits lean administration and Industry 4.0 to be commonly empowering. Thusly its improvement potential is more prominent than the aggregate of the enhancements accomplished by any one methodology autonomously. Common empowering advances benefits past the regular furthest reaches of both of the two methodologies. For instance, utilizing sensors and information to give full straightforwardness into bottlenecks enables the organization to hone the focal point of its lean endeavors to improve generally hardware adequacy [14].

Lean and ICT Performance Management Model

This study included case of a service industry Oracle that implemented Lean Performance Management to great advantage. An Oracle White Paper on 'Lean Performance Management' authored by Patrick M. describes the model in detail [24]. American cities have furthered importantly from the integration of technology with a three figure hotline, 311. The city used to face three specific encounters: service quality, operational efficiency and effectiveness with transparency. As a result of implementing a Lean Performance Organization model – shared services, leveraging Lean, COTS (commercial off-the-shelf) software, and BI (Business Intelligence). Oracle service cloud is a key component in helping USA based cities achieving smart city vision as reported by no other than Rab Lioyed, CIO, City of San Jose who further states, "We are transforming service delivery, enhancing citizen services and enabling data-driven decision making".

5 Conclusions and Scope for Future Work

The study has been carried out to review and find possibility of integration of Lean and Industry 4.0 and their impact on manufacturing excellence. On the other hand challenges being handled separately by each of the systems have been addressed and then effort is made to integrate selected lean tools, technologies and principles with corresponding components of Industy.4. Looking at the role of lean in Industry 4.0, and based upon the survey input, it is concluded that "lean" will stay with Industry 4.0. Lean principles are likely to be more important. Both the systems are customer focused. Smart factories can produce faster with less waste. On the other hand e-kanban will replace physical kanban cards, less whiteboards and other lean practices. In short, Industry 4.technologies may be exactly what is needed to create lean for doing more with less. Future study is required to motivate all and combine lean tools and Industry 4.0.

References

1. Kagermann, H., Hebig, J., Hellinger, A., Washlster, W.: Recommendation for Implementing the Strategic Initiative INDUSTRIES 4.0: Securing the Future of German Manufacturing Industry: Final Report of the Industries 4.0 Working Group, Forschungsunion (2013)
2. Oesterreich, T.D., Teuteberg, F.: Understanding the implication of digitization and automation in the context of Industry 4.0: a triangulation approach and elements of a research agenda for the construction industry. J. Comput. Ind. **83**, 121–139 (2016)
3. Sharma, S., Gandhi, P.J.: Scope of optimizing ICT objectives applying lean principles: an exploratory review. In: International Conference on ICT in Business, Industry & Government (ICTBIG), Catalog No. CFP16F71-POD. IEEE Xplore, IEEE-USA (2016). https://doi.org/10.1109/ICTBIG.2016.7892667. ISBN(e): 978-1-5090-5515-9, ISBN (p): 978-1-5090-5515-6
4. Ing Joachim, M.: A survey report: lean yesterday, today and tomorrow. Staufen AG and PTW of Technical University (TU), Germany (2016)
5. Womack, J., Jones, D., Roos, D.: The Machine that Changed the World. Simon & Schuster, New York (2007). ISBN 13: 978-1-8473-7055-6
6. Liker, J.: The Toyota Way: 14 Management Principles from the World's Greatest Manufacturer. Tata Graw-Hill Publishing Company Limited, Noida (2006). Indian Version. ISBN-0-07-058747-7
7. Shah, R., Ward, P.T.: Defining and developing measures of lean production. J. Oper. Manag. **25**(4), 785–805 (2007). https://doi.org/10.1016/j.jom.2007.01.019
8. Wagner, T., Herrmann, C., Thiede, S.: Industry 4.0 impacts on lean production systems. Procedia CIRP **63**, 125–131 (2017)
9. Kaufmann, T.: Geschäftsmodelle in Industrie 4.0 und dem Internet der Dinge: der Weg vom Anspruch in die Wirklichkeit. Springer (2015). https://doi.org/10.1007/978-3-658-10272-2
10. Lee, J., Bagheri, B., Kao, H.A.: Recent Advances and Trend of Cyber-Physical Systems and Big Data Analysis in Industrial Informatics. SAGE Publishing, New Delhi (2014)
11. Geisberger, E.E., Broy, M.M.: Agenda CPS – Integration Cyber-Physical System. Springer (2012)

12. Muller, E., Chen, X.-L., Riedel, R.: Challenges and requirement for the application of Industry 4.0: a special insight with the usage of cyber-physical system. Chin. J. Mech. Eng. **30**(5), 1050 (2017)
13. Fujino, N., Konno, Y.: Industry 4.0 and significance for Japanese manufacturing. NRI Papers No. 210, September 2016 (2016)
14. Daniel, K., Alike, H., Johannes, S., Daniel, S., Claudio, K.: When Lean Meets Industry 4.0 (2017). https://bcg.com/ publications/ 2016/ learning manufacturing operation
15. Senders, A., Elangeswaran, C., Wulfsberg, J.P.: Industry 4.0 implies lean manufacturing: research activities in Industry 4.0. Research Gate (2016). http://www.researchgate.net/publication/308751310
16. Lewandowski, M., Gath, M., Werthmann, D., Lawo, M.: Agent-based control for material handling system in in-house logistics – towards cyber-physical systems in in-house- logistics utilizing real size, smart objects, systems and technologies (smart system). In: Proceedings of 2013, European Conference, Erlangen/Nuremberg, Germany, pp. 1–5. VDF (2013)
17. Wan, Y.L., Zhu, H.P., Mu, Y.P., Yu, H.C.: Research on IoT–based material delivery system of the mixed-model assembly workshop. In: Proceedings of 2013, 4th International Asia Conference on Industrial Engineering and Management Innovation (IEMI 2013) (2014)
18. Brettel, M., Frienderichen, N., Keller, M., Rosenberg, M.: How virtualization, decentralization and network building change the manufacturing landscape: an Industry 4.0 perspective. Int. J. Mech. Ind. Sci. Eng. **8**(1), 37–44 (2014)
19. Lee, J., Kao, H.A., Yang, S.: Service innovation and smart analytics for Industry 4.0 and big data environment. Procedia CIRP **16**, 3–8 (2014). https://doi.org/10.1016/j.procir.2015.02.001
20. Haddara, M., Elragal, A.: The readiness of ERP systems for the factory of the future. Procedia Comput. Sci. **64**, 721–728 (2015). https://doi.org/10.1016/j.procs.2015.08.598
21. Schuh, G., Gartzen, T., Rodenhauser, T., Marks, A.: Promoting work-based learning through Industry 4.0. Procedia CIRP **32**, 82–87 (2015). https://doi.org/10.1016/j.procir.2015.02.213
22. Nicoletti, B.: Lean and automate manufacturing and logistics. In: Advances in Production Management Systems, Sustainable Production and Service Supply Chains, pp. 278–285. Springer, Heidelberg. http://dx.doi.org/10.1007/978-3-642-41263-9_34
23. Brauner, P., Zietle, M.: Human factor in production systems. In: Advances in Production Technology, pp. 187–199. Springer (2015). http://dx.doi.org/10.1007/978-3-319-12304-2-14
24. Mungovan, P.: "Lean Performance Management" for Public Sector: An Oracle White Paper on Lean Performance Management" USA: Oracle Corporation (2009)

User Authentication in VANET Using SGSK (Self Generated Session Key)

Lal Singh(✉) and Ram Bahadur Patel

National Institute of Technology, Srinagar, Uttarakhand, India
{lal.singh.cse17,rbpatel}@nituk.ac.in

Abstract. Vehicular Ad-Hoc Network (VANET) is a network of smart moving vehicles in which vehicles can communicate with each other and form a Mobile Ad-Hoc Network (MANET). This network is formed by smart vehicles, fixed Road Side Units (RSUs) and Trusted Authority (TA). This setup of the network is called a smart network because it is integrated with Smart traffic management system to monitor the traffic movement in the network. In this system data exchange between the vehicles in VANET is very crucial. It means any type of security failure may affect system performance. There is a need for security in the network in such a way that unauthorized user cannot access the network. To achieve security of the communication channel is very important for isolating intruders/attackers. This paper proposes a scheme which uses the concept of Self Generated Session Key (SGSK). This scheme protects the secret key which is shared over the network channel. A comparative is also made with some existing schemes.

Keywords: Authentication · Password · Smart card · UID · Security analysis

1 Introduction

VANET shows the significant role in the intelligent transportation system by providing safety for drivers. It provides a much safer, efficient, and comfortable driving experience. It can be used to manage the traffic in metropolitan cities caused by the thousands of moving vehicles. VANET provides many attractive safety services to drivers, such as traffic information, weather information, broadcasting warning message as emergency and sign-violation warnings in traffic, curve speed warning, the current position of the road, turn in the road, intersections. Hence shared information must fulfill the cryptographic security requirements such as authentication, integrity, privacy, non-repudiation, traceability, anonymity, of which authentication is very essential for effective security. Although VANET promises several exciting services, there must be carefully resolved the issues of privacy and communication trust before applying in real lives, otherwise, it would bring many problems. For example, a vehicle can forge wrong messages by created traffic situations to get road free.

N. Nain and S. K. Vipparthi (Eds.): ICIoTCT 2019, AISC 1122, pp. 196–203, 2020.
https://doi.org/10.1007/978-3-030-39875-0_21

This paper proposed a user authentication scheme using UID (Unique Identificatioon Number). SGSK is a cryptographic mechanism which ensures authentication in VANETs. This scheme using SGSK ensures identity authentication, privacy, non-repudiation in VANETs.

2 Why VANET?

The main motivation for VANET is to increase safety, as hundreds of thousands of people are killed worldwide and Many more are injured due to road accident yearly. This problem can be solved by giving timely information to the drivers. This can be achieved with the help of VANET. The other motivation for VANET is traffic optimization, as due to Congestion and traffic jams there is a waste of cost, time and fuel. This problem can be avoided if the driver has prior information about traffic jams. So VANET can help in traffic optimization. There are many identity-based authentication schemes for secure exchange information in VANET. In this proposed scheme no any secret key is shared on the network and the secret key is generated by user and TA separately.

The major components of VANET are OBU (On Board Unit), RSU (Road Side Unit), TA (Trusted Authority). The vehicles can communicate with each other through OBUs which will be loaded in vehicles by manufacturers to provide wireless communication capability. Traffic information can be provided by the RSU in their particular area, it is responsible for providing information or access to the internet for the vehicle within range. The TA is responsible for installing the secure parameter in the OBUs to authenticate the users in VANET

Generally, two kinds of communication take place in VANET, first one is communication among vehicles and second one is communication between vehicles and RSUs. To established communication between vehicles initially, each vehicle has to communicate with RSU within its communication range. Hence in order to establish communication among vehicles firstly, Vehicle to RSU (V2R) and RSU to Vehicle (R2V) communication have to be established before the communication establishment among vehicles. Figure 1 shows the main components and types of communication in VANET. V2V (Vehicle to Vehicle) and V2I (Vehicle to Infrastructure) communications in VANET has been indicated by the circle.

3 Literature Survey

Many security schemes have been proposed to ensure user authentication in VANET. Some public key infrastructure (PKI) authentication schemes have been proposed. Digital signature schemes in conventional PKI provides authentication and integrity, maintenance of certificates for vehicles public keys require a lot of computation and communication overhead. To overcome these difficulties in conventional PKI many identity-based (ID-based) authentication schemes are proposed [1,4]. In [1] an identity-based security system for user privacy in vehicular ad-hoc networks. In this paper, there is an approaching concern about the privacy of the user in VANET. The security of the system is fully based on user identity protocol. In [4], the author considered different types of attacks (Brute force attacks, Illusion attack, Traffic analysis attacks, etc.) in VANET and also provides some solutions to avoid these attacks. The author provides a detailed study of different privacy-preserving authentication approaches in VANET [5]. A self-healing key distribution approach with a certificateless signature, where the environment is semi-trusted. In this scheme receiver do not need to ask certificate revocation lists (CRLs) [6].

4 Proposed Method

The Proposed approach SGSK (Self-generated session key) uses UID to provide the user (vehicle) authentication on the Trusted Authority (TA). In this approach, many users are connected with TA, this TA serves as a server. In this approach, the session key is generated by vehicles and TA separately, less number of keys is generated, so there is no headache of maintaining a large number of keys as we have seen in previously existing techniques for user authentication in VANET. Table 1 gives the notations used in the proposed scheme. Brief of the scheme is as follows.

Table 1. Notations

SNO	Notations	Meaning
1	TA	Trusted Authority
2	RSU	Road Side Unit
3	OBU	On Board Unit
4	ID_V	Vehicle identity
5	ID_{TA}	TA identity
6	ID_{RSU}	RSU identity
7	U_i	User identity
8	$pswd_i$	User password
9	$pswd_i^*$	User's new password
10	MD	Message Digest
11	$SESK_{VTA}$	Session key between user and Trusted Authority (TA)
12	S-AES	Simplified advance encryption standard

4.1 User Registration Phase

A user (vehicle) who want to take part in VANET should have to register with the server (TA) only once (Fig. 2). Each user has its own attributes as VIN (vehicle identification number), UID (Unique Identification Number), user ID and password. These entries are generated by the user at his/her own vehicle.

These entries are stored on the server's (TA) database. Now, these values are available at both side as user side as well as server side.

Step1: User sends U_i and $MD_1(U_i,\ pswd_i,\ \text{VIN},\ \text{UID})$ to TA via a secure channel.

Step2: TA stores U_i, $pswd_i$, VIN, UID into U_i's smart card and sends it to user via a secure channel.

4.2 User Login Phase

The user who wants to login inserts the smart ID card into a panel of ZigBee integrated keyboard panel and input user identity U_i and $pswd_i$. The User Login process is shown in Fig. 3.

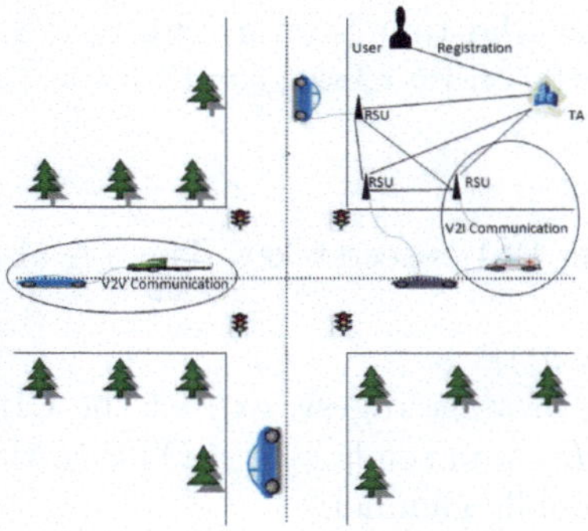

Fig. 1. VANET architecture

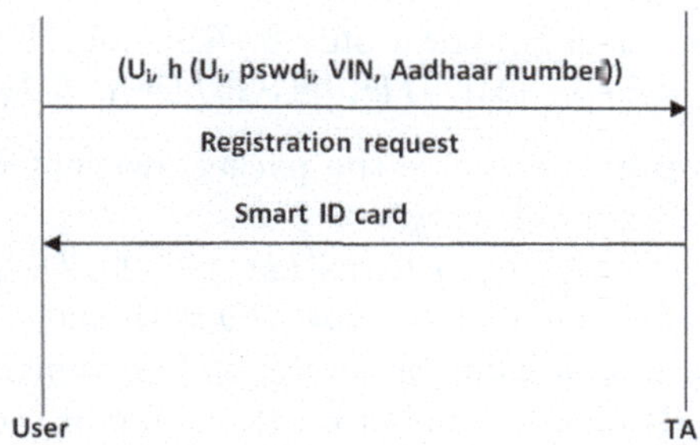

Fig. 2. User registration

4.3 User Authentication on TA Phase

Authentication of the Vehicle on the Trusted Authority (TA):

Step1: Insert the ignition key into the key holder of the vehicle.

Step2: Swap the smart ID card on the panel of Zigbee integrated keyboard panel. This process will read ID card number of the owner of the vehicle and get authenticated with the Trusted Authority (TA) and send a four character OTP (One Time Password) on the device of the owner of the vehicle.

Step3: Now the driver will authenticate the vehicle with the OTP received on the owner's device. If the authentication is done. The vehicle starts, otherwise, the ZigBee system attached to the system responds accordingly.

Proposed Protocol:

The Proposed Protocol Consists of the General Outline of the Proposed Algorithms. Which Described as Following:

Step1: The vehicle generates message digest MD using user id U_i and password $pswd_i$.

MD = SHA-256 ($U_i||pswd_i$).

Step2: The vehicle broadcasts the MD along with vehicle id ID_V, ID_{TA} and Time Stamp TS1. This shows the authentication request.

$<\mathrm{MD}||ID_V||\mathrm{TS1}||ID_{TA}>$.

Here ID_V, ID_{RSU}, ID_{TA} are the dummy identities which are generated at the time of its registration in the following manner. The TA choose two random number a_1, b_1 such that $a_1, b_1 \in Z_q^*$ and computes $ID_V = g_1^{a_1} * g_2^{b_1} modq$. Here g_1 and g_2 are the generators of Z_q^*.

Step3: The RSU receives the packet, increase the time stamp TS1 to TS2, append its identity ID_{RSU}. The RSU encrypt the entire message with its secret key which is known only to RSU and TA.

$<E_{RSU}\ (\mathrm{MD}||ID_V||\mathrm{TS2}||ID_{TA}||ID_{RSU})>$.

Step4: TA receive the packet, decrypt it using RSU's secret key. The TA also verifies its identity.

$<D_{RSU}(E_{RSU}(\mathrm{MD}||ID_V||\mathrm{TS2}||ID_{TA}||ID_{RSU}))>$.

The TA matches the MD with the existing message digest, extract the UID and mobile number corresponding vehicle id ID_V and sends 4-digit OTP to the user's mobile number on the different secure mobile channel.

Step5: The TA generates session key SESKvta between vehicle and TA using UID and 4-digit OTP. $SESK_{VTA}$ = S-AES (UID||OTP).

Step6: The TA sends a random number N along with lifetime LT encrypted with session key $SESK_{VTA}$ between vehicle and TA. This shows authentication response.

$<E_{RSU}(SESK_{VTA}\ (\mathrm{N}||ID_V||\mathrm{LT}||\mathrm{TS3})||ID_{TA})>$.

Step7: The RSU receives packet and decrypt it using its secret key.

$D_{RSU}\ (E_{RSU}(SESK_{VTA}\ (\mathrm{N}||ID_V||\mathrm{LT}||\mathrm{TS3})||ID_{TA}))>$.

The RSU is able to check the identity of TA, verifies it whether it is sent by the authorized TA or malicious node. After verifying RSU sends the packet to the vehicle user.

$<E_{SESK_{VTA}}(\mathrm{N}||\mathrm{IDv}||\mathrm{LT}||\mathrm{TS3})||ID_{TA}>$.

Step8: The vehicle receives the packet and decrypt it with the session key $SESK_{VTA}$ which is generated by the vehicle user itself on receiving the 4-digit OTP on vehicle user's mobile number from TA. $SESK_{VTA}$ = S-AES (UID||OTP).

$D_{SESK_{VTA}}(E_{SESK_{VTA}}(\mathrm{N}||ID_V||\mathrm{LT}||\mathrm{TS3}||ID_{TA})$.

Step9: The vehicle user broadcasts random number N, LT, in encrypted form with $SESK_{VTA}$, along with ID_V and ID_{TA}.

$<E_{SESK_{VTA}}(\text{N}||\text{LT})||ID_V||ID_{TA}>$.

Step10: The RSU receive packet, append its id ID_{RSU}, encrypt this entire packet with its secret key and send to TA.

$<E_{RSU}(E_{SESK_{VTA}}(\text{N}||\text{LT})||ID_V||ID_{TA}||ID_{RSU})>$.

Step11: The TA receives packet, decrypt it with RSU's secret key.

$D_{RSU}(E_{RSU}(E_{SESK_{VTA}}(\text{N}||\text{LT})||ID_V||ID_{TA}||ID_{RSU})$.

The TA verifies its identity ID_{TA}, decrypt the packet with session key $SESK_{VTA}$.

$D_{SESK_{VTA}}(E_{SESK_{VTA}}(\text{N}||\text{LT}))$.

The TA matches the random number N, if matches, the user are authorized user, otherwise unauthorized user. The User Authentication on TA process is shown in Fig. 4.

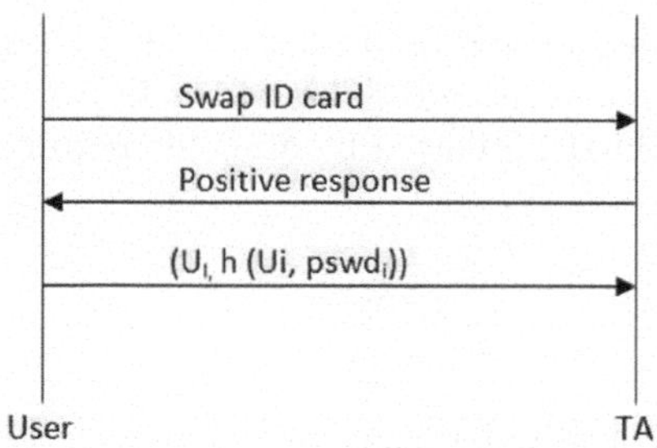

Fig. 3. User login

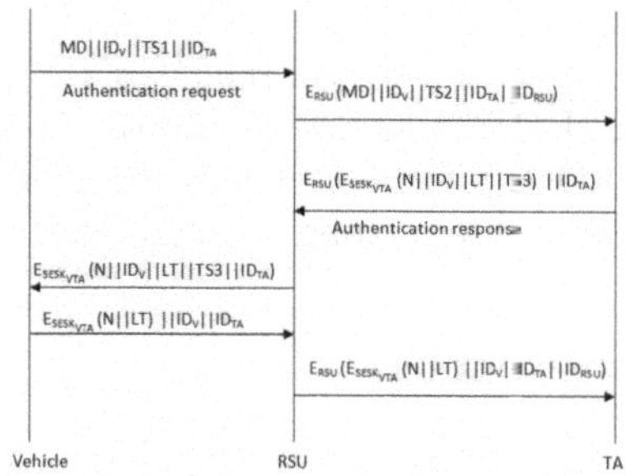

Fig. 4. User authentication

4.4 Password Change Phase

The user U_i freely can change the original password $pswd_i$ to a new password $pswd_i^*$. The password change phase of this scheme is described as follows.

Step1: User U_i insert his/her smart ID card into a panel of Zigbee integrated keyboard panel and inputs ID and password.

Step2: The smart card replaces $pswd_i$ with $pswd_i*$, now the password has been changed.

5 Security Analysis

The proposed technique is mainly used to provide the user authentication by Self Generated Session Key (SGSK) mechanism using the UID. An unauthorized user could not take part in VANET and also can't communicate with other authorized user without knowing the session key. The session key does not share over a network channel, So difficult to threat on user authentication in VANET. The session key between user and server generated separately, which reduces the threat on user authentication in VANET.

6 Conclusions

The proposed SGSK scheme does not share session key over a network channel. This scheme separately generates the session key at the user side as well as server side (TA). It is more reliable than existing approaches/schemes. An SGSK using Aadhaar number approach makes it more reliable because each user (vehicle) generate the session key and do not share it over the network channel. The number of requests is less compared to the existing schemes. In this approach, there is no headache of maintaining a large number of keys. Generating the session key between users taking fewer requests from a user to the server (TA) while existing approaches take more request for establishing the session between users in VANET.

In future we will replace UID with Unique International Identification Number (UIID) which will comprise 16 byte alphanumeric code. This code will permit to identify the location of the user (place of birth) followed by all fields required in an address along with country code.

References

1. Sun, J., Zhang, C., Zhang, Y., Fang, Y.: An identity-based security system for user privacy in vehicular ad hoc networks. IEEE Trans. Parallel Distrib. Syst. **21**(9), 1227–1239 (2010)
2. Zhang, C., Ho, P.H., Tapolcai, J.: On batch verification with group testing for vehicular communications. Wireless Netw. **17**, 1851–1865 (2011)
3. Shim, K.A.: CPAS: an efficient conditional privacy-preserving authentication scheme for vehicular sensor networks. IEEE Trans. Veh. Technol. **61**(4), 1874–1883 (2012)
4. Isaac, J.T., Zeadally, S., Camara, J.S.: Security attacks and solution for vehicular ad hoc networks (2009)
5. Greeshma, T.P., Roshini, T.V.: A review on privacy preserving authentication in VANETs. In: International Conference on Control, Power, Communication and Computing Technologies (ICCPCCT) (2018). https://doi.org/10.1109/iccpcct.2018.8574315
6. Cui, J., Wu, D., Zhang, J., Xu, Y., Zhong, H.: An efficient authentication scheme based on semi-trusted authority in VANETs. IEEE Trans. Veh. Technol. **68**, 2972–2986 (2019). https://doi.org/10.1109/tvt.2019.2896018
7. Horng, S.J., Tzeng, S.F., Pan, Y., Fan, P., Wang, X., Li, T., Khan, M.K.: b-SPECS+: batch verification for secure pseudonymous authentication in VANET. IEEE Trans. Inf. Forensics Secur. **8**(11), 1860–1875 (2013)
8. Zhang, J., Xu, M., Liu, L.: On the security of a secure batch verification with group testing for VANET. Int. J. Netw. Secur. **16**(5), 355–362 (2014)

9. He, D., Zeadally, S., Xu, B., Huang, X.: An efficient identity based conditional privacy-preserving authentication scheme for vehicular ad hoc networks. IEEE Trans. Inf. Forensics Secur. **10**(12), 2681–2691 (2015)
10. Lo, N.W., Tsai, J.L.: An efficient conditional privacy-preserving authentication scheme for vehicular sensor networks without pairings. IEEE Trans. Intell. Transp. Syst. **17**(5), 1319–1328 (2016)
11. Lu, H., Jie, L.: Privacy-preserving authentication schemes for vehicular ad hoc networks: a survey. Wirel. Commun. Mob. Comput. **16**(6), 643–655 (2016)
12. Tzeng, S.F., Horng, S.J., Li, T., Wang, X., Huang, P.H., Khan, M.K.: Enhancing security and privacy for identity-based batch verification scheme in VANET. IEEE Trans. Veh. Technol. **66**(4), 3235–3248 (2017)
13. Pointcheval, D., Stern, J.: Security arguments for digital signatures and blind signatures. J. Cryptol. **13**(3), 361–369 (2000)
14. Horng, S.J., Tzeng, S.F., Huang, P.H., Wang, X., Li, T., Khan, M.K.: An efficient certificateless aggregate signature with conditional privacy-preserving for vehicular sensor networks. Inf. Sci. **317**, 48–66 (2015)

Grid Scrutinize Based Heuristic Sensor Node Scheduling Protocol with Partial Coverage Constraint in WSN

Anamika Sharma(✉) and Siddhartha Chauhan

Computer Science and Engineering Department, National Institute of Technology Hamirpur, Hamirpur, Himachal Pradesh, India
anamika@nith.ac.in

Abstract. Random deployment of wireless sensor nodes' with high density to enhance the coverage area and lifetime imposes many challenges in wireless sensor networks such as coverage redundancy, redundancy in packet generation, radio channel contention, high network cost and network congestion. This paper presents a distributed Grid Scrutinize based Heuristic Sensor Node Scheduling (GS-HSS) coverage protocol to design the efficient node duty schedules that maximize the area coverage and minimize the energy expenditure of sensor nodes. GS-HSS also detects and eliminates the coverage redundancy to mitigate the problem of data and packet generation redundancy. Simulation results show that GS-HSS outperforms in terms of average scheduling rounds and energy expenditure.

Keywords: Coverage area · Customize sensing range · Node scheduling · Partial coverage · Random deployment

1 Introduction

The necessity for the surveillance of sensitive remote hostile regions provide motivation for the deployment of wireless sensor nodes [1]. A set of spatially dispersed sensor nodes having the ability of self-configuration to accomplish its desired task in a distributed manner. A sensor node is a miniature and resource constrained device. The emphasis on miniaturization limits battery power, processing unit and memory. These nodes coordinate and communicate among themselves and base station to either transmit or process the sensory data. Distributed sensing and integration of sensing results to produce important information through data aggregation is one of the main objectives of WSN [2]. The data is transmitted in the form of bits through a wireless channel using ZigBee an IEEE 802.15.4 communication protocol [3]. This communication and coordination among nodes form a wireless sensor network (WSN). Energy conservation is eminent QoS metric for resource constrained wireless sensor network [4–6]. Therefore, the entire task done by a WSN is according to its energy consumption rate. The size of a WSN is relied on sensor count and their deployment strategy. Sensor count is the number of sensor nodes required to fulfil the application requirements.

N. Nain and S. K. Vipparthi (Eds.): ICIoTCT 2019, AISC 1122, pp. 204–213, 2020.
https://doi.org/10.1007/978-3-030-39875-0_22

WSN configurations entirely rely on the real-time application. It is extraneous to propose a protocol without considering the real-time application and its environment. However, the sensing results are vague if the sensing region is not optimally covered. Therefore, coverage is a quality of service metric of WSN which represents the amount of area covered under the sensing range of deployed sensor nodes. This definition is also entitled as coverage rate or coverage ratio. Maximum coverage rate signifies the efficiency and performance of WSN. It is also proved in the previous research that coverage reflects the connectivity among nodes [4, 6].

Deterministic and random are the two well-known node deployment strategies in WSN. In deterministic deployment strategy, the coverage area or coverage rate is known to the network designer because of pre-computed node's location in a specific topology. Therefore, the deployed sensor nodes can provide complete coverage to the sensing region. The node's locations are pre-optimized through extensive simulation on any network simulation (NS-2, NS-3, TOSSIM or OMNeT++) [7]. These simulators provide a virtual environment to optimize the proposed coverage protocols. The virtual environment is very much similar to the real world. Coverage protocols' simulation examines the complete coverage area and energy expenditure of deployed sensor nodes. These results assist network designers in improving the performance of their network.

Deterministic deployment strategy is irrelevant in hostile or environmentally challenged regions where the nodes are deployed with the help of aircrafts and form random topology. It is impossible to have a knowledge about the exact coverage rate provided by deployed sensor nodes in such region [8]. Because of the random dispersion of nodes some regions get overcrowded with nodes cause high node density and others have lower density of nodes. Sensor node deployment in a hostile region with limited battery power requires its energy to be conserved for maximizing the network's lifetime. Random deployment always provides partial coverage and does not assure complete coverage [9]. Therefore, the main objective of a coverage protocol in such scenario is to maximize the coverage with minimum energy expenditure. In such case, there is a need of a coverage protocol that can efficiently schedule the duty time of deployed nodes to get maximum coverage and minimum energy expenditure.

In this paper, we have proposed a distributed Grid Scrutinize based Heuristic Sensor Node Scheduling (GS-HSS) coverage protocol that optimally schedule the duty time of each sensor node in each scheduling round. The active sensor nodes in each scheduling round provide maximum area coverage with minimum energy expenditure. There are many constraints in random deployment strategy such as coverage redundancy, coverage holes, data redundancy, redundant packet generation, radio channel contention and packet traffic cause congestion. This paper has solved the issue of coverage redundancy by optimally schedule the duty time of sensor nodes. Coverage redundancy caused due to the coverage area overlapping among nodes. Due to this redundant sensory data is generated for the same spatial points leads to wastage of precious resources such as battery power. Therefore, the main objective of this paper is to design an efficient nodes' duty scheduling protocol that can maximize the area coverage and minimize the energy expenditure.

The rest of the paper is organized as follows. Section 2 reviews the existed research on node scheduling protocols based on their performance metrics. Section 3 covers the detailed description of the Grid Scrutinize based Heuristic Sensor Node Scheduling (GS-HSS) coverage protocol along with its network architecture. The performance analysis of the proposed protocol is explained in Sect. 4. The entire paper is concluded in Sect. 5.

2 Literature Review

The node scheduling problem is competently solved by various researchers working in the field of wireless sensor networks. The objective of node scheduling protocols is to design a sleep/wake-up mechanism for each sensor node in each scheduling round. This working schedule for sensor nodes maximize the lifetime of a network. Coverage Configuration Protocol (CCP) [4] is the first protocol in this field that dynamically assigns the coverage degree to each spatial point by scheduling the working time of sensor nodes. Each spatial point must be covered by at least one sensor node. The sensor nodes can be in any of one sleep, listen or active state to conserve the energy. Coverage degree defines the number of sensor nodes that covers each spatial point. Along with this they proposed that if the communication range is twice to the of sensing range ($Rc \geq 2Rs$) then the deployed network is fully connected. PEAS [5] activated the necessary set of sensor nodes and deactivated rest of the nodes for energy conservation. The main disadvantage of PEAS is that the nodes are concerned about its local observation and did not incorporate the state of neighboring sensor nodes.

Optimal Geographic Density Control (OGDC) [6] protocol put the unnecessary set of nodes in sleep state while preserving the criteria of optimal node density, complete coverage and connectivity. Randomized Connected k-Coverage ($RACC_k$) [10] proposed duty-cycling coverage connectivity protocol for random deployment of sensor nodes. According to ($RACC_k$) each spatial point is covered by at least k-active sensors and guaranteed connectivity between all sensor nodes in each scheduling round. It also derived an expression for the required minimum spatial density for active sensor nodes that can provide k-coverage to an AoI. Scheduling with Predictive Coverage Redundancy (SPRC) [11] protocol addressed the issue of coverage redundancy and solved it by performing redundancy check at the initial stage of each scheduling round. Further, to remove the repetitive redundancy check in each scheduling round it predicted its requirement and then schedule the working time of nodes. Coverage Contribution Area (CCA) [12] is a centralized and distributed coverage computation protocol. At first, CCA divided entire sensing region in the form of Reuleaux Triangles and then analyzed the contribution of a number of sensor nodes towards each Reuleaux Triangle. Each Reuleaux Triangle must be covered with at least k sensor nodes. It formed two sets of sensor nodes in each scheduling round. One set comprised of active sensor nodes and another comprised of unused inactive sensor nodes. The simulation results showed formation of maximum number of cover sets with the deployed sensor nodes.

Grid Scrutinize based Heuristic Sensor Node Scheduling (GS-HSS) protocol is a node scheduling based coverage protocol. GS-HSS have addressed the issue of coverage redundancy. To eliminate the coverage redundancy entire sensing region is divided into grids and perform redundancy check in each scheduling round after activating at least k ($k \geq 1$) sensor node in each grid. The coverage redundancy check in each scheduling round is eliminated by either customized the sensing range of sensor nodes or deactivated one sensor node based on their overlapped coverage area and number of neighbors.

3 Distributed Coverage Protocol

This section explains about the network architecture and assumptions used for the designing of protocol. It also explains about the working procedure of the proposed protocol and presents its pseudo code in Algorithm 1.

3.1 Network Architecture

A remote hostile region S_R of area A_s comprises of $\mathcal{N}$ abundant sensor nodes. These $\mathcal{N}$ sensor nodes are randomly distributed inside the sensing region S_R and form random topology. Each sensor node S_i $(i = 1, 2, 3 \ldots, \mathcal{N}))$ has uniform sensing range $R_{initial}$ during initialization of the sensor network. The deployed sensor nodes are aware of its location ((x, y) coordinates) via existing localization algorithm proposed in [13]. The communication range is considered to be twice of the sensing range ($R_{com} \geq 2R_{initial}$) for better connectivity and communication among nodes. S_i $(i = 1, 2, 3 \ldots, \mathcal{N})$ perform sensing, computation and communication in distributed manner. CA_i is the coverage provided by individual sensor node.

3.2 Grid Scrutinize Based Heuristic Sensor Node Scheduling (GS-HSS) Coverage Protocol

GS-HSS is a distributed sensor node's duty scheduling protocol that maximizes the area coverage and minimizes the energy expenditure of $\mathcal{N}$ randomly deployed abundant sensor nodes inside sensing region S_R. The step by step working procedure of GS-HSS is represented in Algorithm 1. The sensor nodes can be in any of one state 'active' or 'inactive'. During the network initialization, each sensor node has state active. For exact coverage computations and duty schedules, S_R is divided into G number of grids. The side length of each grid is $2 * R_{initial}$. The center coordinates of each grid are computed. Scan each grid from left to right to identify the number of sensor nodes lying inside each grid. Store these sensor nodes in ascending order of their distance from the center of the belonging grid ($S_i \in g$) in a list. The number of scheduling rounds is equal to the average number of sensor nodes lying inside each grid g. Each scheduling round is of t interval.

For each scheduling round select one sensor in ascending order of their distance keep its state as active and rest as inactive. Again, scan each grid from left to right to identify the grids whose coverage level is below the threshold level ($C_g < \tau$) and mark them as partial covered grids C_{Pg}. If partially covered grids exist, then identify the sensor nodes in the neighboring grids which can provide maximum coverage to C_{Pg}. Update the state of these sensor nodes as active. Afterward the algorithm performs coverage redundancy check.

Coverage Redundancy Check

The random distribution causes coverage redundancy due to the uneven location of sensor nodes. Coverage redundancy is defined as the total amount of the sensing area of any sensor node is covered under the sensing range of m sensor nodes. Coverage redundancy is derived from coverage area overlapping among nodes [14]. After activating all desired node in each scheduling round, GS-HSS performs a coverage redundancy check. If there exists coverage redundancy for any pairs of sensor nodes, then it customizes the sensing range of sensor nodes by R_{reduce} units and computes the new coverage area provided by the sensor nodes after customization of sensing range. The computation of R_{reduce} is present in the Algorithm 1. This Algorithm 1 also represents the step by step procedure of coverage redundancy check and elimination from line 26 to 42.

Algorithm 1: GS-HSS Protocol.

```
1.  Initialize:
      𝒩 ← number of deploying sensor nodes
      R_initial ← sensing range of deployed sensor nodes
      S_active ← 𝒩          // number of active sensor
      nodes
      state [i] = 'active'    //i=1 to 𝒩
      sensing_range [i] = R_initial //i=1 to 𝒩
      grid_sidelength = 2 * R_initial
      total_coverage_area = 0
2.  begin:
3.    divide area of sensing region A_s into G grids
      (G = A_S / grid_area)
4.    randomly deploy 𝒩 sensor nodes inside sens-
      ing region S_R
5.    for each grid g ∈ G
6.      compute center coordinates as g(cx,cy)
7.      S_N^g ← arrange sensor nodes lying inside g
        in ascending order of their distance from
        g(cx,cy)
8.    compute number of scheduling rounds N_SR
      N_SR = average number sensor nodes lying inside
      g
9.    for i^th scheduling round in N_SR  // i= 0 to N_SR
10.     S_active = 0
11.     for each grid g ∈ G
```

```
    if g contains k sensor nodes   //(k>0)
      activate i^th sensor node and deactivate
      remaining sensor nodes S_remain
      state [S_N^g[i]] = 'active'
      S_remain = S_N^g - S_N^g[i]
      state [S_remain] = 'inactive'
      S_active = S_active +1
  scan each grid g ∈ G
    C_g ← coverage provided by S_active to g
    if (C_g < τ) then  // τ is coverage threshold
      check all the neighboring grids of g
      and identify the eligible m inactive
      sensor nodes to increase the sensing
      coverage of g
      State [m] = 'active'
      S_active = S_active + m
    check coverage redundancy among active
    sensor nodes
    for each pair (i,j) of active sensor
    nodes
    if (d(i,j)<2* R_initial) then
      compute reduction in sensing range R_reduce, to remove redundancy
      R_reduce = (2*R_initial - d(i,j)) / 2
      compute the sensing coverage of i and j
      by reducing their sensing range with R_reduce units
      CA_i,j = 2π(R_initial - R_reduce)^2
      if (CA_i,j < CA_i) then
        State [j] = 'inactive'
        total_coverage_area+=CA_i
      else
        reduce the sensing range of i and j by
        R_reduce units
        sensing_range [i] -= R_reduce
        sensing_range [j] -= R_reduce
        total_coverage_area += CA_i + CA_j
    else
      total_coverage_area+=CA_i + CA_j
end
```

4 Performance Analysis

This section evaluates proposed coverage protocol, Grid Scrutinize based Heuristic Sensor Node Scheduling (GS-HSS) through extensive simulation on OMNET++ and Python to ensure the accuracy of the algorithm. The list of parameters is presented in Table 1.

Table 1. List of parameters

Parameter	Value
Number of sensor nodes ($\mathcal{N}$)	50
Sensing area (A_S)	2500 m^2
Sensing range ($R_{initial}$)	15 m
Communication range (R_{com})	30 m
Total energy ($E_{initial}$)	100 J
Power consumed by active sensor node	54 mA
Power consumed by sleep sensor node	0.49 μA
Time of each round (T)	15 min
Threshold coverage area (τ)	0.65

A set of 50 sensor nodes with 15-m initial sensing range is deployed in 2500 m^2 sensing area. The sensing region is divided into 25 grids. Figure 1 shows the random deployment of 50 sensor nodes in a square sensing region. The x-axis and y-axis represents the coordinates of deployed sensor nodes.

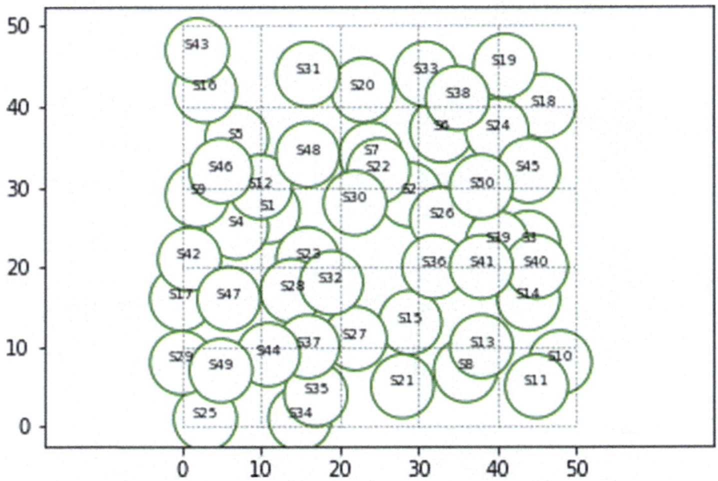

Fig. 1. Random deployment of sensor nodes in 2500 m^2 sensing area

After the implementation of the Algorithm 1 on deployed sensor nodes in Fig. 1, the resultant active sensor nodes with a customized sensing range is shown in Fig. 2. The figure shows that the sensing range of S3, S17, S18, S19, S24, S25, S29, S41 and S47 has been customized after applying GS-HSS algorithm. Afterward, the total coverage area (T_{CA}) is computed. The total coverage area is the sum of the coverage area of individual sensor nodes (CA_i). Equation 1 represents the computation of total coverage area ($T_{CA} \geq \tau$). The threshold value τ is decided after 50 simulations of the GS-HSS protocol.

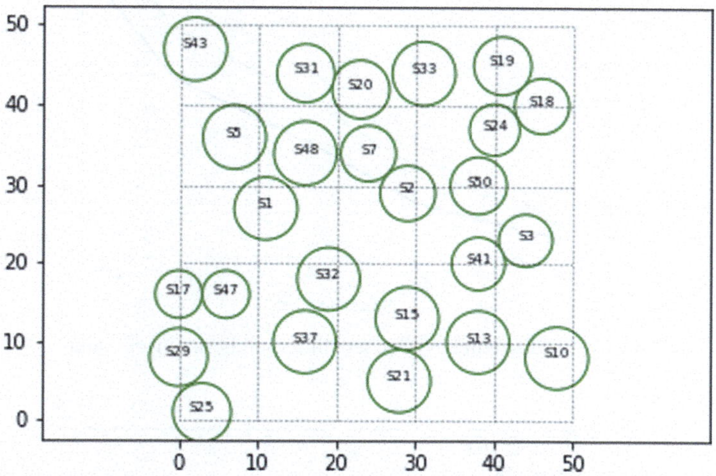

Fig. 2. Resultant active sensor nodes after implementing Algorithm 1 on the deployed sensor nodes of Fig. 1

$$T_{CA} = CA_1 \cup CA_2 \cup CA_3 \cup \ldots\ldots \cup CA_i \geq \tau \tag{1}$$

The performance of GS-HSS protocol is compared with CCP [4], SPRC [11] and CCA [12] in terms of average number of scheduling rounds and average energy consumption for different sensing regions. The sensor count in each sensing region is different and the initial sensing range is kept fixed, i.e. 25 m. The sensing region area is 500 * 500 m^2, 1000 * 1000 m^2, 1500 * 1000 m^2, 1500 * 1500 m^2 and 2000 * 1500 m^2. The sensor counts corresponding to these sensing regions is 128, 510, 765, 1147 and 1318. The sensor counts are computed using uniform random distribution. Sensor nodes have the same amount of energy at the beginning of first duty scheduling round. The energy consumption model reflects the network lifetime. Zhu et al. [15] developed an analytical energy consumption model for randomly deployed multi-hop wireless sensor networks. In this paper, the energy expenditure is modelled according to Zhu et al. [15] energy model. Figures 3 and 4 represent that GS-HSS outperformed as compare to CCP, SPRC and CCA. Figure 3 represents that GS-HSS generates more scheduling rounds as compare to the other protocols. This enables the functioning of the network for long duration. Figure 4 represents that GS-HSS consumed the least amount of energy as compared to the SPRC and CCP. Hence, least amount of energy consumption leads to the maximum network lifetime. The increase in the number of duty scheduling rounds and minimum energy expenditure enhance the performance of wireless sensor network.

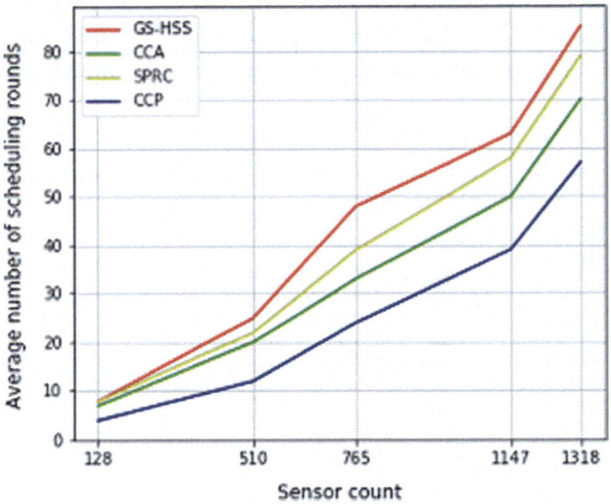

Fig. 3. Average number of scheduling rounds vs sensor count

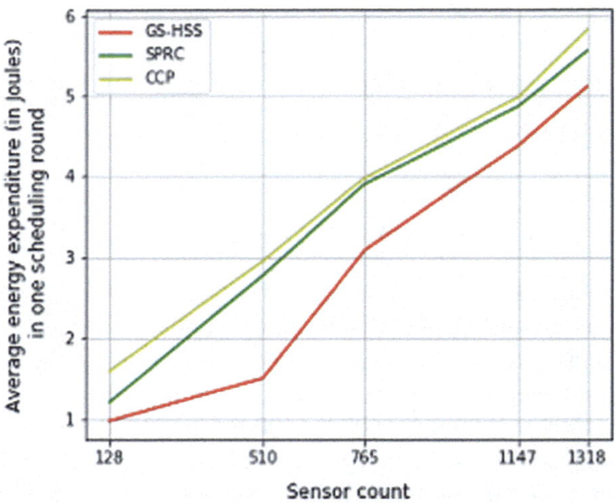

Fig. 4. Average energy expenditure in one scheduling round vs sensor count

5 Conclusion

Network wide coverage and lifetime are crucial WSN performance metrics. This paper has incorporated both the metrics for analyzing the network performance in random deployment strategy and proposed GS-HSS coverage protocol. Grid Scrutinize based Heuristic Sensor Node Scheduling (GS-HSS) protocol is a sensor node scheduling based coverage protocol. GS-HSS have addressed the issue of coverage redundancy to maximize the coverage in each scheduling round and minimize the energy expenditure.

References

1. Onur, E., Ersoy, C., Deliç, H., Akarun, L.: Surveillance wireless sensor networks: deployment quality analysis. IEEE Netw. **21**, 48–53 (2007). https://doi.org/10.1109/MNET.2007.4395110
2. Randhawa, S., Jain, S.: Data aggregation in wireless sensor networks: previous research, current status and future directions. Wirel. Pers. Commun. **97**, 3355–3425 (2017). https://doi.org/10.1007/s11277-017-4674-5
3. Baronti, P., Pillai, P., Chook, V.W.C., Chessa, S., Gotta, A., Hu, Y.F.: Wireless sensor networks: a survey on the state of the art and the 802.15.4 and ZigBee standards. Comput. Commun. **30**, 1655–1695 (2007). https://doi.org/10.1016/j.comcom.2006.12.020
4. Wang, X., Xing, G., Zhang, Y., Lu, C., Pless, R., Gill, C.: Integrated coverage and connectivity configuration in wireless sensor networks. In: Proceedings of the 1st International Conference on Embedded Networked Sensor Systems (SenSys 2003), pp. 28–39. ACM, New York (2003). https://doi.org/10.1145/958491.958496
5. Ye, F., Zhong, G., Cheng, J., Lu, S., Zhang, L.: PEAS: a robust energy conserving protocol for long-lived sensor networks. In: 23rd International Conference on Distributed Computing Systems, pp. 28–37. IEEE (2003). https://doi.org/10.1109/icdcs.2003.1203449
6. Zhang, H., Hou, J.C.: Maintaining sensing coverage and connectivity in large sensor networks. Ad Hoc Sens. Wirel. Netw. **1**, 89–124 (2005). https://doi.org/10.1201/9780203323687.ch28
7. Yu, F., Jain, R.: A survey of wireless sensor network simulation tools. Washington University in St. Louis, Department of Science and Engineering (2011)
8. Kashi, S.S., Sharifi, M.: Coverage rate calculation in wireless sensor networks. Computing **94**, 833–856 (2012). https://doi.org/10.1007/s00607-012-0192-1
9. Wang, B.: Coverage problems in sensor networks: a survey. ACM Comput. Surv. (CSUR) **43**, 32 (2011)
10. Ammari, H.M., Das, S.K.: Centralized and clustered k-coverage protocols for wireless sensor networks. IEEE Trans. Comput. **61**, 118–133 (2012). https://doi.org/10.1109/TC.2011.82
11. Chenait, M., Zebbane, B., Benzaid, C., Badache, N.: Sleep scheduling with predictive coverage redundancy check in wireless sensor networks. In: 2015 International Symposium on Wireless Communication Systems (ISWCS), pp. 366–370. IEEE (2015). https://doi.org/10.1109/iswcs.2015.7454365
12. Yu, J., Wan, S., Cheng, X., Yu, D.: Coverage contribution area based k-coverage for wireless sensor networks. IEEE Trans. Veh. Technol. **66**, 8510–8523 (2017). https://doi.org/10.1109/TVT.2017.2681692
13. Ji, X., Zha, H.: Sensor positioning in wireless ad-hoc sensor networks using multidimensional scaling. In: IEEE INFOCOM 2004, vol. 4, pp. 2652–2661. IEEE (2004). https://doi.org/10.1109/infcom.2004.1354684
14. Sharma, A., Chauhan, S.: Optimal threshold coverage area (OTCA) algorithm for random deployment of sensor nodes in large asymmetrical terrain. In: International Conference on Advances in Computing and Data Sciences, pp. 34–42. Springer, Singapore (2018). https://doi.org/10.1007/978-981-13-1813-9_4
15. Zhu, J., Papavassiliou, S.: On the energy-efficient organization and the lifetime of multi-hop sensor networks. IEEE Commun. Lett. **7**, 537–539 (2003). https://doi.org/10.1109/LCOMM.2003.820097

Intensity Transformation Fusion of Landsat 8 Thermal Infrared (TIR) Imagery

Kul Vaibhav Sharma(✉), Sumit Khandelwal, and Nivedita Kaul

Civil Engineering Department, MNIT Jaipur, JLN Marg, Jaipur 302017, Rajasthan, India
{2015rce9519,nkoul.ce}@mnit.ac.in, khandusa@gmail.com

Abstract. The spatial resolution of panchromatic (PAN) and thermal infrared (TIR) band is 15 m and 100 m respectively in Landsat-8 satellite dataset. The current research proposes an Intensity transformation based fusion method (ITFM) of PAN and TIR imagery. The proposed fusion method introduces unscented spatial filtering of input TIR and PAN images and component based fusion to downscale coarse resolution thermal data. The proposed algorithm has been examined with three thermal image downscaling methods, i.e., DisTrad, TsHARP and Local model. The relative comparison of fusion algorithms results has shown that the proposed ITFM fusion method has outperformed the other conventional methods. The proposed ITFM fusion method has merged edge details from PAN band and earth surface thermal information from TIR band precisely.

Keywords: Data fusion · Land surface temperature · Thermal infrared (TIR) · Panchromatic (PAN)

1 Introduction

A general definition of image fusion is the cumulative presentation of two or greater exclusive images by using definite set of rules to form a brand new image [1]. The spatial and temporal adaptive reflectance fusion model (STARFM) has been used to blend high-frequency temporal information from MODIS and high-resolution spatial information from Landsat 7 ETM+. The author has mentioned the importance of landscape and mask size selection in the downscaling process when operating in urban area [2]. The STARFM fusion method can be used to generate time series images of high spatial resolution thermal data. Further new improvements has been incorporated in STARFM by including viewing angular correction on the MODIS daily bidirectional reflectance. The author has proposed automated co-registration on and automatic selection of Landsat and MODIS paired dates [3]. The advanced area-to-point regression kriging (ATPRK) approach has been employed to enhance the Landsat 8 30 m spatial resolution bands to 10 m using 10 m Sentinel-2 bands as covariates [4]. The modified Kriging methodology helped author to produce frequent observations for continuous monitoring to a finer spatial resolution of 10 m [4]. The image fusion by High Past Division Index (HPDI) method has been proposed to improve the spatial resolution by calculating the spatial-frequency of the edge regions of the image [5]. Statistical Cokriging has been introduced

N. Nain and S. K. Vipparthi (Eds.): ICIoTCT 2019, AISC 1122, pp. 214–220, 2020.
https://doi.org/10.1007/978-3-030-39875-0_23

for enhancing the spatial resolution of thermal band by daily available the coarser band and other spectral indices at higher spatial resolution [6]. The Cokriging algorithm performed better when compared with TsHARP model as it utilized autocorrelation of the images with the cross-correlation among different sensors [7].

Sparse fusion has been in practice to enhance hyper spectral and, multispectral image fusion. The sparse coding has not assumed any arbitrary spectral model of the PAN images [8]. The sparse coding has been based on training samples inputs which gave higher spatial resolution and less spectral distortion [9]. The image quality parameters are meant to be used for assessing the accuracy of different image fusion algorithms. The various spectral and spatial quality assessment parameters has been detailed out for sharpened multispectral imagery [10]. The Pan sharpened Landsat-8 images has been further used to calculates vegetation greenness (i.e., normalized difference vegetation index (NDVI)), canopy structure (i.e., enhanced vegetation index (EVI)), and canopy water content (i.e., normalized difference water index (NDWI))-related variables [11]. Contourlet, non-subsample contourlet and sharp frequency localization contourlet transform fusion has been used for Landsat 7 satellite [12]. A Landsat-MODIS fusion model has given 30-m resolution evapotranspiration (ET) maps for the 2010 growing season over heterogeneous vegetation, agricultural region using the METRIC (mapping evapotranspiration with internalized calibration) algorithm [13]. The study proposed using guided shortwave infrared (SWIR) band imagery as guided image filter in the disaggregation algorithm of TIR imagery in wildfire regions [14].

We can see from literature that the spatial frequency information has not been highly adaptable to local image content since it applies to the image as a whole. The previously attempted fusion methods enhance spatial information of thermal imagery using multispectral data and vegetation indices at lower resolutions. The fundamental image fusion methodologies have been unable to merge PAN band's low and high frequency inputs with the point Land surface temperature (LST) values precisely. Since past studies have limitedly applicable on improving the thermal images along with urban classes and climatic pattern detection, we have used intensity transformation of the spatial features in Pan and TIR images over fix number of pixel selected. The different band property information based on local image content has been pre rectify in Fourier domain by Kalman filter. The noise and bias between both image sate variables has been removed via Kalman filter. A moving window operator has been applied in fusion schemes to identify similar pixels available in both images. The proposed fusion process incorporated PCA transformation for modeling input bands data. We have computed weights for PCA transformed image pairs of different dates of calendar year with components for ITFM fusion algorithm. The ITFM algorithm has estimated earth surface thermal values at finer resolution as well as contains the terrain feature and edges precisely.

2 Study Area

The study area selected is the Jaipur urban and its surroundings rural area in Rajasthan state of India. In perspective of area the Rajasthan state is largest state of India in which its capital city Jaipur is 10th biggest city of India. The study area is located geographically between 26 ∘ 30′18″ to 27 ∘ 00′54″ North latitudes and 75 ∘ 20′16″ to 76 ∘ 10′40″ East longitudes. The topographical positioning of the study area is shown Fig. 1. Landsat 8 satellite covers study area by 147/041 (path/row) in the world reference system-2 (WRS-2).

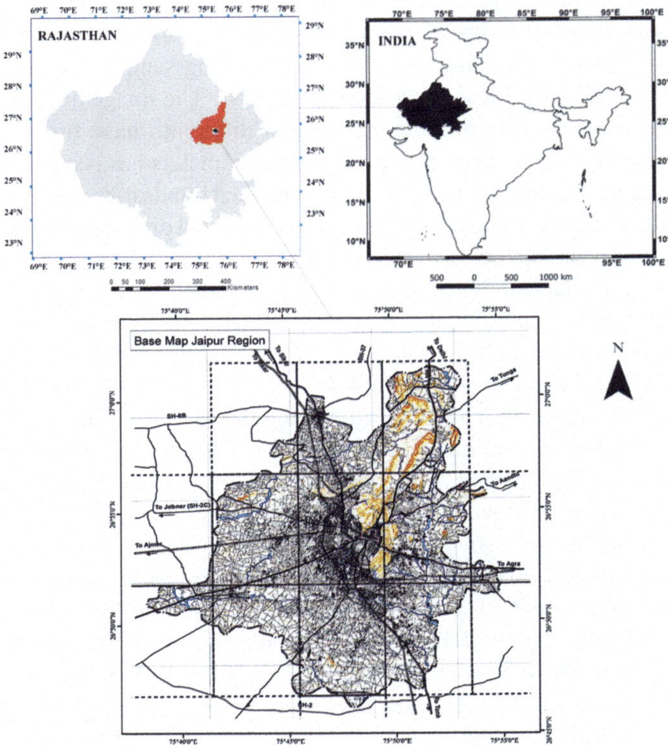

Fig. 1. Study area

3 Methodology

The proposed fusion algorithm processing has been undertaken following initial preprocessing and LST estimation of PAN and TIR bands as their respective resolution. The selected scenes have been first geometrically rectified in the UTM projection system with datum WGS 84 and Zone 43 N. The Landsat 8 Level-1A products have been georectify using nearest neighbor resampling method and the second-order polynomial equation. The proposed algorithms data set pair has been co registered at nearest date available having cloud cover less than 1%. Initially Landsat 8 satellite TIR band LST values at 30 m resolution has been resample and co registered to the PAN band data pixel by pixel. The noise and bias between both image sate variables has been removed via Kalman filter. The resample bands at same resolution have been assigned to make sure the comparison of all data set likely. Sensor data pre processing has been done by using ENVI version 5.3 and ERDAS Imagine 2014 software.

$$L_\lambda = GR_{scale} \times DN + offset \tag{1}$$

At first we calculate sensor gain (L_λ) using grey scale (GR_{scale}) assessment and digital number (DN) of selected pixels of TIR band pixels. Equation (2) has been used to calculate offset.

$$Offset = L_{\min\lambda} - GR_{scale} DN_{\min} \tag{2}$$

Here DNmax and DNmin are maximum and minimum values of digital number of thermal image respectively. The $L_{max\lambda}$ and $L_{min\lambda}$ are maximum and minimum values of spectral radiance values of selected thermal scenes.

To calculate sensor gain we need to estimated GR_{scale} of TIR band that has been calculated using (3). Here DN is reflectance of thermal band of selected data in fusion algorithm.

$$GR_{scale} = \left[\frac{L_{\max\lambda} - L_{\min\lambda}}{DN_{\max} - DN_{\min}} \times \left(DN - DN_{\min}\right)\right] + L_{\min\lambda} \tag{3}$$

At satellite sensor reflected spectral radiance ($L_{sensor\lambda}$) in (w/m^2 * sr * μm) at the top of atmosphere has been calculated by using (4). Here earth surface emissivity ($\in_\lambda$), Sensor brightness temperature (B_λ), downwelling thermal earth radiation gain ($L_{d\lambda}$) and sensor gain (L_λ) has been calculated separately.

$$L_{sensor\lambda} = [\in_\lambda .B_\lambda(L_\lambda) + (1 - \in_\lambda) L_{d\lambda}] \tag{4}$$

NDVI (Normalized difference vegetation index) has been determined for further processing using band 4 and band 5 of Landsat 8 data using (5).

$$NDVI = (B5 - B4) / (B5 + B4) \tag{5}$$

Vegetation proportion (P_v) using above calculated NDVI values using maximum and minimum values of NDVI i.e. $NDVI_{max}$ and $NDVI_{min}$ as shown in (6).

$$P_v = Square\,((NDVI - NDVI_{min}) / (NDVI_{max} - NDVI_{min})) \tag{6}$$

The thermal band earth surface emissivity ($\in_\lambda$) obtained from calculated vegetation proportion (PV) values using (7). Here .004 and .986 are calibration constant taken from Ref. [15].

$$\in_\lambda = .004Pv + .986 \tag{7}$$

The surface brightness temperature (BT) in Kelvin units has been calculated by ($L_{sensor\lambda}$) using (8).

$$BT = (1321.0789 / Ln\,((774.8853 / L_{sensor\lambda}) + 1)) \tag{8}$$

The second last step (9) is to get earth surface temperature in degree centigrade has been calculated using brightness temperature and planks constants ρ.

$$LST = \frac{BT}{1 + \left(\frac{BT}{\rho} * \lambda\right).In \in} - 273.15 \tag{9}$$

The calculated upper and lower limits have been applied in Kalman filter on PAN and TIR image at same resolution to identify similar pixels within limits of good pixels available. The similar pixel moving window's optimum size (W) between PAN and TIR pixels has been identified by "(10)".

$$W = \alpha + (1 - \alpha).\cos(2\pi f / BW_{LR}) \tag{10}$$

Here W stands for bias between bands, f is operating frequency, BW_{LR} is band width at lower resolution of band pairs. The TIR and PAN band values have been modeled in a probabilistic (Gaussian) process. The input bands has been substituted with principal components (PC1–PC4) having maximum relative thermal information.

4 Results and Discussion

The proposed image fusion method operates over the authentic panchromatic picture and the images which have been modified using the proprietary set of rules in Fourier rework. The Table 1 shows the input Landsat 8 data set that has been undertaken for the proposed study. The Fig. 2 shows that proposed ITFM fusion method has outperformed other thermal fusion methods in predicting LST in our study area.

Table 1. Landsat 8 data set (level 1thermal product) used in present study for 2017 and 2018 with day of year (DOY).

DOY	Year	Resolution	Band number
16, 92, 124, 156, 184, 312, 344	2017	30 m	Band 4
90, 154, 186, 202, 218, 282, 314	2018	30 m	Band 5
		15 m	Band 8
		100 * (30) 100 * (30)	Band 10 Band 11

The average annual R^2 values for 2016 and 2017 DOY data has been .89 for ITFM. The average annual R^2 value for DisTrad, TsHARP and Local model is .45, .67 and .61 respectively.

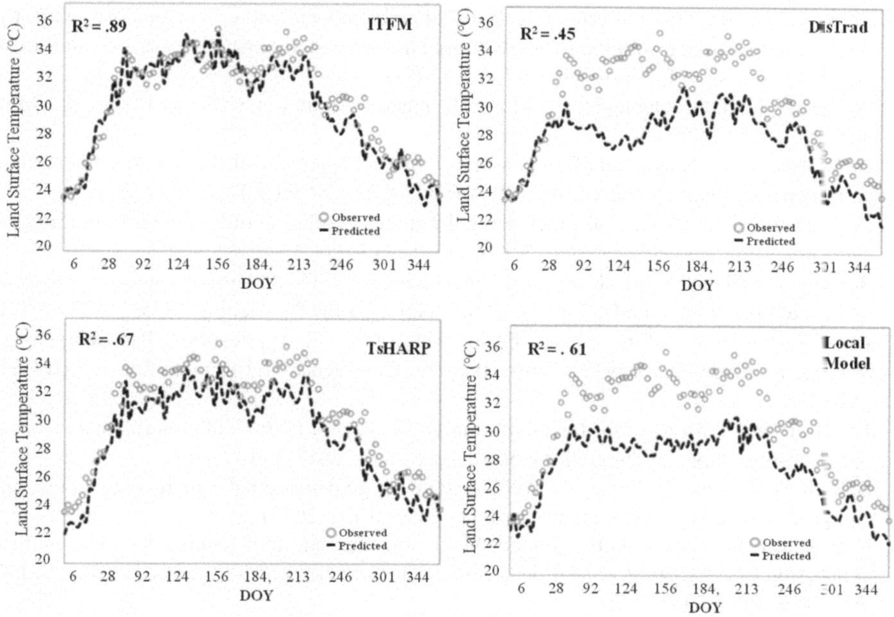

Fig. 2. Scatter plot between observed and predicted LST in °C with DOY for the fusion methods ITFM, DisTrad, TsHARP and Local Model

5 Conclusions

The fusion algorithm generates modeled high-resolution TIR image of Landsat 8 satellite at the more detailed PAN band spatial feature. The proposed fusion algorithm has been able to integrate Landsat-8 PAN image spatial resolution and the temperature information of TIR image without introducing significant distortions. Output image contained both high spatial resolutions with high quality spectral content. The temperature variation at the smaller scale and their pattern of spread can be distinguished by incorporating fusion schemes. For the future work in this area, image fusion schemes can be investigated on different satellite data, in different climatic and topographic conditions. The potential uses can be for various applications like Study of thermal properties of multiple objects in urban areas, analysis of thermal comfort in day and night time, heat losses from buildings, volcanology and forest coal fires monitoring, soil moisture studies and environmental modeling etc.

References

1. Yang, J., Zhang, J., Li, H., et al.: Pixel level fusion methods for remote sensing images: a current review. In: Remote Sensing Symposium (ISPRS), vol. 38, pp. 680–686 (2010)
2. Gao, F., Masek, J., Schwaller, M., Hall, F.: On the blending of the Landsat and MODIS surface reflectance: predicting daily Landsat surface reflectance. IEEE Trans. Geosci. Remote Sens. **44**, 2207–2218 (2006)

3. Li, X., et al.: Generating a series of fine spatial and temporal resolution land cover maps by fusing coarse spatial resolution remotely sensed images and fine spatial resolution land cover maps. Remote Sens. Environ. **196**, 293–311 (2017)
4. Wang, Q., Alan, G., Onojeghuo, A.O., et al.: Fusion of Landsat 8 OLI and sentinel-2 MSI data, pp. 1–24 (2015)
5. Al-Wassai, F.A., Kalyankar, D.N.: A novel metric approach evaluation for the spatial enhancement of pan-sharpened, pp. 479–493. CS & IT-CSCP (2012)
6. Rodríguez-Galiano, V.F., et al.: Increasing the spatial resolution of thermal infrared images using cokriging. Procedia Environ. Sci. **3**, 117–122 (2011)
7. Rodriguez-Galiano, V., et al.: Downscaling Landsat 7 ETM+ thermal imagery using land surface temperature and NDVI images. Int. J. Appl. Earth Obs. Geoinf. **18**, 515–527 (2012)
8. Zhu XX, Bamler R.: A Sparse Image Fusion Algorithm With Application to Pan-Sharpening. Vol. 51, pp. 2827–2836. IEEE Transactions on Geoscience and Remote Sensing (IEEE), (2013)
9. Jin, H., Han, D.: Multisensor fusion of Landsat images for high-resolution thermal infrared images using sparse representations. Math. Probl. Eng. **2017**, 1–10 (2014)
10. Baraldi, A., Despini, F., Teggi, S.: Multi-spectral image panchromatic sharpening – outcome and process quality assessment protocol, pp. 1–39. IEEE (2017)
11. Rahaman, K.R., Hassan, Q.K., Ahmed, M.R.: Pan-sharpening of Landsat-8 images and its application in calculating vegetation greenness and canopy water contents. Int. J. Geo-Inf. (MDPI) **6**, 168 (2017)
12. Farhanj, F., Akhoondzadeh, M.: Fusion of Landsat-8 thermal infrared and visible bands with multi-resolution analysis contourlet methods. Int. Arch. Photogramm Remote Sens. (ISPRS) **42**, 77–81 (2017)
13. Jamshidi, S., et al.: Application of a simple Landsat-MODIS fusion model to estimate evapotranspiration over a heterogeneous sparse vegetation region. Int. J. Geo-Inf. (MDPI) **11**, 741 (2019)
14. Cho, K., Kim, Y., Kim, Y.: Disaggregation of Landsat-8 thermal data using guided SWIR imagery on the scene of awildfire. Int. J. Geo-Inf. (MDPI) **10**, 105 (2018)
15. Wang, F., Qin, Z., Song, C., et al.: An improved mono-window algorithm for land surface temperature retrieval from Landsat 8 thermal infrared sensor data. Int. J. Geo-Inf. (MDPI) **7**, 4268–4289 (2015)

Efficient Spectrum Provisioning in Elastic Optical Networks

Neha Mahala and Jaisingh Thangaraj(✉)

Department of Electronics Engineering, Indian Institute of Technology (Indian School of Mines), Dhanbad 826004, India
mahala.neha@gmail.com, t.jaisingh@gmail.com

Abstract. The emerging technology of Elastic Optical Network (EON) has appeared to satisfy bandwidth hungry demands of today's world. Routing and spectrum assignment is a major problem in EON for efficient spectrum utilization. In this paper we proposed a novel routing and spectrum allocation technique which gives efficient spectrum utilization by reducing blocking probability. This approach guarantees spectrum allocation upon arrival of a connection request but there may be some delay. This is called Request Provisioning with Reservation (RPR). The simulation results shows that this technique enhances spectrum utilization and reduces call blocking.

Keywords: Routing and spectrum assignment · Elastic optical networks · Flexgrid optical networks · Spectrum allocation

1 Introduction

Traditional Wavelength Division Multiplexed (WDM) networks are not able to fulfill the emerging applications. Because of rising demand for high speed communication, cloud computing, huge bandwidth and high-speed networks for data transfer. Although there exists many advantages in WDM networks, like, low bit-error rate, fast connectivity, low propagation delay and low cost re-generator etc. Though the improved version of WDM networks is available and called as Dense Wavelength Division Multiplexed (DWDM) networks but their use is still limited by the fixed-grid channel for wavelength assignment. This causes to spectral fragmentation which occurs because of unused wavelength channel and hence results in degradation in spectrum efficiency. For all these problems a new solution emerges for future data transmission needs is known as the Elastic Optical Network (EON). It has flexible (elastic) optical spectrum and allocates spectrum in slots as per the demand of connection request to get efficient spectrum consumption. The EON is enabled by two important technologies, that are optical Orthogonal Frequency Division Multiplexing (O-OFDM) and another one is Bandwidth-Variable Cross-Connect (BV-OXC) [1]. OFDM allows overlapping of adjacent sub-carriers and parallel transmission of data for a connection request, it leads to better spectrum utilization. Then, BV-OXCs further

N. Nain and S. K. Vipparthi (Eds.): ICIoTCT 2019, AISC 1122, pp. 221–228, 2020.
https://doi.org/10.1007/978-3-030-39875-0_24

enhances flexibility by dynamically adjusting bandwidth capacity by using different modulation format as per the transmission reach and bit rate configurations. Further guard bands between neighboring channels are added to enhance the spectrum overhead and to avoid interference which helps in better spectrum consumption. Spectrum assignment in flex-grid network gives good spectral utilization. The RSA has some Constraints such as: spectrum-continuity and spectrum-contiguity. According to the spectrum-continuity constraint same set of spectrum resources along each link can only be assigned. And next, the spectrum-contiguity says that only a set of neighboring sub-carrier slots can be assigned for a request [2].

The procedure of finding out a suitable path and then assigning spectrum to a connection demand is referred to Routing and Spectrum Assignment (RSA). The RSA concept was first introduced in [3]. Then RSA was applied to the dynamic traffic in [4], and for the static traffic in [5]. In [6], authors considered the transponder assignment for RSA problem. Many routing and spectrum assignment approaches have been proposed in the literature such as fragmentation aware RSA [7] and RSA with traffic grooming [8]. In [9], the authors proposed an ILP and sequential algorithm for RSA, in [10] authors proposed three different RSA algorithms for good blocking performance, ILP and heuristic approach based fragmentation, joint optimization of delay-bandwidth product and link congestion is proposed in [11]. In [12] authors focused on the fragmentation issues in EON and proposed a spectral de-fragmentation algorithm. To get the optimal schemes of different kinds of mapping like virtual links, virtual nodes in virtual optical network a genetic algorithm is proposed [13]. In [14], authors proposed a holding time aware algorithm and considered the impact of fragmentation from the view of occupied and free frequency slots and to measure it. RSA technique must be able to find available contiguous slots for a connection request. By following the spectrum continuity and contiguity constraint, the required set of contiguous slots are not always available on the fiber links along its selected route at a given service time, which causes poor blocking performance.

In this paper, we considers Advance Reservation (AR) for spectrum assignment. Here spectrum is reserved for a request for a known book ahead time and after some predetermined time spectrum is assigned. We have proposed a novel RSA policy called Request Provisioning with Reservation (RPR). This approach reduces the blocking probability by efficiently utilizing spectrum. It is a two-phased approach, first is finding a shortest route between source and destination node and second is spectrum allocation based on resource availability. Our simulation results shows that by delaying the allocation of spectrum, we can utilize spectrum in better way and can serve more number of requests. We present the system model in Sect. 2, explanation of proposed RRP algorithm in Sect. 3. We present the performance of our technique through simulation results in Sects. 4, and 5 concludes the paper.

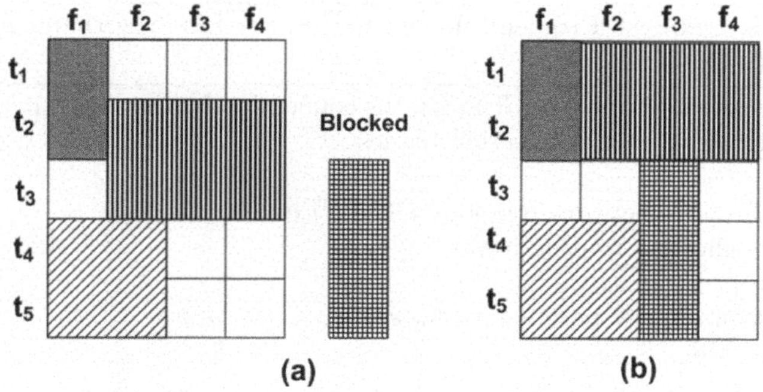

Fig. 1. Spectrum assignment using (a) Shortest Path First Fit and (b) proposed RPR.

2 System Model

EON can be expressed as a graph $G(N, E)$, where N denotes the number of nodes, E denotes the set of links. M_d is the highest modulation level (M_d is 1, 2, 3, 4 for BPSK, QPSK, 8-QAM, and 16-QAM respectively). Modulation technique is assigned to each connection request based on its transmission distance. A connection request is expressed as $R_q(s, d, L_c, t_c, \beta_c, \mu_c)$, where s, d denotes the source and destination node respectively, L_c denotes the capacity of the connection request (Gb/s) and t_c denotes the arrival time. Start time for a connection request is represented by β_c and μ_c denotes end time. In order to complete a request, first, a routing path has to be searched which can accommodate the capacity within provided service duration. The connection request capacity is mapped into N_{cg} number of available contiguous subcarrier slots within provided service duration on a routing path. The number of contiguous slots required to carry a connection request depends on the capacity of the connection request L_c, transmission reach and spectral width of slot L_m.

3 Proposed Algorithm

In this section we will present the proposed technique to assign spectrum considering Advance Reservation (AR) for spectrum assignment as shown in Fig. 1. Here spectrum is reserved for a request based on prediction and after some predetermined time spectrum is assigned. We have proposed a novel RSA policy called Request Provisioning with Reservation (RPR). In the network, shortest path between each source destination pair has been find out using Dijkstra's shortest path algorithm. When a connection request arrives then corresponding shortest paths are chosen. Based on the distance modulation level assigned. Then, required number of contiguous slots assigned following the spectrum continuity and contiguity constraint. If the contiguous slots are not available then reserve slots along that path for reservation window $[t_c, \beta_c + \tau]$ as explained in

Algorithm 1. Request Provisioning with Reservation Algorithm for assigning spectrum in EON

INPUT: The network topology $G(N, E)$, the connection request $R_q(s, d, L_c, t_c, \beta_c, \mu_c)$,
OUTPUT: RSA for each connection request

```
Begin
for all (s, d) pair ϵ network topology G(N, E) do
    determine shortest routing paths
end for
get connection request parameters R_q(s, d, L_c, t_c, β_c, μ_c)
for the start time β_c do
    select P-shortest paths
    for the P-shortest paths do
        find out the modulation level (M_d) based on transmission distance
        calculate the contiguous subcarrier slots N_cg on for service time τ = β_c − μ_c
        end for
    end for
if the contiguous slots are available then
    assign slots
    else
    reserve the slots along that path for reservation window [t_c, β_c + τ]
    plan to serve the connection request from β_c
end if
```

Algorithm 1. This approach reduces the blocking probability by efficiently utilizing spectrum. For assigning spectrum to a request N_{cg} contiguous subcarrier slots should be occupied (due to spectrum continuity constraint). Here, N_{cg} can be determined with the highest modulation level M_d and capacity L_c of a request.

$$N_{cg} = \left\lceil \frac{L_c}{M_d * L_m} \right\rceil + g_b \tag{1}$$

Here N_{cg} denotes the number of contiguous slots, L_c is the capacity of connection request, g_b is the number of slots required for the guard band, M_d indicates modulation level and L_m means slot width. Blocking probability for a call request can be defined as shown in (2).

$$BP = \frac{\sum_{r \epsilon R_b} L_c \tau}{\sum_{r \epsilon R} L_c \tau} \tag{2}$$

Here, r denotes the index of connection requests, R is the set of connection requests, R_b is the set of blocked connection requests and τ is the service duration of the connection request.

4 Performance Analysis

In this section, we have presented the performance evaluation of our proposed technique in terms of initial delay and blocking probability. We simulate the proposed algorithm with NSFNET topology as shown in Fig. 2 and for a large number of connection requests. Later we compared it with Shortest Path First Fit (SPFF) policy through simulation results.

The parameters we have assumed are shown in Table 1. The source and destination nodes for the connection requests are selected randomly from the network nodes. For evaluation of our algorithm we have generated randomly three hundred lightpath requests and the traffic demands. Connection request arrives following the Poisson distribution with arrival rate λ per sec. It stays for holding time $1/\mu$ following an exponential distribution. The traffic load (λ/μ) is varying from 100 to 300 Erlang. We have carried out all simulations in MATLAB R2013a environment on a computer with 1.70 GHz intel core i3 processor and CPU with 4 GB RAM.

Table 1. Simulation parameter.

Parameter	Value
Spectrum width of slot, (L_m) (GHz)	12.5
Link capacity (THz)	1.6
Frequency slots (S) per link	128
Guard band number per request (g_b)	1
Transmission reach of BPSK (km.)	5000
Transmission reach of QPSK (km.)	2500
Transmission reach of 8-QAM (km.)	1250
Transmission reach of 16-QAM (km.)	625
Range of connection request	12.5 Gbps–200 Gbps

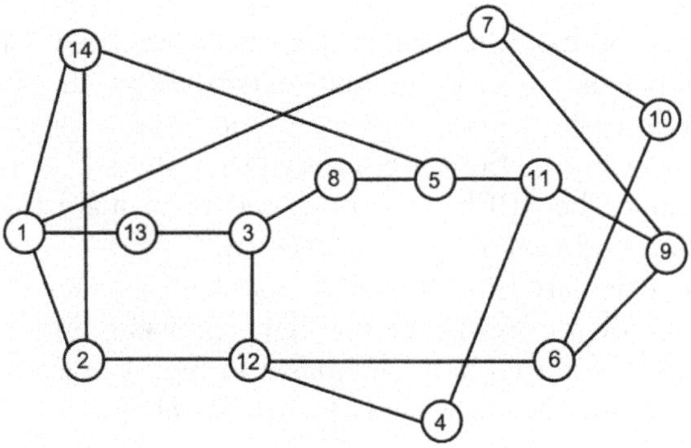

Fig. 2. NSFNET topology which have been used.

Blocking probability (BP) for connection requests is expressed as the number of requests blocked among the total requests arrived. For proper evaluation

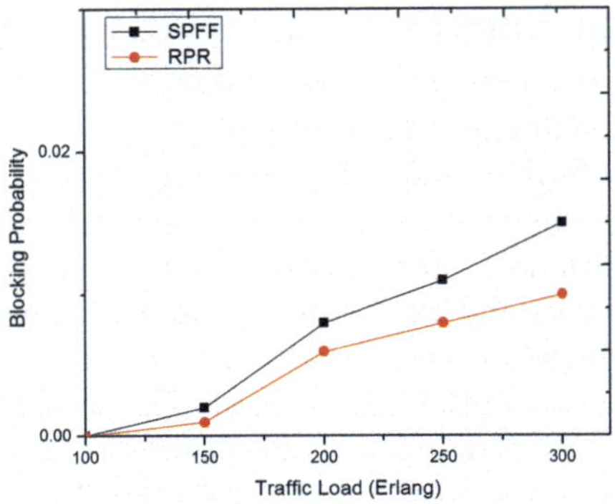

Fig. 3. Blocking probability with respect to traffic load for the Shortest Path First Fit and proposed RPR.

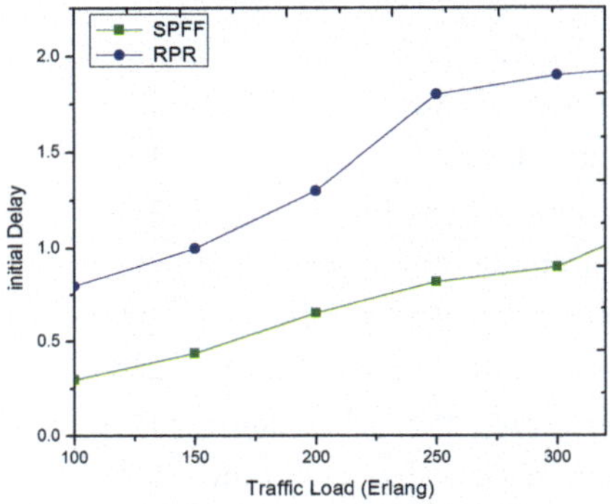

Fig. 4. Initial delay with respect to traffic load for the Shortest Path First Fit and proposed RPR.

purpose, a suitable modulation format have been assigned for the SPFF and proposed algorithm based on the distance according to the Table 1. The NSF network topology have been used for evaluation. The Simulation results were observed as shown in Fig. 3. Here, we observe that the BP is lower for our proposed algorithm than the SPFF. The proposed RPR algorithm can efficiently utilize the spectrum and therefore reduce the BP.

Initial delay is calculated for NSFNET topology as shown in Fig. 4 for the SPFF and RPR algorithm. Note that the higher blocking improvement leads to longer initial delay. The initial delay increases in high loads which is because of more connection requests are in waiting for the spectrum.

5 Conclusion

In this paper, we presented the solution for the problem of routing and spectrum assignment in elastic optical networks by using advance reservation scheme. We evaluated the proposed algorithm on a 14-node NSFNET topology using simulation. These results reveal that our proposed algorithm outperforms the SPFF in terms of BP and initial delay under two network topology. It shows that the proposed algorithm gives better results in compare to traditionally used algorithm.

References

1. Jinno, M., Takara, H., et al.: Spectrum-efficient and scalable elastic optical path network: architecture, benefits, and enabling technologies. IEEE Commun. Mag. **47**(11), 66–73 (2009)
2. Alaskar, R.W., Ahmad, I., Alyatama, A.: Offline routing and spectrum allocation algorithms for elastic optical networks. Opt. Switching Networking **21**, 7992 (2016)
3. Jinno, M., Kozicki, B., et al.: Distance-adaptive spectrum resource allocation in spectrum-sliced elastic optical path network. IEEE Commun. Mag. **48**(8), 138–145 (2010)
4. Wang, X., Kuang, K., et al.: Dynamic routing and spectrum allocation in elastic optical networks with mixed line rates. IEEE/OSA J. Opt. Commun. Networking **6**(12), 1115–1127 (2014)
5. Christodoulopoulos, K., Tomkos, I., Varvarigos, E.: Elastic bandwidth allocation in flexible OFDM-based optical networks. J. Lightwave Technol. **29**(9), 1354–1366 (2011)
6. Christodoulopoulos, K., Soumplis, P., Varvarigos, E.: Planning flexible optical networks under physical layer constraints. IEEE/OSA J. Opt. Commun. Networking **5**(11), 1296–1312 (2013)
7. Kozicki, B., Takara, H., Sone, Y., et al.: Distance-adaptive spectrum allocation in elastic optical path network (slice) with bit per symbol adjustment. In: Conference on (OFC/NFOEC) Optical Fiber Communication (OFC), Collocated National Fiber Optic Engineers Conference, pp. 1–3 (IEEE) (2010)
8. Eira, A., Pedro, J., Pires, J.: Optimized design of shared restoration in flexible-grid transparent optical networks. In: National Fiber Optic Engineers Conference, JTh2A–37. Optical Society of America (2012)
9. Christodoulopoulos, K., Tomkos, I., Varvarigos, E.A.: Routing and spectrum allocation in OFDM-based optical networks with elastic bandwidth allocation 3. In: Global Telecommunications Conference (GLOBECOM 2010), pp. 1–5. IEEE (2010)
10. Wan, X., Wang, L., Hua, N. et al.: Dynamic routing and spectrum assignment in flexible optical path networks. In: Optical Fiber Communication Conference and Exposition (OFC/NFOEC) and the National Fiber Optic Engineers Conference, pp. 1–3. IEEE (2011)
11. Behera, S., George, J., Das, G.: Effect of transmission impairments in CO-OFDM based elastic optical network design. Comput. Netw. **144**, 242253 (2018)
12. Wen, K., Yin, Y., Geisler, D.J. et al.: Dynamic on-demand lightpath provisioning using spectral defragmentation in flexible bandwidth networks. In: European Conference and Exposition on Optical Communications, Mo–2. Optical Society of America (2011)

13. Xuan, H., Wang, Y., Xu, Z., et al.: Virtual optical network mapping and core allocation in elastic optical networks using multi-core fibers. Opt. Commun. **402**, 2635 (2017)
14. Yang, F., Wang, L., Wang, L., et al.: Holding-time-aware spectrum allocation algorithm for elastic optical networks. Opt. Fiber Technol. **41**, 155162 (2018)

Intrusion Detection and Prevention Mechanism Implemented Using NS-2 Based on State Context and Hierarchical Trust in WSNs

Abhishek Vyas[1(✉)] and Satheesh Abimannan[2]

[1] CSE Department, JK Lakshmipat University, Jaipur 302026, Rajasthan, India
abhishek_vyas@msn.com, abhishek.vyas@jklu.edu.in
[2] National Taipei University, New Taipei City, Taiwan
satheesha23@gm.ntpu.edu.tw

Abstract. Wireless Sensor Networks are ubiquitous. The use of WSN saves a lot of man-power and instrumentation for a lot of healthcare and especially industrial applications. This paper describes the work done on securing cluster based WSNs using IDP mechanism. An Intrusion Detection and Prevention Mechanism is proposed here based on State Context and Hierchical Trust in cluster based WSNs. An IDP system will prevent and detect attacks on WSN nodes before they happen and jeopardize the whole network. The paper proposes a system which is flexible and is suitable for the WSNs that are constantly changing, characterized by changes in their perceptual environment, transitions of state of nodes and variation in trust value. A multilevel 2-tier hierarchical trust mechanism at the level of SNs and CHs evaluated on the basis of interactive trust, honesty trust and content trust is put forward, which is a combination of direct evaluation and feed-back based evaluation in the fixed-hop range. The calculation of interactive trust and honesty trust is done based on the interaction and behavior of the sensor nodes during routing. While the calculation of content trust is done based on the deviation of data transferred with respect to the rest of the members in the cluster during data aggregation. CHs evaluates the trust of the cluster member SNs, while the trust of the CH is evaluated by its neighboring SNs. When the trust evaluation process is done, then the data from low trust cluster members are avoided. Forwarding of data through low trust cluster members is avoided. Initially the attacker node tries to communicate to the rest of the nodes via fake control messages for getting itself elected to the level of CH. The mechanism proposed here in this paper will not only prevent the attacker node but also reduce the overall resource overhead while improving network performance. This paper tries to give its small contribution in the security of cluster based WSNs using IDP mechanisms.

Keywords: Sensor nodes (SNs) · Cluster heads (CHs) · Content trust · Honesty trust · Interactive trust

1 Introduction

Wireless Sensor Networks (WSNs) technology is almost 10–15 years old now and a lot of new developments are happening in this field every year. WSN will be as pervasive

N. Nain and S. K. Vipparthi (Eds.): ICIoTCT 2019, AISC 1122, pp. 229–240, 2020.
https://doi.org/10.1007/978-3-030-39875-0_25

in the near future as today we have internet, DTH T.V., smart phones etc. Presently WSNs are used in industrial and commercial applications, because of the technical developments in processor speed, communication protocols, and low power usage of embedded-computing devices.

'A Wireless Sensor Network is one kind of wireless network includes a large number of circulating, self-directed, minute, low powered devices named sensor nodes called motes. These networks certainly cover a huge number of spatially distributed, little, battery-operated, embedded devices that are networked to carefully collect, process, and transfer data to the operators, and it has control over the capabilities of computing & processing. Sensor nodes are tiny computers, which work jointly to form the networks' [1].

1.1 Problem Statement, Motivation and Objectives

To implement an intrusion detection and prevention system based upon state context and hierarchical trust evaluation and estimation of cluster based WSN. The intrusion detection system will work on the trust evaluation of individual nodes and cluster head nodes and a threshold value of the trust. If the nodes trust value is less than the threshold value then that node is detected as malicious. This paper implements different types of attacks; common to cluster based WSNs and then it implements the IDP mechanism to prevent and detect those attacks to improve network overhead and network performance. One of the most important point of the problem statement is to find out a method to prevent an attack by finding that particular malicious node before data transmission begins between nodes. This will be the major contribution of this paper. The author thinks that this will solve the problem of high energy consumption during transmission and reception by nodes, high communication overhead if the malicious nodes are detected way ahead of time.

The motivation for working on the problem of implementing an intrusion detection and prevention system came after reading a lot of literature on WSN and IoT security issues. After reading the research paper published in IEEE Access in 2017 [2]. This paper was about intrusion detection based on state context and hierarchical trust estimation; it enlightened the author of implementing an intrusion prevention and detection system based on 2-tier state context and hierarchical trust estimation on multidimensional data in cluster based WSNs. This idea lead to the development of the implementation of the hypothesis on Network Simulator-2 and the end result is this detailed research paper detailing the whole theoretical and experimental study conducted during the project.

The main objectives for the implementation and getting the required results from this research project paper are – to detect the implemented attacks on WSN & improve the overall network performance, to prevent one of the implemented attacks on WSN & improve the network resource overhead and to implement attack resilient routing and data aggregation in WSN.

Moving ahead; section two of this paper lays out some of the related works in the literature that are similar in nature to this paper's work; section three discusses in detail the theoretical background for the implemented work & how the implemented algorithm works; section four talks in brief about the simulation parameters, experimental design in NS-2 for implementation; section five talks about results & analysis and finally section six is about the conclusion and the future enhancements that can be done to improve upon the proposed work implemented in this paper.

2 Related Works

Apart from the base paper cited in reference number 2. There were many papers referred and studied before the implementation of this research project and the authoring of this paper.

In one of the papers the authors He et al. proposes and implements a new attack resistant and lightweight trust management for MSNs. 'In this paper the authors suggest a 2-tier architecture for an MSN, i.e., a Medical Sensor Network. The proposed system is based on 2-tier architecture, develop an attack-resistant and lightweight trust management protocol named ReTrust which remedies the security and efficiency weaknesses of existing trust systems' [3].

In another of the papers by He et al. talks about, 'a distributed trust evaluation model and its application scenarios for medical sensor networks' [4]. They discuss about the importance of the need for development of medical sensor networks for betterment in healthcare and as an asset to medical professionals. But the authors feel that security remains a main challenge in this field of development of MSNs and the SNs connected to MSN. 'Traditional cryptographic mechanisms are not sufficient given the unique characteristics of MSNs, and the fact that MSNs are susceptible to a variety of node misbehaviors. In such situations, the security and performance of MSNs depend on the cooperative and trust nature of the distributed nodes, and it is important for each node to evaluate the trustworthiness of other nodes. This paper also proposed Collection Tree Protocol model in the network of TelosB motes, the use of which significantly improves the network performance' [4].

In yet another research paper, the authors propose a 'Reputation Based Framework for High Integrity WSNs.' 'A reputation-based intrusion detection system for WSNs, which adopted reputation and trust to gauge the behaviors of nodes. The evaluation considers only communication factors such as correct and incorrect interactions using a beta function, which is executed by each node. Every node has evaluation and detection tasks for other nodes; thus, the overhead is still high for WSNs. This method could decrease the overhead of nodes by executing tasks in CHs and BSs with more resources' [7].

In one of the research papers which is more related to the work done and shown in this paper, the authors propose a 'Distributed Trust based Intrusion Detection Approach in WSNs'. 'Trust models have become very important as far as detection of malicious behavior is concerned. The Distributed Trust based Intrusion Detection (DTBID) model not only considers communication behavior to detect trustworthiness of sensor node but also considers some other factors of trust which are distributed in energy, data trust, reliability, communication trust etc., During DTBID, the calculation of direct trust, recommendation trust & indirect trust have also been discussed in this research paper. The authors in this research paper compare subjective trust derived by the proposed model with the objective trust which is based on ground truth status of sensor node to detect any kind of malicious behavior in wireless sensor networks' [9].

This concludes the related works survey for this research paper. More papers and other sources were referred as one can see in the references, in related works only important referred works are described in brief.

3 Description of the IDPS Based on State Context and Hierarchical Trust in Cluster Based Wireless Sensor Networks

3.1 Description of the Intrusion Detection System

In the implemented system that is theorized and proposed in this paper. 2-tier hierarchical trust mechanism is introduced in the cluster based WSN. Unlike the previous works referred in [2], here the first level of trust is simplified by CH to sensor node SN evaluation, due to the direct communication between SN and CH in that particular cluster, whereas the second level of trust is calculated by CH-to-CH direct evaluation and BS-to-CH direct evaluation or indirect evaluation via the feedback of 1-hop neighbor of CH in the cluster. The trust of sensor nodes is evaluated by the CH of that cluster. This means that, CH-to-SN trust, which considers multidimensional trust, including interactive trust, content trust and honesty trust, during the process of trust calculation. As already mentioned the trust value of the SN is evaluated by their respective CHs, while the trust value of the CHs is evaluated by the BS, reducing the workload on the SNs. The process of finding the malicious SNs is executed by the respective CH of that cluster. After the trust estimation is done, the data from low trust members is filtered at the time of data aggregation. Data forwarding through the low trust members is avoided. The one major drawback of the existing system is overhead cost that will be incurred in detecting an alternate routing path that does not involve malicious nodes (Fig. 1).

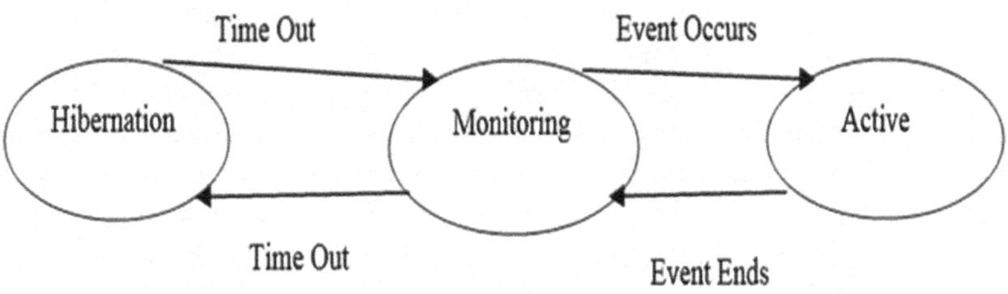

Fig. 1. Schematic diagram of state transition of sensor nodes

3.2 Description of the Intrusion Prevention System

In the implemented system here in this paper, intrusion prevention along with intrusion detection is implemented. In the intrusion prevention system only the cluster dropping attack scenario is implemented and prevented using the SHTIP routing algorithm.

So, here initially the attack is launched by the attacker node so that it can be selected as the CH of that particular cluster of nodes. The attacker node sends fake control information like high energy, high neighbor count etc. in control messages to the other nodes in the network during cluster forming and cluster head selection process. This way the attacker node wants to get selected as the cluster head of a particular cluster. By employing a strategy that can effectively detect these fake information control messages, i.e., SHTIP routing, the IPS can be implemented successfully. This IPS scheme will reduce the WSN overhead due to countermeasures taken after the detection of attacks in IDS system, like for example finding an alternative route to reach the sink node i.e., the base station after detection. Here, in IPS the actual neighbor count information can be collected using the statistics collected from neighbors. And actual energy can be estimated form the time differential (Δt) between previous and current control message information.

Some advantages of the implemented system are as follows - the implemented system detects & prevents the attacker in a cluster based WSN, the implemented system reduces the overall network overhead & energy consumption in the network and last but not the least the implemented system improves the overall network performance which is measured by the stipulated performance metrics like detection time, packet delivery ratio & communication overhead.

Some assumptions made by the authors/implementers of this paper/project while developing & conceptualizing the network model architecture of this projects' implementation are – the WSN in this implementation is cluster based & SNs in a cluster communicate directly with their respective CHs, whereas CHs communicate with BS directly or indirectly via other cluster-heads; each SN possess a unique ID & belongs to a unique cluster & CHs have more energy than SNs; the data transmission model in this simulated WSN is hybrid, i.e. both continuous & event-driven; hibernation, monitoring & active are the 4 states of a SN; there is a transition between monitoring & active, which is taken into consideration during the trust evaluation of sensor nodes and final assumption is that for reasons that consider the reliability of data transfer the SNs are deployed densely & redundantly.

3.3 Implemented System Design

This section describes in detail the module wise design of the implemented system. There are various modules in both the IDS and the IPS part of the implemented system, these modules of the completed implemented IDPS system for cluster based WSN can be explained briefly module-wise in a streamlined manner as follows:

1. MODULE 0: [(Input in both IDS & IPS)]
   Overall Input is Residual Energy and Sensor Nodes
2. MODULE 1: [Clustering Process (for both IDS & IPS)]
   Input here is also Residual Energy and Sensor Nodes
   Output here is selection of CH and cluster formation

Here in this module, the logic as per the implementation methodology dictates that in clustering process based on the initial control signals and information passed along all the nodes in the WSN setup, a cluster head is selected based on the following logic:

For Each SN in the Simulation Setup for WSN these formulae apply.

$$\text{energy\_val} = \text{residual\_energy}/50.0 \tag{1}$$

$$\text{neighbour\_val} = \text{neighbour\_count}/15 \tag{2}$$

$$\text{energy\_weight} = \text{energy\_val} * 0.5 \tag{3}$$

$$\text{neighbour\_weight} = \text{neighbour\_val} * 0.5 \tag{4}$$

$$\text{weight} = \text{energy\_weight} + \text{neighbour\_weight} \tag{5}$$

Now, if weight is greater than max_weight, then we have

$$\text{max\_weight} = \text{weight} \tag{6}$$

so, the cluster head is the node with maximum weight.

In this module, after the CH is selected for all the given input of SNs in the given setup of WSN, then Clusters are formed and then the process of data collection & aggregation and transmission form the SNs to the CHs and form CHs to the BS begins.

3. MODULE 2: [Implementation of the Attacks for both IDS & IPS]
   Input here is collected data transmission from SNs
   Output here will be Malicious CH dropping, false data injector SN identified, & energy exhaustion attacker SN identified.

Here in module 2, the implementation of all the attacks is done, here in this project the implemented attacks against the cluster based WSN are false data injection attack by a malicious SN, energy exhaustion attack by another malicious SN & a malicious attacker SN trying to make itself CH by transmitting fake control messages, i.e., CH dropping attack. The ideal outcome of this module is the successful implementation of the above mentioned attacks and finally in subsequent modules their detection and prevention.

4. MODULE 3: [Trust Estimation & Calculation of SNs by their respective CHs; applicable for both IDS & IPS]
   Input here is respective CHs and SNs
   Output here is the calculation of the OT_SN based on the CT_SN, HT_SN & IT_SN, and based on that detection of malicious attacker SNs.

Logic used for SNs' trust estimation and calculation:

$$\text{IT\_SN} = 1 \tag{7}$$

if, the number of arrived data packets is greater than maximum data packets. Otherwise,

$$\text{IT\_SN} = \{1/(\text{arrived data packets}/\text{maximum data packets})\} \tag{8}$$

$$\text{HT\_SN} = \{\text{forwarded data packets to CHs}/\text{received data packets from SNs}\} \tag{9}$$

$$\text{CT\_SN} = \{\text{deviation}/\text{average of sensing data}\} \tag{10}$$

$$\text{OT\_SN} = \{w_1 * \text{IT\_CH} + w_2 * \text{HT\_CH} + w_3 * \text{HT\_CH}\} \tag{11}$$

Here, w_1, w_2 and w_3 represents the weight factor

If, the overall trust value is less than the threshold then the respective CH is an attacker node; otherwise it is not an attacker node.

5. MODULE 4: [Trust Estimation & Calculation of CHs by their cluster member SNs & BS; applicable for both IDS & IPS]
   Input here is respective SNs, CHs & the BS.
   Output here is the calculation of OT_CH based on IT_CH, HT_CH & CT_CH and based on that detection of malicious attacker CHs

The logic used in this module is similar to the one used in the module 3 described in detail above, here the trust calculation is based on cluster heads and base stations.

6. MODULE 5: [Overall Output of the implemented IDS & IPS]
   Input here is all of Module 0 to Module 5;
   Output here is successful detection of the implemented attacks and prevention of the cluster dropping attack.

4 Design of Experimental Simulation for NS-2

This section briefly discusses about the experimental simulation parameters and the NS-2 setup and design used for the implementation of the simulation logic for the IDPS for cluster based WSN based on state context and hierarchical trust estimation.

4.1 Experimental Simulation Design

The following files that were pre-existing that were modified for the experimental simulation design: agent.h, cmu-trace.h, cmu-trace.cc, ip.h, packets.h, agent.cc and ns-packets.tcl file.

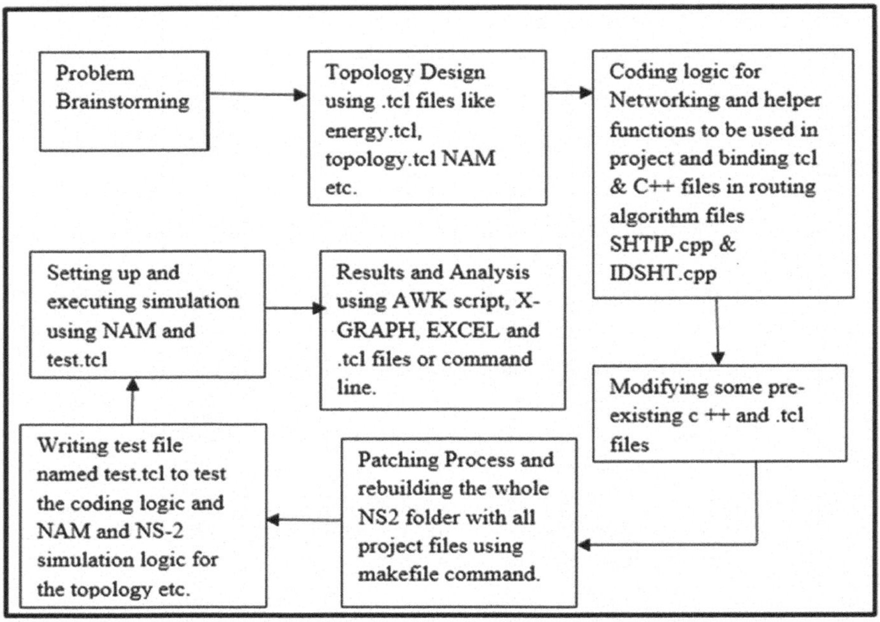

Fig. 2. Functional experimental design for simulation of the experiment in NS-2

The above Fig. 2 shows the functional experimental design for the NS-2 simulation for the implementation of the research project detailed in this research paper. This gives an idea about how the logic and experimentation for the project implementation was done using NS-2 simulation. The Simulation was executed for varying input of nodes to get data for the comparison parameters mentioned in results section for analysis of the results and to compare false positive results with false negative results.

4.2 Simulation Parameters Used in NS-2 for Experimentation

As already known, the implementation of the logic and the entire scheme of this research is done via simulation and the simulator used for this purpose is NS-2, specifically network simulator version 2.35. For the purposes of data collection for performance evaluation during experimentation maximum number of nodes used was 60 and the minimum was 20. Simulation area setup was (500 m^2). The communication area setup was 250 m. Interface type used was Phy/Wireless Phy. MAC type used was IEEE 802.11 standard for wireless networks. Queuing type used was droptail/priority queue. Queuing length was set at 50 packets. Omni antennae type was used during simulation setup and execution. Propagation method used was tworayground based propagation. New secure routing protocols for WSN were implemented in this research project namely IDSHT & SHTIP, both of them were modified from the AODV routing protocol given in NS-2 for the ad-hoc networks. UDP was used as a data transport agent for the NS-2 simulation while CBR was used as an application agent in this simulation in NS-2. Last but not the least the initial residual energy for each SN was set at 50 J and the simulation time was set at 50 s. This concludes the simulation parameters description for this research project.

5 Results and Analysis

This section discusses the results obtained from the simulation of the proposed implementation of the IDP mechanism based on the state context and hierarchical trust estimation in cluster based WSNs. The discussion is based on the performance evaluation between the two algorithms used in this simulation implementation of the proposed scheme, i.e. IDSHT and SHTIP. Here, three performance evaluation parameters for the comparison between the IDSHT and SHTIP are used namely detection time, packet delivery ratio and communication overhead. Lastly a graph displaying false positives and false negatives for both the algorithms is shown and discussed upon for the number of total number of simulations executed for different number of inputs of sensor nodes.

5.1 Results Obtained and Performance Evaluation

Due to the prevention mechanism employed to prevent the attacker node getting selected in the CH selection process, attacker node is detected at the earlier time during the control information broadcast stage. Hence, it is completely eliminated during data transmission phase in the proposed system. But the attacker CH is detected during the data transmission phase in the existing system. Hence, attack detection time is less in proposed system, when it is compared to the existing system (Figs. 3 and 4).

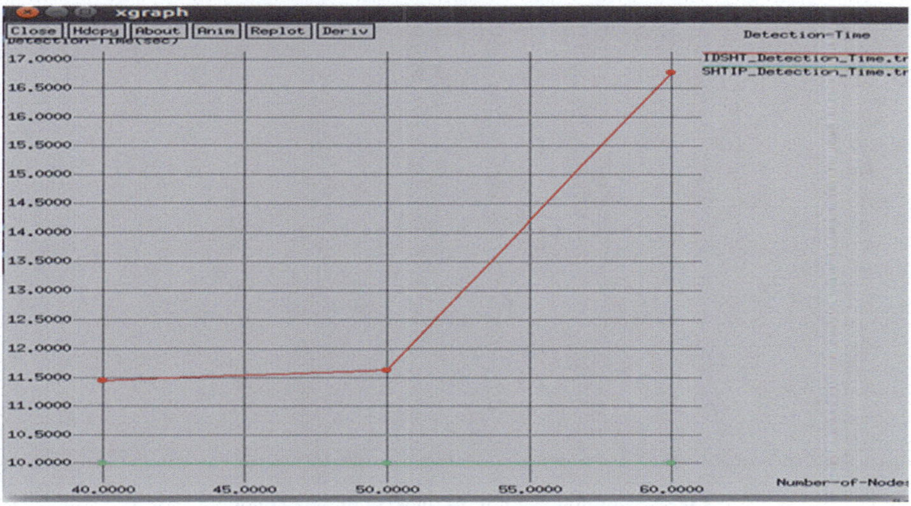

Fig. 3. Detection time comparison graph

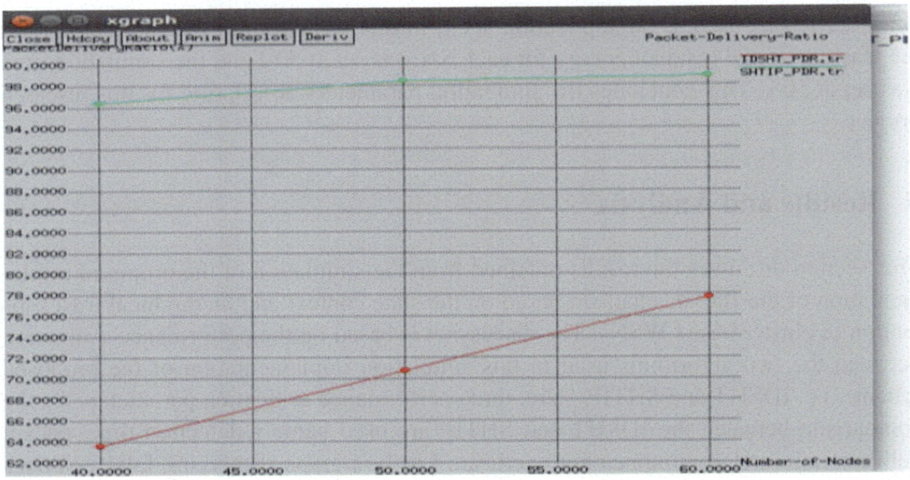

Fig. 4. PDR (%) comparison graph

Due to the prevention mechanism employed to prevent the selection of the attacker node as CH in CH selection process, attacker node for cluster-dropping attack is detected at an earlier time during the stage of control information broadcast. Hence, it is completely eliminated during data transmission phase in proposed system SHTIP. Hence, the packet drop is prevented. Although, the attacker node CH is detected during the data transmission phase in existing system-IDSHT. Attacker node is detected only after packet dropping is done. Hence, the PDR (%) is more in proposed system-SHTIP when comparing it to the existing system-IDSHT (Fig. 5).

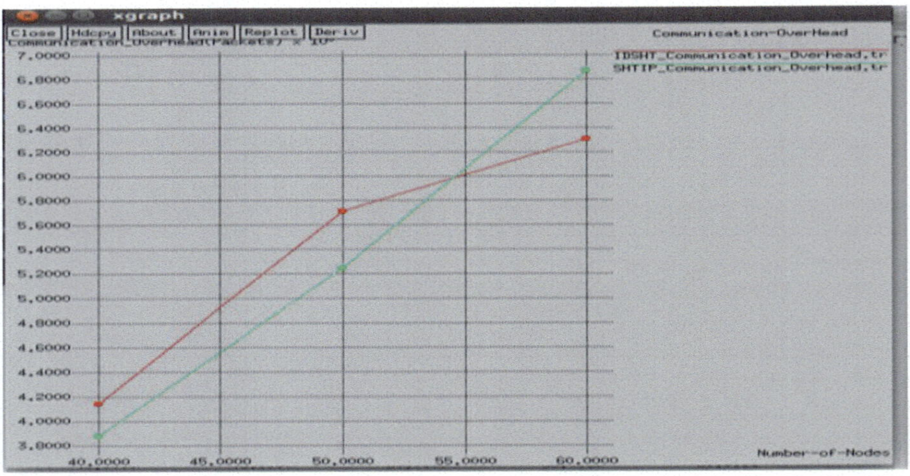

Fig. 5. Communication overhead comparison graph

In the existing system-IDSHT the attacker node, the attacker SN attracts more neighbours so that it can get itself selected as a CH by broadcasting fake control information to the other sensor nodes. Although, in the proposed system-SHTIP this thing is prevented. Hence, the control packets and thus the communication overhead is more in existing system than the proposed system-SHTIP. But in some cases, at some points in the comparative analysis graph as shown in Fig. 6 the communication overhead is the same for both IDSHT and SHTIP.

The following graph displays the false positives and false negatives for the results obtained during the simulation execution for a variable number of input nodes. This graph is self-explanatory. But this shows that simulation results for SHTIP are pretty even for false positives & false negatives obtained.

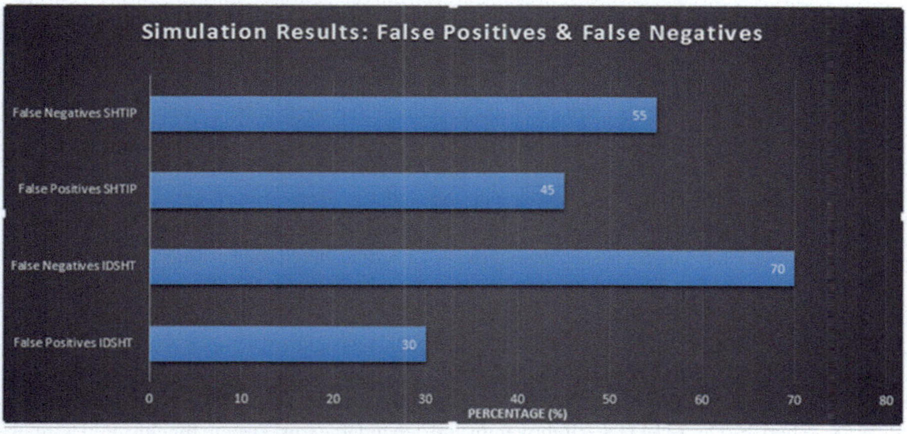

Fig. 6. Graph depicting both the false positive & false negative results for both IDHST & SHTIP implemented algorithms.

6 Conclusion and Future Enhancements for the System

Some conclusions that can be drawn from this paper are as follows - as already know in this paper an Intrusion Detection and Prevention Mechanism Based on State Context and Hierarchical Trust is implemented. This was successfully implemented using NS-2 simulation software package. A self-adaption detection trust threshold was used, as the cluster based WSN is constantly changing. Self-adaption detection trust threshold adapts itself to the overall trust value calculated for the ever-changing cluster based WSN. A two-tier trust calculation mechanism is used here, between SN-to-CH and CH-to-CH & CH-to-BS via 1-hop neighbour. During the process of trust calculation, the state and the transition between the states was taken into account for employing a level-playing field in trust calculation. During trust calculation process different weights were assigned to the nodes to find out their exact neighbour count and energy value for cluster head selection process; this way the CLUSTER HEAD DROPPING ATTACK was prevented (using IPS). All in all the Implementation of the Project can be considered as successful; as shown by the performance evaluation graph results.

In any future work, other interested researchers can use more than one attack to implement an IPS for the cluster based WSN; as cluster based WSN have more attacks in the network layer that needs a prevention mechanism. Other researchers can try and develop an IDPS for cluster based WSN directly at hardware level, i.e. at chip level or microprocessor level. This can be done through light weight algorithms that can be booted into the sensor nodes hardware itself. Last but not the least the proposed mechanism of intrusion detection and prevention described here in this paper can be combined with a software based firewall so that it can also prevent and detect malicious sensor nodes and cluster heads in cluster based WSNs.

References

1. Agarwal, T.: (Web-Blog). Wireless sensor network architecture and its application. https://www.elprocus.com/architecture-of-wireless-sensor-network-and-applications/. Accessed 4 April 2018
2. Zhang, Z., Zhu, H., Luo, S., Xin, Y., Liu, X.: Intrusion detection based on state context and hierarchical trust in wireless sensor networks. IEEE Access **5**, 12088–12102 (2017)
3. He, D., Chen, C., Chan, S., Bu, J., Vasilakos, A.V.: ReTrust: attack resistant and lightweight trust management for medical sensor networks. IEEE Trans. Inf. Technol. Biomed. Eng. **16**(4), 622–632 (2012)
4. He, D., Chen, C., Chan, S., Bu, J., Vasilakos, A.V.: A distributed trust evaluation model and its application scenarios for medical sensor networks. IEEE Trans. Inf. Technol. Biomed. Eng. **16**(6), 1164–1175 (2012)
5. Han, G., Jiang, J., Shu, L., Guizani, M.: An attack-resistant trust model based on multidimensional trust metrics in underwater acoustic sensor network. IEEE Trans. Mob. Comput. **14**(12), 2447–2459 (2015)
6. Li, X., Zhou, F., Du, J.: LDTS: a lightweight and dependable trust system for clustered wireless sensor networks. IEEE Trans. Inf. Forensics Secur. **8**, 924–935 (2013)
7. Ganeriwal, S., Srivastava, M.: Reputation-based framework for high integrity sensor networks. ACM Trans. Sens. Netw. **4**(4), 66–77 (2004)
8. Yao, Z., Kim, D., Doh, Y.: PLUS: Parameterized and localized trust management scheme for sensor networks security. In: Proceedings of IEEE International Conference on Mobile Adhoc Sensor Systems, pp. 437–446 (2006)
9. Dhakne, A., Chatur, P.: Distributed trust based intrusion detection approach in wireless sensor network. In: Proceedings of IEEE International Conference on Communication Control of Intelligent Systems, pp. 96–101 (2015)
10. Gautam, G., Sen, B.: Design and simulation of wireless sensor network in NS2. Int. J. Comput. Appl. **113**(16), 14–16 (2015)
11. Butun, I., Morgera, S.D., Sankar, R.: A survey of intrusion detection systems in wireless sensor networks. IEEE Commun. Surv. Tutor. **16**(1), 234–241 (2014)

Hardware Design of 8 × 8 and 16 × 16 2D Discrete Cosine Transform with N/2 Equations for Image Compression

Nikhil C. Bichwe[1(✉)] and Rahul Kumar Chaurasiya[2]

[1] Department of Electronics and Telecommunication, NIT, Raipur, India
nikhilbichwe@gmail.com
[2] Department of Electronics and Communication Engineering, MNIT, Jaipur, India
rkchaurasiya.39@gmail.com

Abstract. The discrete cosine transform (DCT) has remarkable significance in the area of image and video compression due to its energy compaction property. This paper propose a distinct architecture for 8 × 8 and 16 × 16 2D-DCT for compression of images. A method is proposed to reduce the number of equations with the help of high energy compaction property of DCT. In the conventional N point, 1D-DCT requires N number of equations for the transformation, by using high energy compaction property, only N/2 equations are required to perform the task. Proposed architecture reduces the arithmetic complexity. Further, it exhibits low power consumption with less area requirement. The proposed 2D-DCT is synthesized in XC3S700AN and XC6VLX75T devices, and the simulation results are compared with conventional 2D-DCT. The quality of the reconstructed image is evaluated by peak signal to noise ratio (PSNR). The obtained result shows the 75% reduction in the number of pixels required to store the image.

Keywords: Discrete cosine transform (DCT) · Joint photographic experts group (JPEG) · Field programmable gate array (FPGA) · Peak Signal to Noise Ratio (PSNR) · Xilinx

1 Introduction

The discrete cosine transform (DCT) is a basic building block for applications of multiple image and video processing. It is also used for the applications for various standards for contemporary image and video compression, such as the JPEG. DCT has turned out to be an essential tool in signal and image processing. It's popularity is fundamental because of its high vitality compaction properties. Traditional DCT makes exact calculation and it leads to higher complexity and power consumption results. The utilization of digital cameras has increase the provision for storage, control, and exchange of digital images dangerously. Usually the image files can be huge and it can involve plenty of memory. In the case of images with a high correlation coefficient, DCT-based coding can lead to a decent execution. The calculation of 2D-DCT needs large number of operations and huge

N. Nain and S. K. Vipparthi (Eds.): ICIoTCT 2019, AISC 1122, pp. 241–250, 2020.
https://doi.org/10.1007/978-3-030-39875-0_26

quantity of data. Usually the calculation of 2D-DCT is broken down into 1D-DCT. The algorithms are used for this calculations depends on the row-column decomposition and hence row by row and column by column calculations are done for 1D-DCT operation.

This paper introduces a new 8 × 8 and 16 × 16 2D-DCT architecture using N/2 equations for image compression which possesses an extremely low arithmetic complexity, low power consumption and require less area than the conventional DCT architecture. The conventional architecture requires 28 additions and 22 multiplication for 8-point 1D-DCT and 92 additions and 86 multiplications for 16-point 1D-DCT operation. And the proposed architecture requires 20 additions and 11 multiplication for 8-point 1D-DCT and 64 additions and 43 multiplications for 16-point 1D-DCT operation. The architecture is synthesized in XC3S700AN and XC6VLX75T devices. The performance of the image compression was assessed using the Peak Signal to Noise Ratio (PSNR) values.

2 Literature Survey

Potluri et al. [1] introduced a novel DCT (N = 8) approximation which requires only 14 addition operations and no multiplications. The proposed DCT approximation is used for the reshape of video measures, for example, HEVC. Here DCT approximation are mapped to systolic-array and by using Xilinx FPGA tool implemented the digital prototype circuit and also mapped to 45 nm CMOS technology. Jridi et al. [2, 16] proposed a reconfigurable architecture for orthogonal approximation of DCT. Architecture of DCT performance for one N, two N/2, four N/4 length DCT where N is the input. It calculates multiplication results after m clock cycles which give the advantage of flexibility and low power consumption features for the architecture.

Brahimi et al. [3] proposed 8-point DCT with 8 × 8 orthogonal transform matrix and also introduces some zeros in 8 × 8 Signed DCT (SDCT). By introducing some zeros in SDCT the performance of orthogonal transform matrix is increases and the proposed matrix has orthogonality property. Author compared conventional quantization matrix with new proposed transform matrix and its PSNR is improved by 0.7 dB. Wang [4] present a two types of circuit architecture and a new recursive algorithm for calculation of 2D-DCT and used few number of multiplication and addition operation for the calculation of recursive kernel. For the proposed algorithm, author implemented two version of VLSI architecture to improve the hardware usage and speed of operation. And also deployed a circuit blocks for recursive kernel which requires only four multiplier and three adder and hence it requires less clock cycles. The FPGA result shows that 25% reduction in hardware and 17% increase in clock speed.

Sateesh et al. [5] implemented a new architecture for image compression and encryption technique which is suitable for real-time applications. Here for random number generation LFSR is used for the encryption process and then it added to some output of DCT. Architecture requires 4 multiplier and 14 adders. For integer multiplication author used high performance multiplier (HPM). PSNR for architecture is about 24 dB and the simulation results are around 66%. This architecture requires 78 MHz clock frequency and throughput is 624 M samples/s. Madanayake et al. [6] was proposed an algebraic integer (AI) encoded 2-D Arai DCT with help of a time-multiplexed systolic-array hardware architecture and final reconstruction step (FRS) algorithms. The proposed FRS utilizes two methodologies dependent on (1) optimized Dempster – Macleod multipliers, and (2) development factor scaling. It empowers the high-dynamic low-noise extend applications in computerized video preparing which control the calculation of the 2-D DCT. The frame rate for an image size of 1920 × 1080 is 1187.35 Hz, the maximum clock rate is 307.787 MHz and block rate is 38.47 MHz.

Masera et al. [7] proposed an efficient 2D DCT for the high efficiency video coding (HEVC) standard. Walsh-Hadamarad transform (WHT) is used for the implementation of DCT. Here two architectures are implemented where one is the 1D DCT in complete folded manner and another one is the selected folding degree. Dhandapani et al. [8] proposed 8 × 8 transform matrix which needs only 12 addition by using only zeros and ones and by escaping multiplication and bit shift operations. The proposed transform matrix is implemented on Xilinx Virtex 7 device and synthesized with Cadence RTL compiler using UMC 90 nm.

3 DCT Algorithm

DCT converts an image in to its corresponding frequency domain by dividing image pixels matrix into n × n size blocks where n relies on the image type. Two dimensional N × N order of DCT defined as [9]

$$z_{(u,v)} = \alpha(u)\alpha(v)\sum_{i=0}^{N-1}\sum_{j=0}^{N-1} x_{(i,j)}\, cos\frac{(2i+1)\pi u}{2N} cos\frac{(2j+1)\pi v}{2N} \tag{1}$$

$$\alpha(u) = \alpha(v) = \begin{cases} \sqrt{1/N}\; for\, u = 0,\, 0 \le v \le N-1 \\ \sqrt{2/N}\; otherwise \end{cases}$$

Where $z_{(u,\,v)}$ represent the transform output block of image with dimension u × v, $x_{(i,\,j)}$ represent input block of image with dimension i × j. Utilizing matrix notation, the 8 × 8 point 2D DCT can be communicated as a matrix vector calculation as given below in Eq. (2). where c represents the cosine transform matrix [12].

$$[z] = [c].[x].[c^T] \tag{2}$$

4 Conventional 2D-DCT

2D-DCT is achieved by the mathematical use of two 1D-DCT. 2D-DCT is evaluated by first apply the inputs matrix to 1D-DCT, transpose the output of 1D-DCT and then again apply this transpose matrix to 1D-DCT. Figure 1 shows a 2D-DCT architecture which uses two 1D-DCT blocks, in which x represent the input N × N block of image, y is the output of 1D-DCT block, y^T is transposed output of y and z is the output of

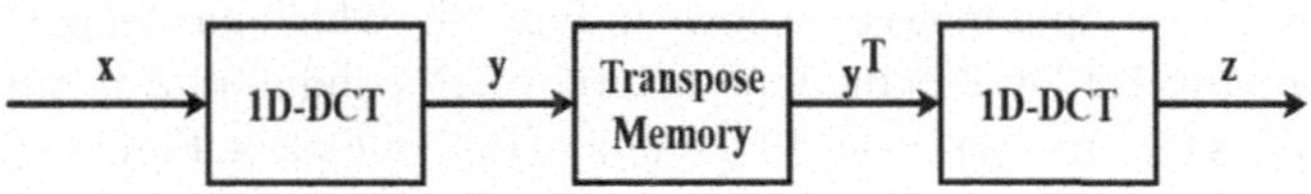

Fig. 1. 2D DCT architecture

second 1D-DCT block. Here z represents 2D-DCT of input N × N block of image. The operation of 2D-DCT is broken down in two 1D-DCT. 1D-DCT of sequence N is defined as [10, 11].

$$y_u = \alpha(u) \sum_{j=0}^{N-1} x_i \, cos \frac{(2i + 1)\pi u}{2N} \tag{3}$$

$$\alpha(u) = \begin{cases} \sqrt{1/N} \; for \, u = 0, \\ \sqrt{1/N} \; otherwise \end{cases}$$

Above Eq. (3) shows the calculation of 1D-DCT for the compression of an image. Consider N = 8, for evaluating the 1D-DCT of input 8 × 8 block of image take 1st column of 8 × 8 block of input image. The simplified form of Eq. (3) shown in the Eq. (4) in which1D-DCT of N samples. Where y represents 1D-DCT matrix of input x matrix and c represents the cosine transform matrix.

$$[y] = [c].[x] \tag{4}$$

Equation (5) shows the 1D-DCT of 1st column of 8 × 8 block of an input image. Similarly calculate for the remaining column of 8 × 8 block of an input image. Where y_0 to y_7 are the output of 1D-DCT of 1st column of 8 × 8 block of an input image and a, b, c, d, e, f, g are the coefficient of cosine transform matrix [12–14].

$$\begin{bmatrix} y_0 \\ y_1 \\ y_2 \\ y_3 \\ y_4 \\ y_5 \\ y_6 \\ y_7 \end{bmatrix} = \begin{bmatrix} a & a & a & a & a & a & a & a \\ b & d & e & g & -g & -e & -d & -b \\ c & f & -f & -c & -c & -f & f & c \\ d & -g & -b & -e & e & b & g & -d \\ a & -a & -a & a & a & -a & -a & a \\ e & -b & g & d & -d & -g & b & -e \\ f & -c & c & -f & -f & c & -c & f \\ g & -e & d & -b & b & b & e & -g \end{bmatrix} \begin{bmatrix} x_0 \\ x_1 \\ x_2 \\ x_3 \\ x_4 \\ x_5 \\ x_6 \\ x_7 \end{bmatrix} \tag{5}$$

5 Proposed Method and DCT Architecture

DCT has high energy compaction property and it converts the signal or image from spatial domain to frequency domain. In the frequency domain 90% of energy is present at the top left corner of block of image [15]. It is along these lines directly to state that the majority of the changed vitality is contained in the initial couple of coefficients and this property is taken favourable position in the proposed design, where the most vital coefficients are discovered less mistaken than the least vital coefficients.

Figure 2(a) shows the input matrix of 8 × 8 block of an image (x_{00} to x_{63}) which include total 64 pixels and further 1D-DCT is apply on each column of input matrix. The output of 1D-DCT of input matrix is shows in Fig. 2(b) which has total 32 values (y_{00} to y_{31}), by applying 1D-DCT to the input matrix of 1st column (x_{00} to x_{07}) we get the four output values (y_{00} to y_{03}) as the 1st column of output matrix. Figure 2(c) shows the transposed of output matrix and Fig. 2(d) shows the 2D-DCT matrix which is calculated by applying 1D-DCT on transposed matrix, it includes total 16 values (z_{00} to z_{16}).

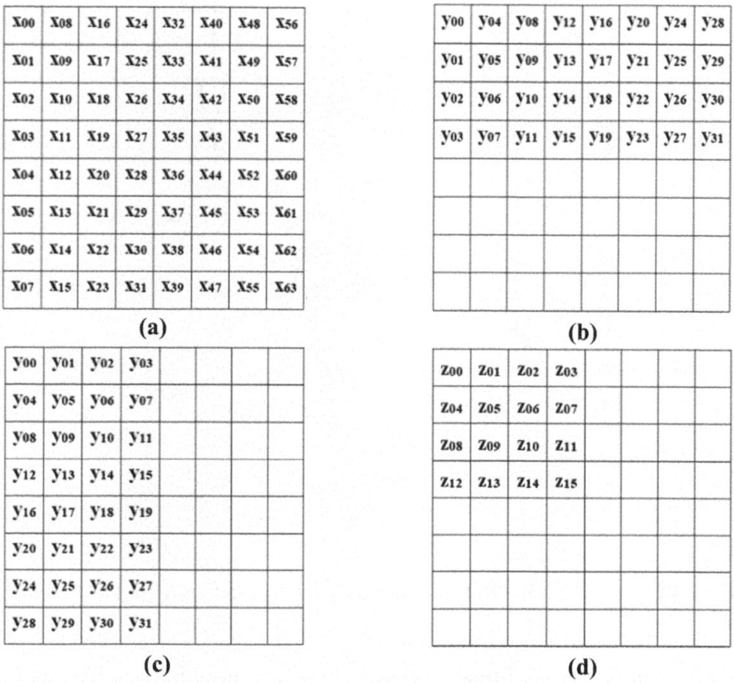

Fig. 2. Proposed method (a) Input matrix (b) 1D-DCT output (c) Transposed matrix (d) 2D-DCT ouput.

5.1 Proposed Architecture for 8-Point 1D-DCT

The proposed architecture for 8-point DCT is shown in Fig. 3(a). For the calculation of 1D-DCT x_0 to x_7 are the inputs pixel values of block of image and y_0 to y_3 are the output values for 1D-DCT. In the architecture doted lines shows the multiplication with coefficients cosine transform matrix and this matrix consists of seven coefficients. The coefficient value of cosine transform matrix is shown in Table 1.

Table 1. Coefficient values for 8-point DCT

Coefficients	a	b	c	d	e	f	g
Values	0.3536	0.4904	0.4657	0.4157	0.2778	0.1913	0.0975

5.2 Proposed Architecture for 16-Point 1D-DCT

The Eq. (6) shows output matrix of 16-point 1D-DCT, where y_0 to y_{15} are the output values, x_0 to x_{16} are input pixel values and a, b, c, d, e, f, g, h, i, j, k, l, m, n, and o are the constant coefficient value of cosine transform matrix. The proposed architecture for 16-point 1D-DCT is shown in Fig. 3(b). For the calculation of 1D-DCT x_0 to x_{15} are

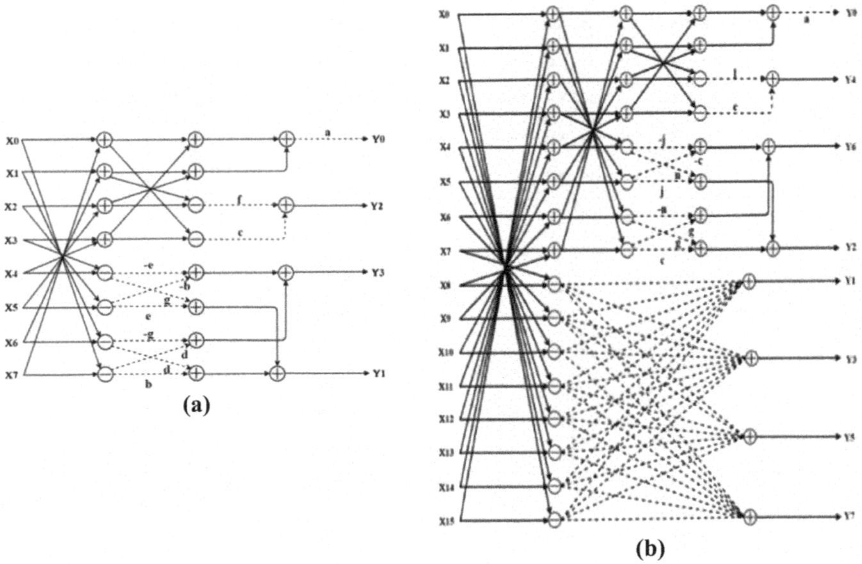

Fig. 3. (a) Proposed 8-point DCT architecture (b) Proposed 16-point DCT architecture

the inputs pixel values of an image and y_0 to y_{15} are the output values. As shown in Fig. 3(b). The operation for 16 point is perform in four stages. In the architecture doted lines shows the multiplication with the coefficients of cosine transform matrix and this matrix consists of 15 coefficients. The coefficient value of cosine transform matrix is shown in Table 2.

$$\begin{bmatrix} y_0 \\ y_1 \\ y_2 \\ y_3 \\ y_4 \\ y_5 \\ y_6 \\ y_7 \\ y_8 \\ y_9 \\ y_{10} \\ y_{11} \\ y_{12} \\ y_{13} \\ y_{14} \\ y_{15} \end{bmatrix} = \begin{bmatrix} a & a & a & a & a & a & a & a & a & a & a & a & a & a & a & a \\ b & d & f & h & i & k & m & o & -o & -n & -k & -i & -h & -f & -d & -b \\ c & g & j & n & -n & -j & -g & -c & -c & -j & -j & -n & n & j & g & c \\ d & i & o & -k & -f & -b & -h & -m & m & h & b & f & k & -o & -i & -d \\ e & l & -l & -e & -e & -l & l & e & e & l & -l & -e & -e & -l & l & e \\ f & o & -h & -d & -m & i & b & k & -k & -b & -i & m & d & h & -o & -f \\ g & -n & -c & -j & j & c & n & -g & -j & n & c & j & -j & -c & -n & g \\ h & -k & -d & o & b & m & -f & -i & i & f & -m & -b & -o & d & k & -h \\ a & -a & -a & a & a & -a & -a & a & a & -a & -a & a & a & -a & -a & a \\ i & -f & -n & b & -o & -d & k & h & -h & -k & d & o & -b & m & f & -i \\ j & -c & n & g & -g & -n & c & -j & -j & c & -n & -g & g & n & -c & j \\ k & -b & i & m & -d & h & o & -f & f & -o & -h & d & -n & -i & b & -k \\ l & -e & e & -l & -l & e & -e & l & l & -e & e & -l & -l & e & -c & l \\ m & -h & b & -f & k & o & -i & d & -d & i & -o & -k & f & -b & h & -m \\ n & -j & g & -c & c & -g & j & -n & -n & j & -g & c & -c & g & -j & n \\ o & -m & k & -i & h & -f & d & -b & b & -d & f & -h & i & -k & m & -o \end{bmatrix} \begin{bmatrix} x_0 \\ x_1 \\ x_2 \\ x_3 \\ x_4 \\ x_5 \\ x_6 \\ x_7 \\ x_8 \\ x_9 \\ x_{10} \\ x_{11} \\ x_{12} \\ x_{13} \\ x_{14} \\ x_{15} \end{bmatrix} \tag{6}$$

Table 2. Coefficient values for 16-point DCT

Coefficients	a	b	c	d	e	f	g	h
Values	0.2500	0.3519	0.3468	0.3383	0.3266	0.3118	0.2940	0.2733
Coefficients	i	j	k	l	m	n	o	
Values	0.2243	0.1964	0.1667	0.1353	0.1026	0.0689	0.0347	

5.3 2D DCT Architecture

The proposed architecture consists of input memory, 1D-DCT block which is explained in Fig. 3(a), intermediate memory, output memory, registers, counters, reset and clock pulses. The proposed 8 × 8 2D-DCT architecture is shown in Fig. 4, the reset signal HIGH initiates the architecture by giving initial address to input memory. The corresponding data of the address is given to 1D-DCT input. In first clock pulse the 1st column of the 8 × 8 block of image is fed to 1D-DCT, 2nd column of block is fed in 2nd clock pulse and similarly the whole 8 × 8 block of image will be fed to 1D-DCT in 8 clock pulses. For the initial 8 clock pulses the output of 1D-DCT is stored in intermediate memory according to the addresses specified by the output initial address block. After the 8 clock pulses the input of 1D-DCT is taken from intermediate memory for the next 4 clock pulses. Therefore within 12 clock pulses the output of 1D-DCT is stored in output memory according to the addresses specified by the output initial address block.

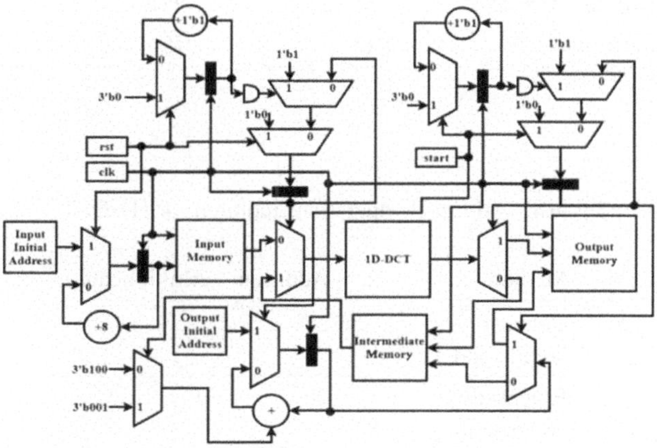

Fig. 4. Proposed architecture for 2D-DCT

For 16 × 16 2D-DCT same architecture is used with one alteration like two 3bit counters is replaced by 4bit counters. For the initial 16 clock pulses the output of 1D-DCT is stored in intermediate memory according to the addresses specified by the output initial address block. After the 16 clock pulses the input of 1D-DCT is taken from intermediate memory for the next 8 clock pulses. Therefore within 24 clock pulses the output of 1D-DCT is stored in output memory according to the addresses specified by the output initial address block.

6 Performance Analysis

The proposed architecture achieves the maximum accuracy as we compare the input matrix data with the reconstructed matrix data for 8 × 8 2D-DCT and similarly for 16 × 16 2D-DCT. For the performance analysis of hardware lena [9] image is used as a input with the resolution of 512 × 512. ThePSNR values for 8 × 8 2D-DCT is 53.1869 dB and 16 × 162D-DCT is 37.7400 dB. The performance analysis of architecture shown in the Table 3. For 8-point 2D-DCT and Table 4. For 16-point 2D-DCT which includes number of 4 input LUTs, number of slice flip-flop, number of occupied slices, maximum frequency and power for the devices XC3S700AN and XC6VLX75T.

Table 3. Performance analysis of proposed architecture for 8 × 8 2D-DCT and 16 × 16 2D-DCT

Device	XC3S700AN		XC6VLX75T	
Logic utilization	Used	Available	Used	Available
Number of 4 input LUTs	5,060	11,776	16,432	46,560
Number of slice flip-flop	788	5,888	3,849	16,432
Number of occupied slices	2,650	11,776	5,284	11,640
Maximum frequency	31.474		97.53	
Power (mW)	45		75	

Table 4. Arithmetic operations required for 1D-DCT

N	Method	Addition	Multiplication
8	Conventional architecture	28	22
	Proposed architecture	20	11
	Percentage reduction	28.57%	50%
16	Conventional architecture	92	86
	Proposed architecture	64	43
	Percentage reduction	30.43%	50%

The proposed 16 × 16 2D-DCT architecture requires more area than 8 × 8 2D-DCT architecture as shown in Tables 3 and 4. But the proposed 8 × 8 2D-DCT architecture can process 64 pixel in 12 clock pulses where 16 × 16 2D-DCT can process 256 pixel in 24 clock pulses. Thus 16 × 16 2D-DCT architecture is faster than 8 × 8 2D-DCT architecture. The proposed architecture requires less number of addition and multiplication operation than the conventional architecture. Table 4 shows the comparison of arithmetic operations required for conventional and proposed architecture for 8 × 8 and 16 × 16 1D-DCT.

7 Conclusion

Efficient 2D-DCT architecture has been proposed for the compression of image by using N/2 equations. For 8-point 1D-DCT, only four number of equations and similarly for 16-point 1D-DCT only eight equations are required. The output of 2D-DCT is evaluated in MATLAB by performing inverse discrete cosine transform. This architecture exhibit low arithmetic complexity, low power consumption and require less area. From the comparison shown in Table 4 the percentage reduction in the arithmetic operation for 8 point DCT is 28.57% and 50% and for 16 point DCT is 30.43% and 50% for addition and multiplication respectively. The proposed architecture required less number of arithmetic operations than the conventional architecture and gives the reduction of 75% in the number of pixels to store the image, hence this architecture is used for the high resolution video compression.

As we implemented the 8 × 8 and 16 × 16 2D-DCT for image compression Further for high resolution of image compression we can implement 32 × 32 and 64 × 64 2D-DCT.

References

1. Potluri, U.S., Madanayake, A., Cintra, R.J., Bayer, F.M., Kulasekera, S., Edirisuriya, A.: Improved 8-point approximate DCT for image and video compression requiring only 14 additions. IEEE Trans. Circuits Syst. I Regul. Pap. **61**(6), 1727–1740 (2014)
2. Jridi, M., Alfalou, A., Meher, P.K.: A generalized algorithm and reconfigurable architecture for efficient and scalable orthogonal approximation of DCT. IEEE Trans. Circuits Syst. I Regul. Pap. **62**(2), 449–457 (2015)
3. Brahimi, N., Bouguezel, S.: An efficient fast integer DCT transform for images compression with 16 additions only. In: International Workshop on Systems, Signal Processing and their Applications, WOSSPA, pp. 71–74. IEEE (2011)
4. An, S., Wang, C.: Recursive algorithm, architectures and FPGA implementation of the two-dimensional discrete cosine transform. IET Image Process. **2**(6), 286–294 (2008)
5. Sateesh, S.V.V., Sakthivel, R., Nirosha, K., Kittur, H.M.: An optimized architecture to perform image compression and encryption simultaneously using modified DCT algorithm. In: 2011 International Conference on Signal Processing, Communication, Computing and Networking Technologies, pp. 442–447. IEEE (2011)
6. Madanayake, A., Cintra, R.J., Onen, D., Dimitrov, V.S., Rajapaksha, N., Bruton, L.T., Edirisuriya, A.: A row-parallel 8 × 8 2-D DCT architecture using algebraic integer-based exact computation. IEEE Trans. Circuits Syst. Video Technol. **22**(6), 915–929 (2012)

7. Masera, M., Martina, M., Masera, G.: Adaptive approximated DCT architectures for HEVC. IEEE Trans. Circuits Syst. Video Technol. **27**(12), 2714–2725 (2017)
8. Dhandapani, V.A., Ramachandran, S.: Area and power efficient DCT architecture for image compression. EURASIP J. Adv. Signal Process. **180**(1), 1–9 (2014)
9. Gonzalez, R.C., Woods, R.E., Masters, B.R.: Digital Image Processing 3rd edn. (2009). J. Biomed. Opt. **14**(2), 029901
10. El-Banna, H., El-Fattah, A.A., Fakhr, W.: An efficient implementation of the 1D DCT using FPGA technology. In: Proceedings of the 12th IEEE International Conference on Fuzzy Systems (Cat. No. 03CH37442), pp. 278–281. IEEE (2003)
11. Patiño, A.M., Peiró, M.M., Ballester, F., Paya, G.: 2D-DCT on FPGA by polynomial transformation in two-dimensions. In: 2004 IEEE International Symposium on Circuits and Systems (IEEE Cat. No. 04CH37512), vol. 3, pp. III-365. IEEE (2004)
12. Trainor, D.W., Heron, J.P., Woods, R.F.: Implementation of the 2D DCT using a Xilinx XC6264 FPGA. In: 1997 IEEE Workshop on Signal Processing Systems. SiPS 1997 Design and Implementation formerly VLSI Signal Processing, pp. 541–550. IEEE (1997)
13. Edirisuriya, A., Madanayake, A., Dimitrov, V.S., Cintra, R.J., Adikari, J.: VLSI architecture for 8-point AI-based Arai DCT having low area-time complexity and power at improved accuracy. J. Low Power Electron. Appl. **2**(2), 127–142 (2012)
14. Kitsos, P., Voros, N.S., Dagiuklas, T., Skodras, A.N.: A high speed FPGA implementation of the 2D DCT for ultra high definition video coding. In: 2013 18th International Conference on Digital Signal Processing (DSP), pp. 1–5. IEEE (2013)
15. Khayam, S.A.: The discrete cosine transform (DCT): theory and application, Department of Electrical and Computer Engineering, Michigan State University, Lansing, MI, USA, Technical report, ECE, pp. 602–802 (2003)
16. Jridi, M., Meher, P.K.: Scalable approximate DCT architectures for efficient HEVC-compliant video coding. IEEE Trans. Circuits Syst. Video Technol. **27**(8), 1815–1825 (2017)

Security and Efficiency Analysis of Anti-jamming Techniques

S. Kshipra Prasadh and Sumit Kumar Jindal(✉)

School of Electronics Engineering, Vellore Institute of Technology, Vellore, Tamil Nadu, India
sumitjindal08@gmail.com

Abstract. Internet of Things has resulted in ubiquitous computing, where all IoT devices are connected almost all the time, to provide continuous services. This makes the network prone to several attacks, one such attack being the Direct Denial of Service (DDoS) attack. Jamming, which is the use of malicious nodes to deliberately lower Signal to Noise ratio (SNR), is a subset of DDoS attacks, which affects physical layer devices and channels, and can cause errors in the upper layers. Anti-jamming techniques, which are used to mitigate the effects of jamming, must be analyzed from the security and efficiency perspective. In this paper, anti-jamming techniques and protocols, viz. JAM- Jammed Area Mapping, Channel surfing and spatial retreat, channel hopping, reactive jamming detection and trigger node detection; are compared with respect to security and efficiency parameters. The suitable techniques are finally selected for specific use cases.

Keywords: Jamming · Device layer · Signal to noise ratio · Nodes · Bit error rate · Collision

1 Introduction

With the advent of Internet of Things, several functions have been automated. However, for achieving real time operations and continuous automation, the devices have to be connected to the internet all the time. This makes IoT devices prone to many security threats and attacks [1]. One such security threat is called the Direct Denial of service (DDoS) attack. Here, the communication of packets from the source to the destination is prevented. This could be done without the user's knowledge, deliberately, to reduce the performance of the wireless network. Jamming is a type of DDoS attack, which can cause security threats to the upper layers too.

Jamming can be intentional or unintentional. Sometimes, the nodes present in a network may jam the network. This is reduced by reducing the traffic or by reducing the number of nodes. Jamming is intentionally done by hackers, who wish to prevent communication in wireless networks, by reducing the signal to noise ratio at the transmitting or receiving end. Jamming can be done either by adding malicious nodes, to disrupt communication, or using multiple wireless signals to confuse the channel. Jamming is also done by distorting packet information, like the headers, trailers, message, checker bits, etc. so the packet either contains wrong information or it is routed to the wrong

N. Nain and S. K. Vipparthi (Eds.): ICIoTCT 2019, AISC 1122, pp. 251–259, 2020.
https://doi.org/10.1007/978-3-030-39875-0_27

node. Jamming attacks are hard to detect, and can be detected only after considerable number of packets have been lost or damaged [2]. This makes usage of anti-jamming techniques a necessity in almost every IoT system.

The working of jammers can be realized by estimating the types of jammers, the protocols used in jammers, and the anti-jamming techniques available to counter the effects of jamming. Anti-jamming techniques need to be analyzed based on the efficiency of the technique. The algorithms must have the least execution time and must provide real-time protection against jamming. They must also be able to detect the presence of malicious nodes and signals within a very short period of time. Another important aspect of anti-jamming algorithms is security. Anti-jamming algorithms must be secure themselves, and should not be accessible to the intruder, lest they would become ineffective. There are several techniques for the detection and prevention of jamming. They include channel surfing and spatial retreat, channel hopping, reactive jamming detection, trigger control detection and the use of Hermes nodes. In this paper, all the anti-jamming algorithms are compared with respect to efficiency and security.

The wireless network scenario was deployed in NETSIM ACADEMIC V2.0. For each of the techniques, the MATLAB code was interfaced with NETSIM, to analyse the functioning of the protocol. The application and network metrics were analyzed to determine the throughput, number of packets collided and number of packets transmitted. The security analysis was done based on the MATLAB code.

The paper is divided as follows. Section 2 describes the different jamming techniques and protocols. Section 3 describes the anti-jamming techniques. Section 4 highlights the simulation and results.

2 Jamming Techniques

Jammers are always seen as a threat to physical layer security. However, it has been proved that jammers can also be useful. A node can jam itself, so that it can prevent external attacks. As a threat, jammers are used to send wireless signals continuously over a radio channel, and this utilizes the complete capacity of the channel. Hence, any transmitter that wants to transmit legitimate data is forced to back off, and the transmission of meaningful data is prevented. There are different types of jammers. Jammers are classified based on their point of attack. Some jammers attack nodes, by preventing them from transmitting data. Some attack the wireless channel, by flooding it with unnecessary signals. Some attack the packets, modifying the packet information [3–7].

Jammers are classified as simple or complex, based on their operation and usage. Simple jammers do not use complex jamming algorithms. They just send packets over the wireless channel. Simple jammers are further classified into proactive and reactive jammers.

Proactive jammers- These jammers keep sending data on the radio channel, until it dies out. The effect of continuous transmission is that the other nodes are unable to transmit their packets, as the radio channel is already engaged. There are three types of proactive jammers- Constant jammer, deceptive jammer and random jammer [8].

Constant jammers send random bits of data across the channel. The other nodes in the network are made to believe that transmission of data is taking place, and hence, they are switched off. This technique is easy to deploy, but is also easy to detect. Deceptive jammers, on the other hand, send continuous streams of data, thus masquerading legitimate communication. The other nodes are in their 'receive state' till the jammer sends data and is drained of energy. They are harder to deploy and harder to detect, as continuous streams of data are sent, instead of random packets. A random jammer combines the principles of the deceptive and constant jammers. It either sends random packets, or continuous steams of data, or does not send any data, depending on the network situation. It makes random decisions on the type of packets to be sent.

Reactive Jammer- Unlike a proactive jammer, which sends data even when no nodes communicate, a reactive jammer starts its operation only when it senses some form of communication in the network. The receiving nodes are blocked, and thus a reactive jammer prevents packets from being received. There are many ways a reactive jammer operates. One such method is by preventing the 'send request' from the sender. Then, the receiver does not send the 'Send' command. So, the sender assumes the receiver is busy and switches off operation. Another technique is to damage the ACK packets, which are sent from the receiver to the sender on the successful receipt of data packets. When the acknowledgement packets are damaged, the sender re-sends the packet again and again till it receives a positive acknowledgement. This causes the network to jam [9].

Complex jammers work on multiple channels, and use concepts of single jammers, in order to produce a complex jamming technique, which makes it very hard to detect the kind of jamming technique that is being used. Some complex jammers are function specific. These jammers make use of a predefined algorithm, so as to perform a specific function, without the unnecessary wastage of packets. This helps the jammer to extend its battery life. Such jammers can be follow- on type, channel hopping type or pulsed noise type. In follow on type, the jammer switches between multiple channel randomly, whereas in channel hopping type, the jammer follows an algorithm to jump between channels [10–12].

Certain jammers are given intelligence, and are able to take real- time decisions as to which channel has to be attacked. These jammers are the hardest to detect. Some jammers may, in addition to distorting packets, damage some nodes, thus switching off the node permanently. Some jammers specifically damage the control channels in a system, thus destroying all the important transmission related information. Others may cause Denial of service attacks [13].

Thus, jamming techniques are becoming more complex by the day and detection of jamming plays a very important role in any IoT system.

3 Anti-jamming Techniques

To counteract the given jamming techniques, several anti-jamming algorithms are used. First, the network scenario is deployed using NETSIM ACADEMIC v2.0. Then the algorithms are coded in MATLAB and interfaced with NETSIM. The throughput, number of packets transmitted and collided are determined from NETSIM. Based on the above information, the efficiency of the algorithm is determined. The anti-jamming techniques studied in this paper are as follows.

3.1 JAM- Jammed Area Mapping (JAM)

This protocol is used to map the jammed area. The mapping is used to determine the jammed and non-jammed paths. Once the paths are determined, the packets are routed between non-jammed paths, around the jammed paths. This protocol helps in successful transmission of the packets, but results in a lot of packet collisions and reduces the throughput of the network. This method does not counter jamming, instead avoids it. Hence, the physical layer is still prone to attacks [13].

3.2 Channel Surfing and Spatial Retreat (CSSR)

Channel surfing moves to another channel when a jammer attacks a channel. The next channel is determined by considering all the orthogonal channels in the wireless network and identifying the free or available channel among all the channels. Spatial retreat involves transferring the nodes from a jammed region to a safer region. Channel surfing requires a quite complex algorithm and additional security needs to be implemented in the algorithm. Without security, the channel surfing algorithm can be accessed and modified by the intruder. In spatial retreat, the entire network has to be reconfigured, as the nodes need to be moved from one part of the network to another.

3.3 Channel Hopping (CH)

Channel hopping involves shifting from one channel to another. While channel hopping may seem a suitable counteractive measure, handoff requests from users, especially in ad hoc networks, severely limit the performance of this technique. Hence, channel hopping is not very suitable for ad-hoc networks. One type of channel hopping involves shifting from one channel to another irrespective of whether there is jamming or not. Another type of channel hopping sets a jamming threshold ‘a’, and if the jamming is more than ‘a’, the next channel is utilized.

3.4 Reactive Jamming Detection (RJD)

This technique uses the strength of the signal of the received bit to determine the presence of jamming. If the strength is low and the bit is distorted, then it is due to internal packet collisions. If the strength is high, but the packet is distorted, it is due to jamming. Reactive jamming detection is easier to implement and reduces the amount of packet collisions [14].

Another technique followed here is the trigger node detection. Here, the nodes which trigger the reactive jamming are identified. First, the entire network is scanned and the harmless nodes are first detected. The information about the harmless nodes are sent to the base station. Second, the triggered nodes which are controlled by the jammer are detected and separated from the other nodes. Then, all the triggered nodes are switched off [15].

4 Results and Discussion

The wireless network scenario was deployed in NETSIM. The IoT scenario is deployed using wireless sensor networks. 16 nodes are deployed in an area of 100 * 100 m^2. Out of these, Node 5 is defined as the jamming (malicious) node. The MATLAB codes for the anti-jamming techniques are interfaced with the NETSIM code.

The block diagram of the above system is given in Fig. 1. A constant simulation time of 100 s was used for all the cases. The default link speeds of 10 m/s uplink and downlink was used and the nodes were arranged in a random order.

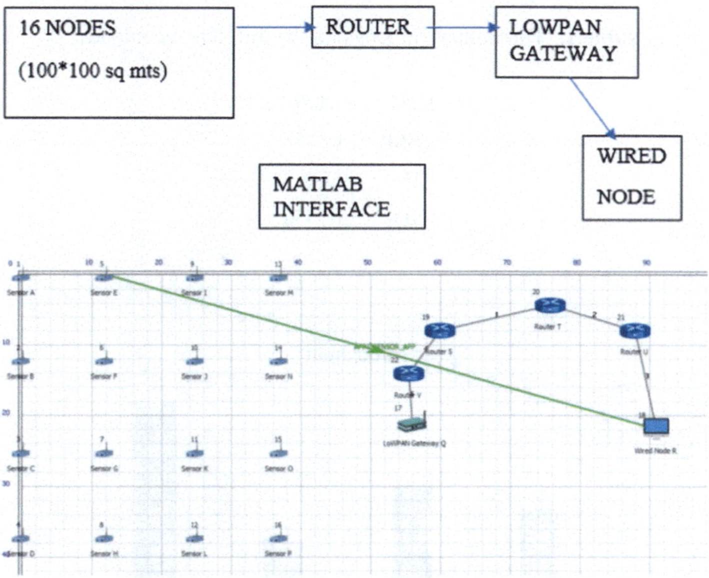

Fig. 1. Network diagram of the WSN system

The router is a network layer device, which transmits packets to the LOWPAN gateway. The MATLAB code is interfaced with the wireless channel between the 16 nodes and the router. It performs the desired modifications to the channel and to the entire NETSIM environment, to eliminate jamming.

The three main parameters which determine the efficiency of any protocol are the throughput, the number of packets collided and the number of packets transmitted. Throughput of any network is the number of packets transmitted successfully from the sender to the receiver. It is measured in Mbps or Kbps. It denotes how effective a protocol is, in terms of routing of packets.

4.1 Throughput

Reactive Jamming Detection technique has the highest throughput, as the trigger node is identified and turned off. Channel spacing and Spatial retreat (CSSR) has the second highest throughput, as it effectively removes a node from the jammed area and places it in an non-jammed region. JAM has the third highest throughput, as it only maps the area, without effectively preventing jamming. Channel hopping causes much more packet collisions, especially in band-limited environments, hence, it has the least throughput. Table 1 shows the throughput values obtained from the network metrics and Fig. 2 shows the graph obtained from the NETSIM plot.

Table 1. Throughput (in Kbps) of the different techniques

JAM	12.8667
CSSR	15.2456
CH	9.8767
RJD	23.4656

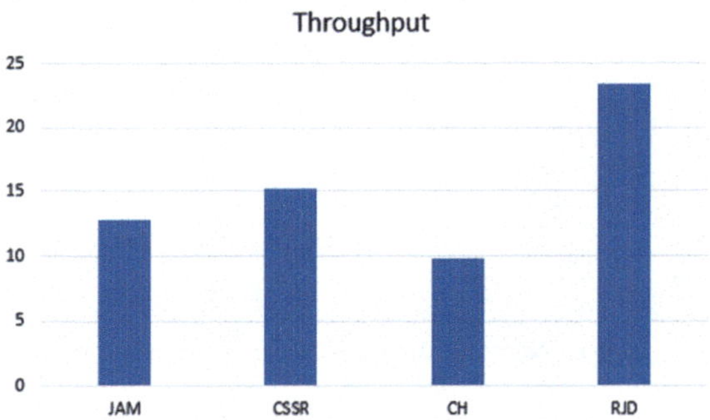

Fig. 2. Throughput (in Kbps) of the different techniques

4.2 Number of Packets Transmitted

Reactive Jamming Detection technique has the highest number of packets transmitted, as the jamming is prevented at an early stage itself. Channel spacing and Spatial retreat (CSSR) has the second highest packets transmitted, as packets are routed to a different area. JAM has the third highest throughput, as it only maps the area, but causes collision of packets, as it leads to congestion of packets in areas around the jammer. Channel hopping causes much more packet collisions, especially in band-limited environments, hence, it transmits the least number of packets. Table 2 shows the number of packets transmitted obtained from the network metrics and Fig. 3 shows the graph obtained from the NETSIM plot [16].

Table 2. Number of packets transmitted for the different techniques

JAM	2864
CSSR	3678
CH	1345
RJD	3989

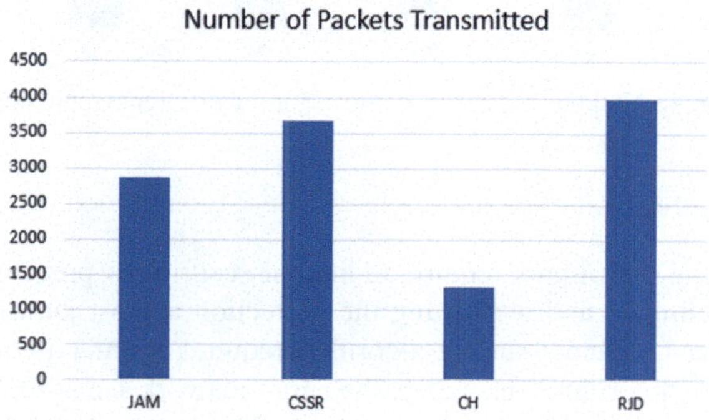

Fig. 3. Number of packets transmitted for the different techniques

4.3 Number of Packets Collided

Reactive jamming detection has the least number of packets collided, as the technique prevents jamming. Jammed area mapping has the maximum number of packets collided, as the protocol leads to congestion in the network. Channel hopping also causes packet collision, due to excess traffic in neighbouring channels. CSSR reduces the amount of collisions, due to selective channel hopping. Table 3 shows the number of packets collided, obtained from the network metrics and Fig. 4 shows the graph obtained from the NETSIM plot.

Table 3. Number of packets collided for the different techniques

JAM	258
CSSR	125
CH	212
RJD	67

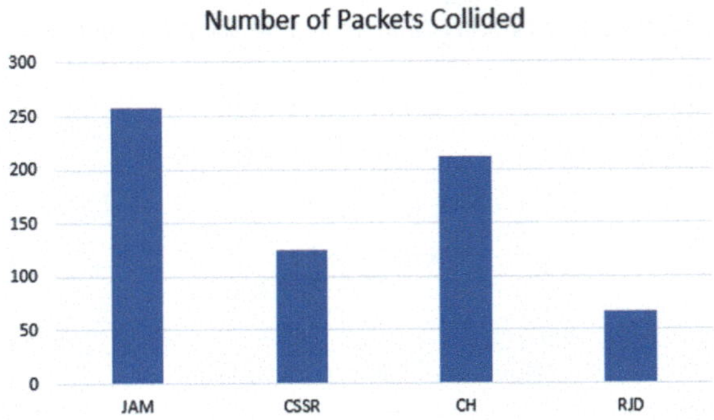

Fig. 4. Number of packets collided for the different techniques

4.4 Security Analysis

Some anti-jamming algorithms require additional security to prevent intruders from modifying the technique and weakening the protection against jamming. Among the algorithms analyzed, channel surfing algorithm requires the maximum security, as a modification in the algorithm can change the entire network scenario. Encryption keys need to be implemented to allow access to the code. If not properly implemented, intruders can change the channel surfing methodology and increase traffic in all channels. Channel hopping does not require much security as it is proactive, and does not follow a specific algorithm. However, if reactive jamming technique is applied, high security has to be provided, as the intruder can turn off victim nodes, if the algorithm is accessed. Hence, the reactive jamming technique algorithm must be accompanied by strong RSA encryption or elliptical curve encryption techniques to avoid turning off the victim nodes. If the JAM protocol is secure in a network, there can be denial of service attacks in the network, where require packets are lost in their path to the destination.

5 Conclusions

In this work, the efficiency and security of the anti-jamming techniques are analyzed and compared. The efficiency is analyzed in terms of throughput, number of packets transmitted and collided. It is found that the reactive jamming protocol has the maximum throughput, and transmits the maximum number of packets in a given time and causes least packet collisions. However, excessive security needs to be provided to this technique, as if the intruder has access, the harmless victim nodes can be turned off instead of the malicious trigger node. The channel hopping technique is very easy to implement, but has the least throughput and causes lots of packet collisions, as the technique just involves proactive switching between adjacent channels. Channel surfing and spatial retreat algorithm is quite complex, and has a run-time of $O(n^2)$. However, the algorithm is effective against most of the jamming techniques available, including smart jamming. It performs reactive channel hopping and hence reduces the number of packet collisions. However, due to the problem of multiple handoff requests, this technique cannot be used for ad hoc networks.

References

1. Alnifie, G., Simon, R.: A multi-channel defense against jamming attacks in wireless sensor networks. In: Proceedings of the 3rd ACM Workshop on QoS and Security for Wireless and Mobile Networks, pp. 95–104 (2007)
2. Bayraktaroglu, E., King, C., Liu, X., Noubir, G., Rajaraman, R., Thapa, B.: On the performance of IEEE 802.11 under jamming. In: IEEE the 27th Conference on Computer Communications, pp. 1265–1273 (2008)
3. Bellardo, J., Savage, S.: 802.11 denial-of-service attacks: real vulnerabilities and practical solutions. In: Proceedings of the 12th Conference on USENIX Security Symposium, pp. 15–28 (2003)
4. Broustis, I., Pelechrinis, K., Syrivelis, D., Krishnamurthy, S.V., Tassiulas, L.: FIJI: fighting implicit jamming in 802.11 WLANs. In: Security and Privacy in Communication Networks, vol. 19, pp. 21–40 (2009)
5. Chiang, J.T., Hu, Y.C.: Cross-layer jamming detection and mitigation in wireless broadcast networks. IEEE/ACM Trans. Netw. **19**(1), 286–298 (2011)
6. Commander, C.W., Pardalos, P.M., Ryabchenko, V., Shylo, O.V., Uryasev, S., Zrazhevsky, G.: Jamming communication networks under complete uncertainty. Optimization Letters **2**(1), 53–70 (2008)
7. Gencer, C., Aydogan, E.K., Celik, C.: A decision support system for locating VHF/UHF radio jammer systems on the terrain. Information Systems Frontiers **10**(1), 111–124 (2008)
8. Gollakota, S., Katabi, D.: iJam: jamming oneself for secure wireless communication. Technical report, Massachusetts Institute of Technology (2010)
9. Andrea, I., Chrysostomou, C., Hadjichristofi, G.: Internet of Things: security vulnerabilities and challenges. In: 2015 IEEE Symposium on Computers and Communication (ISCC), 6 July 2015, pp. 180–187. IEEE (2015)
10. Aman, M.N., Chua, K.C., Sikdar, B.: Position paper: physical unclonable functions for IoT security. In: Proceedings of the 2nd ACM International Workshop on IoT Privacy, Trust, and Security, 30 May 2016, pp. 10–13. ACM (2016)
11. Fang, H., Xu, L., Zou, Y., Wang, X., Choo, K.K.: Three-stage Stackelberg game for defending against full-duplex active eavesdropping attacks in cooperative communication. IEEE Trans. Veh. Technol. **67**(11), 10788–10799 (2018)
12. Bharathi, S., Kumar, D., Ram, D.: Defence against responsive and non-responsive jamming attack in cognitive radio networks: an evolutionary game theoretical approach. J. Eng. **2018**(2), 68–75 (2018)
13. Chen, Y., Li, Y., Xu, D., Xiao, L.: Dqn-based power control for IoT transmission against jamming. In: 2018 IEEE 87th Vehicular Technology Conference (VTC Spring), 3 June 2018, pp. 1–5. IEEE (2018)
14. Bany Salameh, H., Almajali, S., Ayyash, M., Elgala, H.: Spectrum assignment in cognitive radio networks for Internet-of-Things delay-sensitive applications under jamming attacks, 2327–4662. https://doi.org/10.1109/JIOT.2018.2817339
15. Garnaev, A., Trappe, W., Petropulu, A.: An anti-jamming strategy when it is unknown which receivers will face with smart interference. In: International Conference on Wired/Wireless Internet Communication 18 June 2018, pp. 195–206. Springer, Cham (2018)
16. Bany Salameh, H., Almajali, S., Ayyash, M., Elgala, H.: Batch-based security-aware spectrum sharing with simultaneous assignment decisions in time-critical IoT networks with cognitive radio capabilities. Trans. Emerg. Telecommun. Technol. **29**(11), e3317 (2018)

Generalized Dynamic Multilayer Fog Computing Architecture

K. P. Arjun(✉) and S. Mary Saira Bhanu

National Institute of Technology, Tiruchirappalli,
Tiruchirappalli 620015, Tamil Nadu, India
arjunthedragon@gmail.com, msb@nitt.edu

Abstract. The concept of Smart city and Smart Home is a promising and challenging one which defines the future of urban infrastructure development by integrating the Internet of Things (IoT) with Cloud Computing. Fog Computing with its latest framework on multilayering where each layer focuses on the different aspect of the architecture has proven useful in its goal to reduce space, time and computation overhead from cloud perspectives, due to its traffic offloading and edge device proximity for computation. But, a need for platform generalization and dynamic behavior still exists as most of the smart home needs to operate in multiple environments whereas the existing system mainly works on a rigid model with single infrastructure support. In this paper, a modified version of multilayer fog architecture has been proposed with scope for generalization and dynamic operation in terms of data management and placement. The proposed system consists of five sections - the Lower Fog Layer for end device collection operation and prioritization, Dockerized Middle Layer for node-specific operations like event detection and filtered data propagation, ICFN (Interconnecting Fog Node) Layer for interconnecting fog nodes, Cloud Server for data analytics and Fog Maintenance Remote Server for platform customization. The additional platform customization and multiple service support in the proposed architecture have also made significant improvements with regards to data distribution, request bandwidth, and request failure rate.

Keywords: Multilayer fog computing · Dockers and Containers · Cloud computing · Data distribution · Fog generalization · Fog multi-service

1 Introduction

Based on an analysis made in 2016, it is estimated that there are around 4.6 billion connected devices excluding phones, tablets, and laptops. It is expected that this may increase to 15.3 billion devices in the next 5 years. To manage the huge outburst of data with a prime focus on reduced cloud storage, less event latency, reduced request bandwidth and less request failure rate, the concept

N. Nain and S. K. Vipparthi (Eds.): ICIoTCT 2019, AISC 1122, pp. 260–268, 2020.
https://doi.org/10.1007/978-3-030-39875-0_28

of Edge Computing or Fog Computing or Cloudlets are introduced whose basic idea is to offload data traffic and computation to localized computation resources within the edge devices. This reduces computation overhead from remote cloud server end perspective and most of the resource-intensive operations occur at edge devices which are at close proximity to the end devices. Tian wang et al. [1], Deng et al. [2], Singh et al. [3] and Rafat et al. [4] present an architecture based on standard fog system and the key focus is emphasized on latency and request bandwidth, and how it has positive implications on IoT like framework. But, there always exists a need to go for a generalized system with dynamic functionalities, which paves way for a system which could be easily customizable and has provision to include more infrastructure in the long run. The solution to design such a framework is more challenging and interesting as it is expected to perform perfectly in an environment taking all critical aspects of a real-time infrastructure into account, with least probability of failure.

The first section gives the overview of a standard fog system along with its advantage over a cloud system and some scope of improvisation over the standard fog system is also highlighted. The second section gives the literature survey and some gaps associated to the existing work. The third section presents the proposed fog architecture with description on each of its modules. The fourth section highlights the experimentation, results and analysis of measure of improvement of our proposed architecture over the standard fog. The fifth section highlights the conclusion and future scope.

2 Related Work

Computer scientists and researchers especially from companies like Cisco [5], IBM, Intel, GE etc. are working to formulate a better approach in formulating a system which promotes a much hierarchical and organized form of data management and data placement by collecting sensor data from IoT end devices distributed geospatially covering wide span of regions and using it to actuate end components in real time with efficient usage of bandwidth and memory usage with less probability of task failures. Li et al. [6] present a hierarchical multi-layer fog based architecture consisting of three layers - fog lower layer on data sensing, middle layer on event-based detection on anomaly occurrence, upper layer on interconnecting computing nodes. Lower the layer, less the latency and less the data scaling. Bachmann et al. [7] present information regarding the standard implementation of a fog-based system with master and slave nodes which includes necessary components and services as presented in Yu et al. [8], Mayer et al. [9] and Yin et al. [10] in the form of Dockers and Containers. Jordi Fonollosa et al. [11], Okay et al. [12] and Wang et al. [13] illustrate a fire detection scheme relevant to fire alarm service infrastructure and smart grid analysis within IoT-Fog-Cloud Environment ensuring fewer cases of false alarm detection.

However, the existing frameworks have various drawbacks as most of these fog based architecture is single infrastructure service oriented. Supporting multiple environments is quite complex as each of these infrastructures tends to possess their own device gateways and edge devices as their scheme of functionalities are different. A basic fog system is a localized version of cloud and it is not accessible from external system when it comes to introducing updates on workflow and meta-data and lacks access to manual entities for the kind of infrastructure preferred i.e prioritizing requests from Fog, making it more predetermined and rigid. A need to generalize multiple service infrastructure exists as the typical smart home system supports multiple IoT frameworks preventing wastage of a complete fog-cloud space for a single infrastructure. It is also preferred to create a system which would be more dynamic in nature towards data manipulation and operation access, in a more specific user-friendly manner with more control over the kind of requests being sent to the cloud. A software defined multilayer fog system is proposed with an additional remote maintenance server to enable platform customization and support within fog layer to handle multiple service infrastructure.

3 Proposed Work

This section presents a modified version of multilayer fog computing architecture with the support of rich customization, data management and multiple service infrastructure management. As shown in Fig. 1, the proposed architecture consists of three modules as fog layers:

- Fog Lower Layer - The preprocessed data received from simulated end device is converted into service task instances and prioritized for multiple service infrastructure and forwarded to dockerized controller services
- Dockerized Middle Layer - It consists of dockerized controller nodes with services meant for event detection and propagating filtered data to an ICFN Node
- ICFN Layer - An interconnecting fog node performs data aggregation from multiple houses and forwards processed infra readings periodically to cloud

and two modules as remote servers:

- Remote Cloud Server - contains application specific data with analytics features. Ex: geospatial searching
- Fog Maintenance Remote Server - A remote server with platform customization support based information which is periodically retrieved by all the modules within the system

The network communication among modules occurs by means of TCP/IP protocol in terms of HTTP/JSON due to its simplicity and REST support. To prevent loss of network requests, the requests are queued before sending till the retry attempts on failures reaches maximum retry limit.

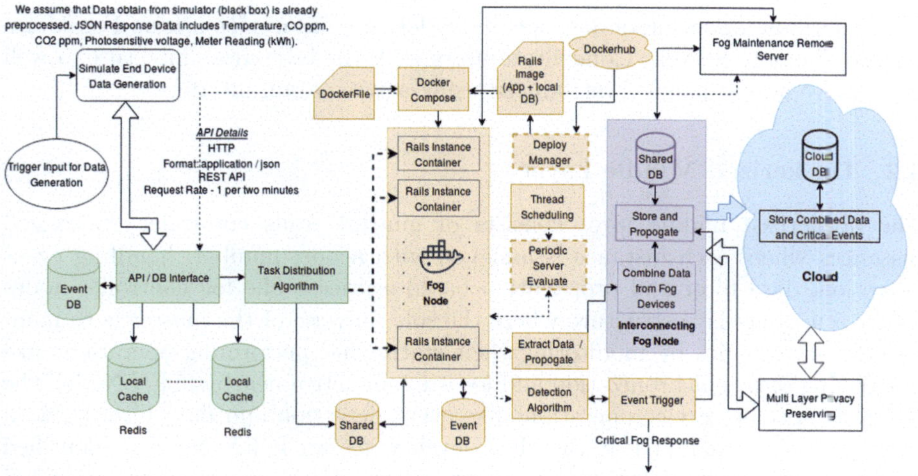

Fig. 1. Basic block diagram of the proposed fog architecture

3.1 Fog Lower Layer

In fog lower layer module, the end device preprocessed data generated in JSON format is collected and parsed, as shown in Fig. 2. The data is cached in redis and converted into service task instances which are then enqueued on the basis of timestamp and service infra priority, for service infrastructure prioritization as the proposed system supports multiple service infrastructure. The enqueued entry is then dequeued and converted to route requests, and forwarded to the dockerized rails controller instances. Every service task requests is given a maximum retry limit based on its associated service infrastructure so as to ensure the request is not lost. On failure of any instance, the request will be then automatically forwarded to any other available instance. This layer also takes part

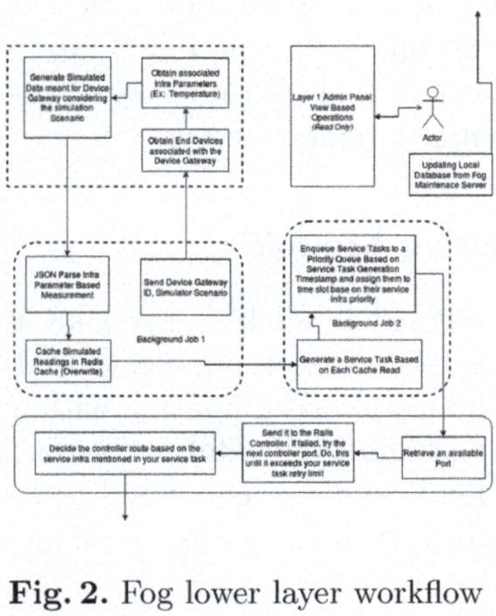

Fig. 2. Fog lower layer workflow

in the periodic updating of data locally by fetching information from Fog Maintenance Remote Server. To facilitate users with the layered architecture, a well descriptive admin panel is introduced using active admin interface.

3.2 Dockerized Middle Layer

The dockerized middle layer consists of multiple rails controller dockerized instances where each instance is assigned with a port number, handling event detection, data filter and propagation based services. The fog controller nodes follow a master-slave structure where the sole purpose of the master is to manage the slave nodes by monitoring their status and performing operations like underlying data synchronization and locating an active nearby ICFN Node. The fog node specific service operations like event detection and data filtering with propagation are handled at the slave level. Once an ICFN Node is identified and connection established for data propagation, the data is sent to the ICFN Node by means of an ICFN Agent. This is done since the dockerized controller layer's version can't communicate with the outside module with the standard IP address. The data is forwarded from controllers to ICFN Agent using Docker IP to communicate with the machine localhost and then it is forwarded to the connected ICFN Node.

3.3 Interconnecting Fog Node (ICFN)

An ICFN Node is a data processing node which is located within a region with its own location coordinates defined in Fog Maintenance Remote Server. Every ICFN node is associated to its icfn id, location, associated cloud id and url, which is required for communicating with remote maintenance server in order to seek information to communicate with cloud. Once an ICFN Node is identified for forwarding data from the preceding layer, the data is received at that particular node and then it is filtered and propagated to cloud with its schedule interval much higher than its corresponding lower layers. Within the ICFN Node, it informs its neighbor about any critical events since more than one dockerized controller services might be linked to this ICFN Node due to its geographical proximity and running status. The ICFN node synchronize itself by periodically contacting the fog maintenance remote server.

3.4 Remote Cloud Server Module

The Cloud Backend is designed in such a way that it is currently meant for long term analytics and persistent storage. Geospatial operations like fetching house event detection with location coordination within a radius, listing average temperature or other infra parameter reading captured for the day at an hourly basis, are added. Every cloud server is associated to its cloud id and url which is required for communicating with remote maintenance server for periodic synchronization.

3.5 Fog Maintenance Remote Server

There are entities within the current framework whose values need to be synchronized among all the modules. Some of these entities include the meta information which needs to be communicated among the modules. For eg: the list of end device info and service infrastructures information needs to communicate within all the modules to prevent inconsistency in data manipulation. With the key focus to enable software defined platform customization within the entire system, a maintenance server is thereby introduced to communicate information and changes, which could be managed by root level administrators and is not accessible to the local system administrators. Almost all details could be manipulated at this level other than admin user details which could only be done at the local system level. The system is maintained in read-write mode unlike the read-only admin panels of lower fog layer and the cloud server. The maintenance server basically contains information on the fog specific infra parameters, houses within the system, service infrastructures supported, icfn nodes, cloud servers, end device generic information etc.

4 Experiments and Results

The current experimental setup consists of a local ubuntu LTS 16.04 with processor Intel Core-i5, 8 GB RAM, clock speed 2.5 GHz, 976 GB disk size machine to incorporate fog layer services and components. In real time scenario, the ICFN layer services should be deployed in a nearby geographical location to our localized fog system. The Ruby on Rails framework with Postgres database and Redis cache support is included along with docker version 18.06.1-ce and docker compose version 1.22.0 to program fog specific operations and services. The remote cloud server is deployed in an amazon ec2 instance of a t2.micro type with hvm virtualization, ebs root device type, 1 GB RAM, 1GB disk size and 1 virtual CPUs. Ruby on Rails along with Postgres framework is included for backend specific operations for data analytics. The fog maintenance remote server also uses the same specification of remote cloud server excluding the disk size of 20 GB and the number of virtual CPUs as 2. To analyze the performance of the proposed framework, four evaluation metrics are considered: data distribution, response latency, number of incoming requests to cloud and request failure rate. Three environments were introduced to compare the performance of these metrics: traditional cloud environment, standard fog system and the proposed fog architecture. This comparison approach is to evaluate the speed up incurred by standard fog system with respect to a traditional cloud environment(A). Then, the speed up incurred by the proposed architecture is evaluated for the same with respect to traditional cloud environment(B). Then, the improvement factor is assessed by comparing A and B. To evaluate performance, twelve parameters are considered: traditional cloud storage size(S_{TC}), fog based cloud storage size(S_{FC}), cloud storage improved ratio($S_{imp} \rightarrow S_{TC}/S_{FC}$), traditional cloud latency response(T_{TC}), fog based cloud storage(T_{FC}), latency improved over fog architecture($T_{imp} \rightarrow T_{TC}/T_{FC}$), traditional cloud incoming requests count(R_{TC}), fog system incoming requests count(R_{FC}), incoming requests minimization over fog architecture($R_{imp} \rightarrow R_{TC}/R_{FC}$), traditional cloud request failure percent(Rf_{TC}), fog system request failure percent(Rf_{FC}), request failure minimization over fog architecture($Rf_{imp} \rightarrow Rf_{TC}/Rf_{FC}$).

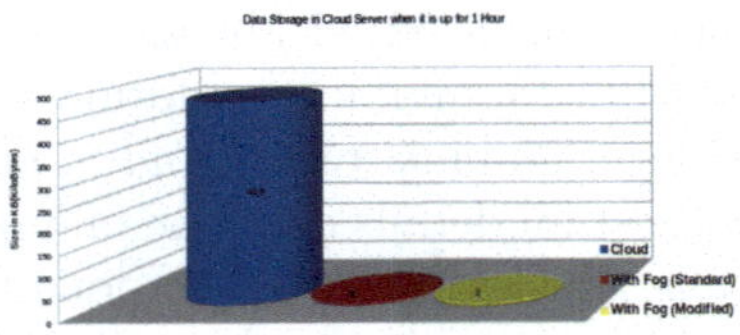

Fig. 3. Data distribution graph

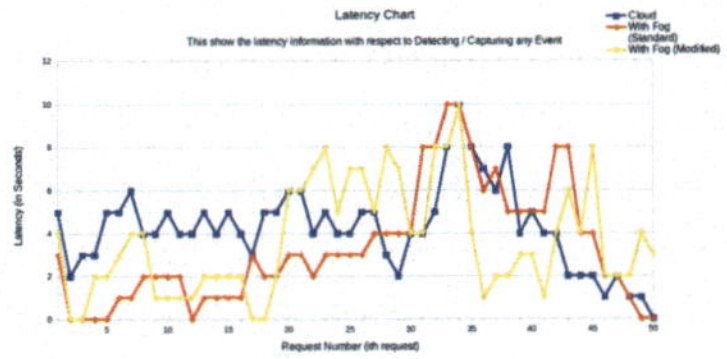

Fig. 4. Response latency graph

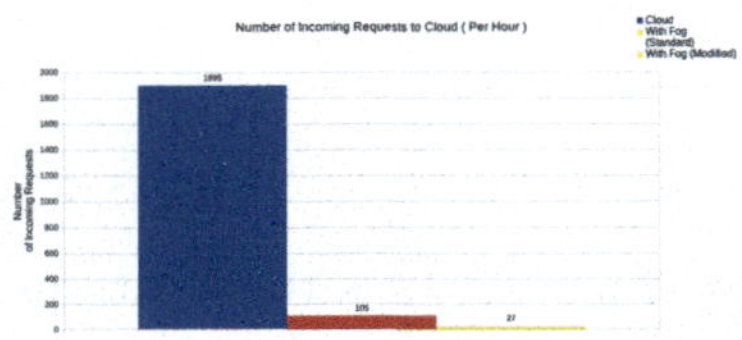

Fig. 5. Number of incoming requests graph

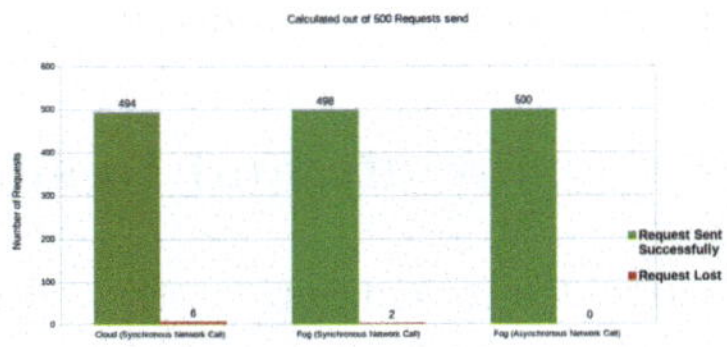

Fig. 6. Request failure rate graph

The assessment made in Fig. 3 and Table 1 shows significant improvement in data distribution made compared to traditional cloud infrastructure due to storage on three layers which does not burden cloud for storing huge amount of data. But, the improvement made by the modified system is nearly the same which standard fog systems made over traditional cloud systems. There is no significant improvement in latency as shown in Fig. 4 and Table 1 even though the proposed architecture gives less latency compared to traditional cloud system. This might be the presence of slot based prioritizing of requests in the modified framework since it works on a generalized approach focussing on more than one service infrastructure. The bandwidth efficiency in terms of usage has improved approximately four times the standard fog achieved as shown in Fig. 5. This is expected as the major reduction for incoming requests for the modified architecture happens at dockerized controllers layer and ICFN layer unlike a standard fog system where it happens only at fog services layer. It is not possible to possible to provide a mathematical inference for request failure rate as per the assessment made in 6 and Table 1. The number of requests failed per 1000 requests is zero, which is due to the delayed job services as it ensures that any request which is expensive, is enqueued and then send to the server for execution, along with maximum number of attempts to be allowed on request failure (Fig. 6).

Table 1. Assessment made on the basis of Traditional Cloud, Standard Fog and Proposed Fog Architecture

Evaluation parameter	Readings $C \rightarrow Cloud$ $S \rightarrow Standard\,Fog$ $P \rightarrow Proposed\,Fog$	Improvement by standard fog system $I_{imp}(S) \rightarrow (C/S)$	Improvement by proposed fog system $I_{imp}(P) \rightarrow (C/P)$	Fog improvement measure $I_{imp} \rightarrow I_{imp}(P)/I_{imp}(S)$
Data Distribution – Reduction in Cloud Storage (in KB)	$\mathrm{C} \rightarrow \mathrm{S}_{TC} \rightarrow 468$ $\mathrm{S} \rightarrow \mathrm{S}_{FC}(\mathrm{S}) \rightarrow 2$ $\mathrm{P} \rightarrow \mathrm{S}_{FC}(\mathrm{P}) \rightarrow 2$	$\mathrm{I}_{imp}(\mathrm{S}) \rightarrow \mathrm{S}_{imp}(\mathrm{S}) \rightarrow 468/2 \rightarrow 234$	$\mathrm{I}_{imp}(\mathrm{P}) \rightarrow \mathrm{S}_{imp}(\mathrm{P}) \rightarrow 468/2 \rightarrow 234$	1
Response Latency (in seconds)	$\mathrm{C} \rightarrow \mathrm{T}_{TC} \rightarrow 4.3745$ $\mathrm{S} \rightarrow \mathrm{T}_{FC}(\mathrm{S}) \rightarrow 3.3$ $\mathrm{P} \rightarrow \mathrm{T}_{FC}(\mathrm{P}) \rightarrow 3.6078$	$\mathrm{I}_{imp}(\mathrm{S}) \rightarrow \mathrm{T}_{imp}(\mathrm{S}) \rightarrow 4.375/3.3 \rightarrow 1.33$	$\mathrm{I}_{imp}(\mathrm{P}) \rightarrow \mathrm{T}_{imp}(\mathrm{P}) \rightarrow 4.375/3.61 \rightarrow 1.2125$	0.91468481948
Number of Incoming Requests to Cloud (per hour)	$\mathrm{C} \rightarrow \mathrm{R}_{TC} \rightarrow 1895$ $\mathrm{S} \rightarrow \mathrm{R}_{FC}(\mathrm{S}) \rightarrow 105$ $\mathrm{P} \rightarrow \mathrm{R}_{FC}(\mathrm{P}) \rightarrow 27$	$\mathrm{I}_{imp}(\mathrm{S}) \rightarrow \mathrm{R}_{imp}(\mathrm{S}) \rightarrow 1895/105 \rightarrow 18.05$	$\mathrm{I}_{imp}(\mathrm{S}) \rightarrow \mathrm{R}_{imp}(\mathrm{P}) \rightarrow 1895/27 \rightarrow 70.1852$	3.89
Request Failure Rate (per 500 requests)	$\mathrm{C} \rightarrow \mathrm{Rf}_{TC} \rightarrow 1.2$ $\mathrm{S} \rightarrow \mathrm{Rf}_{FC}(\mathrm{S}) \rightarrow 0.4$ $\mathrm{P} \rightarrow \mathrm{Rf}_{FC}(\mathrm{P}) \rightarrow 0.0$	$\mathrm{I}_{imp}(\mathrm{S}) \rightarrow \mathrm{Rf}_{imp}(\mathrm{S}) \rightarrow 1.2/0.4 \rightarrow 3$	$\mathrm{I}_{imp}(\mathrm{S}) \rightarrow \mathrm{S}_{imp}(\mathrm{S}) \rightarrow 1.2/0 \rightarrow$ NaN (indicates improvement)	NaN (indicates improvement)

5 Conclusion

A generalized multilayer fog-cloud computing system is introduced with modules consisting of fog layers, based on its proximity to the end device, along with an analytics-driven application cloud server and a centralized fog maintenance remote server to handle the entire system in global scale. Other than improvement in various performance metrics, support for multiple service infrastructure and rich customization is also made possible by means of meta information synchronization and docker support. In future work, it would be the major focus to further improve the framework by taking into account its security prospects for secure storage and communication, along with fog intelligence.

References

1. Wang, T., Zhou, J., Chen, X., Wang, G., Liu, A., Liu, Y.: A three-layer privacy preserving cloud storage scheme based on computational intelligence in fog computing. IEEE Trans. Emerg. Top. Comput. Intell. **2**(1), 3–12 (2018)
2. Deng, R., Lu, R., Lai, C., Luan, T.H., Liang, H.: Optimal workload allocation in fog-cloud computing toward balanced delay and power consumption. IEEE Internet Things J. **3**(6), 1171–1181 (2016)
3. Singh, S.P., Nayyar, A., Kumar, R., Sharma, A.: Fog computing: from architecture to edge computing and big data processing. J. Supercomput. **75**, 2070–2105 (2018)
4. Raafat, H.M., Hossain, M.S., Essa, E., Elmougy, S., Tolba, A.S., Muhammad, G., Ghoneim, A.: Fog intelligence for real-time iot sensor data analytics. IEEE Access **5**, 24062–24069 (2017)
5. Cisco: White paper: Fog computing and the internet of things: extend the cloud to where the things are. Technical report, Cisco Systems Inc., San Jose, CA (4 2015)

6. Tang, B., Chen, Z., Hefferman, G., Pei, S., Wei, T., He, H., Yang, Q.: Incorporating intelligence in fog computing for big data analysis in smart cities. IEEE Trans. Ind. Inf. **13**(5), 2140–2150 (2017)
7. Bachman, K.: Design and Implementation of a Fog Computing Framework. Master's thesis, Technical University of Vienna, A-1040 Wien Karlsplatz 13, thesis in Software Engineering and Internet Computing, Technical University of Vienna, Reg no: 1126001, February 2017
8. Yu, R., Xue, G., Zhang, X.: Application provisioning in fog computing-enabled internet-of-things: a network perspective, pp. 783–791, April 2018
9. Mayer, R., Gupta, H., Saurez, E., Ramachandran, U.: Fogstore: toward a distributed data store for fog computing. In: 2017 IEEE Fog World Congress (FWC), pp. 1–6 (2017)
10. Yin, L., Luo, J., Luo, H.: Tasks scheduling and resource allocation in fog computing based on containers for smart manufacturing. IEEE Trans. Ind. Inf. **14**(10), 4712–4721 (2018)
11. Solórzano, A., Fonollosa, J., Marco, S.: Improving calibration of chemical gas sensors for fire detection using small scale setups. In: Multidisciplinary Digital Publishing Institute Proceedings, vol. 1, p. 453, August 2017
12. Okay, F.Y., Özdemir, S.: A fog computing based smart grid model. In: 2016 International Symposium on Networks, Computers and Communications (ISNCC), pp. 1–6 (2016)
13. Wang, Y., Chen, Q., Hong, T., Kang, C.: Review of smart meter data analytics: applications, methodologies, and challenges. IEEE Trans. Smart Grid **10**(3), 3125–3148 (2018)

An Efficient Quantum Key Management Scheme

Vishal[1](✉) and S. Taruna[2]

[1] Banasthali University, Banasthali, India
vishalhim@yahoo.com
[2] JK Lakshmipat University, Jaipur, India

Abstract. The quantum key distribution playing the role of trusted carrier to augment the symmetric key distribution and is competing with public key cryptography in case of reliability and security. It is secure against unlimited computing power. In the implementation front of quantum key distribution, there is an intersection of fields like quantum physics, optics, computer science and electronics make it more complex. In this paper, we proposed Quantum key distribution technique in which traditional cryptographic techniques are not needed for mutually authenticating the keys. We proposed the base distribution center as a third party, in which both communicating parties share their bases before communication and mutually authenticate each other by transmitting a series of bits and deduce the encryption and decryption keys.

Keywords: Quantum key · Quantum cryptography · Qubit · Base distribution center · Public key cryptography · Photon polarization

1 Introduction

The concept of "Quantum cryptography" was first developed by Bennett and Brassard in 1984 [1]. The first experimental demonstration was done in 1992 [2]. Quantum cryptography obtains its security from the basic principles of quantum mechanics [3] i.e. the quantum-mechanical properties of single photons. Quantum cryptography is also defined as quantum key distribution due to its plan is to provide a shared, secret key in between two authorized parties who want to communicate securely. In the classical cryptographic techniques [4], keys and related algorithm are kept secret. The encryption security totally depends on the time needed to decode the key value from a large domain of possible key pools. e.g. a 128-bit length key size needs 2^{128} key check using brute force. Quantum cryptography is not a secret key exchange and transmission, but as a secret key generation, initially neither transmitter nor receiver can predetermine the key, they only know the value of key after completion of agreement protocol. Since it is a symmetric process and both transmitter and receiver perform the similar procedure on their qubits [5], it cannot be said that either transmitter or receiver generates the key. Quantum communication [6] is the way of transmitting a qubit state in between the transmitter and receiver. The quantum state [7] neither copied, nor cloned and provides highly secure key agreement protocol. The Heisenberg Uncertainty Principle [8] states

N. Nain and S. K. Vipparthi (Eds.): ICIoTCT 2019, AISC 1122, pp. 269–277, 2020.
https://doi.org/10.1007/978-3-030-39875-0_29

that the certain pairs of physical properties are linked in such a way that measuring one property prevents the observer from knowing the value of others. Whenever man in middle wants to measure the polarization of a photon, the choice of direction or angle of measurement will disturb all subsequent measurements and transmission which can be easily detected. Certain pairs of physical properties are related in such a way that measuring one property prevents the observer from knowing the value of the other. When measuring the polarization of a photon, the choice of what direction to measure affects all subsequent measurements. If a photon passes through a vertical filter it will have the vertical orientation regardless of its initial direction of polarization. The traditional public key cryptography is susceptible to new algorithmic techniques for factorization, discrete logs as well as to quantum computers. Whereas QKD is often promoted as the alternative to public key cryptography for the future [9]. As it allows the communicating entities to validate that the key has not been intercepted.

2 Quantum Cryptography

In order to achieve high security in cryptographic techniques, the quantum mechanical properties of photons are exploited to perform cryptographic task. Quantum cryptography is completely secure against interception and impossible to decrypt the information. Quantum cryptography is totally dependent on the physical properties of photon (individual light particle/wave-particle) and their inherent quantum properties to create an unbreakable cryptosystem. It is already found that, it is not possible to measure the quantum state of any system without disturbing that system and if state of photon is disturbed then not possible to find the exact state [10]. Photon has all properties for quantum cryptography along with transmission of photon in optical fiber provides high speed, high bandwidth communication [11]. The acceptance of quantum mechanics is greatly influenced by its accurate prediction of the physical behavior of the system. the quantum state are characterized to a specific wave packets (photon spin) and a set of quantum numbers and these number can be measured exactly during the transition from one quantum state to another the wave packets changes these structure continuously [12]. a normal computer uses binary digit 0 or 1,but quantum bit (qubit) is a superposition of two quantum states. in general qubits can be characterized by two states of an electron $|0>$ and $|1>$ the combination of two states of one electron makes one qubit.

$$|0> + |1> = 1 \text{ qubit.}$$

The combination of four states of two electrons

$$|0, 0> + |1, 1> + |0, 1> + |1, 0> = 2 \text{ qubits.}$$

Hence there are combination of 2^n states of n electrons and makes n qubits. The normal computer can process only one combination of n bits, where as the quantum computer can process 2^n combination of two states at the same time, which is equivalent to normal computer with 2^n processor.

3 Photon Polarization

The behavior of photon is governed by the laws of quantum mechanical wave functions follow the principle of superposition. The sum of two wave function describes ar allowed state of the photon. However one cannot add the wave function of two different photons because the right handed and left handed photon are described by two different representations. Suppose a plane wave of monochromatic light of frequency ω propagating in positive z-direction. the electric field $\vec{E}(t)$ at a given point is a vector orthogonal to the direction of propagation . so it lies in orthogonal X, Y plane and has $\{E_x(t), E_y(t), E_z(t) = 0\}$ (Fig. 1).

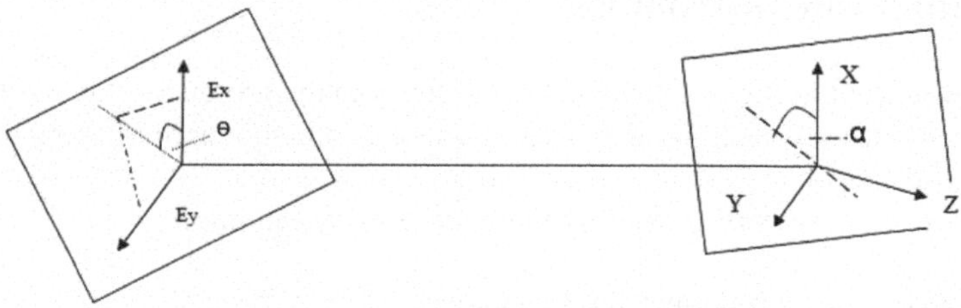

Fig. 1. Photon polarization in X, Y Plane

$$\vec{E}(t) = \begin{Bmatrix} E_x(t) = E_{0x}Cos(\omega t - \delta_x) \\ E_y(t) = E_{0y}Cos(\omega t - \delta_y) \end{Bmatrix} \tag{1}$$

By choice of the original of time, it is always possible to choose $\delta_x = 0, \delta_y = \delta$ The intensity I of light wave is proportional to the square of electric field.

$$I = Ix + Iy = K(E_{0x}^2 + E_{0y}^2) = KE_0^2 \tag{2}$$

Where $\delta = 0$ the polarization is linear and if we take

$$E_{0x} = E_0Cos\theta, E_{0y} = E_0Sin\theta \tag{3}$$

In (1) for $\delta_x = \delta_y = 0$ the electric field oscillates in the η_θ direction of the X, Y plane making an angle θ with X axis. Such a light wave can be obtained using a linear polarizer whose axis is parallel to η_θ. The polarization filter allows only light of a specified polarization direction to pass. in the polarization photon with either be pass or not pass through a polarization filter and if it passes it must be aligned w th filter regardless of its original state and there is not partial photon (Fig. 2).

In the first vertically aligned filter some light is absorbed and remaining light is polarized in the vertical direction. The second filter is tilted at angle θ absorb some of the light and transmit rest resulting it a new polarization.

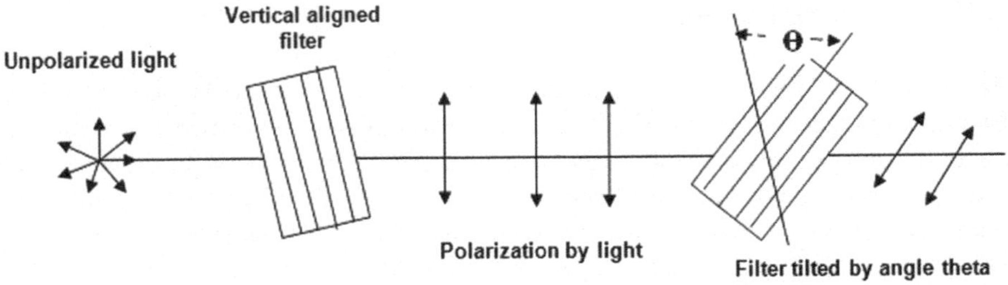

Fig. 2. Photon polarization by filter at different angle

4 Quantum Key Distribution

Alice and Bob want to communicate in total security in the presence of an eavesdropper Eve. In order to achieve security they need to share a common random string of number that is key (Fig. 3).

In order to achieve perfect security there are some assumption.

a. The source must emit perfect single photons.
b. Channel can be noisy but should be lossless. i.e. there should be no absorption in channel.
c. It should be considered that measurement can disturb a quantum state.
d. The detector should have 100% detection efficiency.
e. The alignment of the source and detector should be perfect the angle between X and Y axis should be exactly 45°.

These assumption leads to security in BB84.theoretical assumption for QKD security is that the quantum state cannot be cloned as a result the eavesdropper, Eve cannot have the same information as Bob. Also single photon is secure but multi-photon signal can be

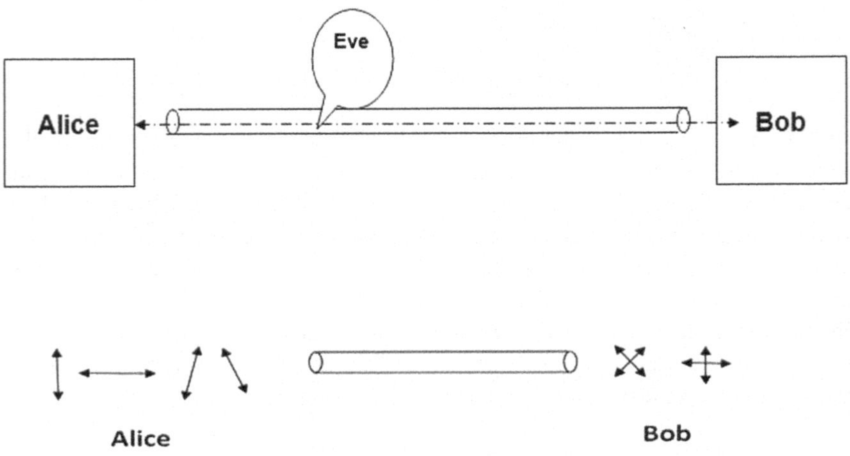

Fig. 3. opticalcommunication in between alice and Bob

split and can make copy for both receiver and eavesdropper and therefore are insecure. Basically the quantum key distribution requires two channel one quantum channel and it may be noisy and other public channel subject to eavesdropping (Fig. 4).

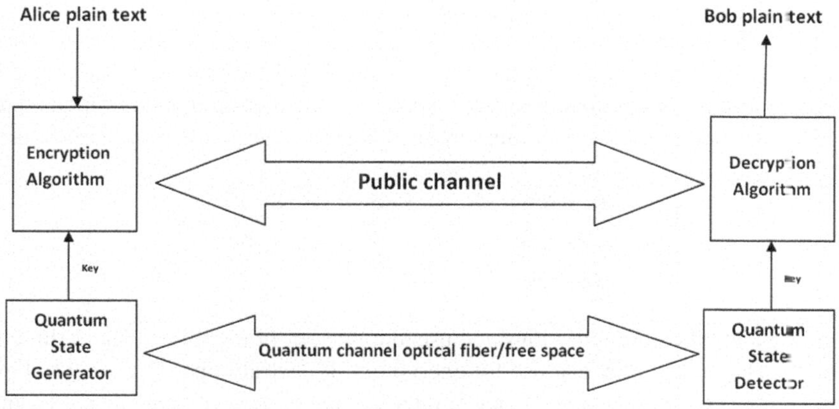

Fig. 4. Quantum cryptographic communication System

BB84 uses the technique in which Alice sends a short light signals and polarization in each signal is randomly modulated to one of four states (horizontal, vertical, left-circular, and right-circular) and Bob measures photon polarization in a random sequence of bases (rectilinear or diagonal) bob tells the sender publically which bases were chosen. And the signal are interpreted using a binary scheme either left-circular or horizontal 0 and right circular and vertical 1. In this scheme Alice encodes her information randomly in one of two base.

Bases A	***Bases B***
|0>= 0	*|+>= 0*
|1>= 1	*|−>= 1*

Alice prepares 8 bits

bits	**01011000**
bases	*BAABAABA*
state	*+10-10 + 0*

Now bob receives the stream of qubits and measure each one in a random basis. ABAABAAA so bob get 1-00-0 + 0

Received bits by bobs analyzer	**+10-10 + 0**
Bob bases	*ABAABAAA*
Actual bits detected	*1-00-010*

Bob selected their own bases

Bases A	*Bases B*
|+>= *1*	|*1*>=–
|–>= *0*	|*0*>= +

A pair of polarization states, such as horizontal and vertical, or left-circular and right-circular that can be reliably distinguished by some measurement is called a basis. The Alice and Bob compare their measurement basis not the result via a public channel

Alice Bases	**B**	**A**	**A**	**B**	**A**	**A**	**B**	**A**
Bobs bases	A	B	A	A	B	A	A	A
Matching	–	–	*0*	–	–	*0*	–	*0*

So bob and Alice are left with three useable bits out of eight bits. These three bits will be the shared key they can use for encryption. If eavesdropper Eve want to tap on Alice and Bob communication. Then eve intercept the bit stream from alice measure it and prepare a new bit stream to Bob based on new measurements. but eve does not know what bases to measure in. so eve would have to measure randomly and 50% of the time eve will be wrong and if the bits Bob measure in the correct bases there is 50% chances that eve had changed the basis of bits and thus the error in bits is detected by Bob and eavesdropping is found in error. Even the error in perfect transmission system can occur but only increase in error rate in transmission will detect the eavesdropper.

5 Quantum Key Distribution Using Base Distribution Center

The peer to peer key agreement protocol is less efficient and is not well managed. The proposal for induction of third party known as base distribution centre is introduced to provide more security as well as management of keys. Since the two parties who want to communicate securely require a secret key to secure their communication over a quantum channel uses a trusted third party, here each communicating entity share their base with key distribution centre. Whereas protocols like BB84 and B92, the sender and receiver does not identify the secret key until the last step when they conclude the comparison of their bases. The induction of classical method in quantum techniques we can enhance the authentication and data integrity. The algorithms will follow the following steps (Fig. 5).

i. Alice requests the bases status of Bobs from base distribution center using ID of Bobs along with sharing its quantum base QBA.

$$Alice : E_{Alice-BDU}(ID_{Bob}||QBA)$$

ii. Base distribution center send the request to Bob for sharing its bases along with its ID and quantum base of Alice.

$$BDU : E_{BDU-Bob}(ID_{BDU}||QBA)$$

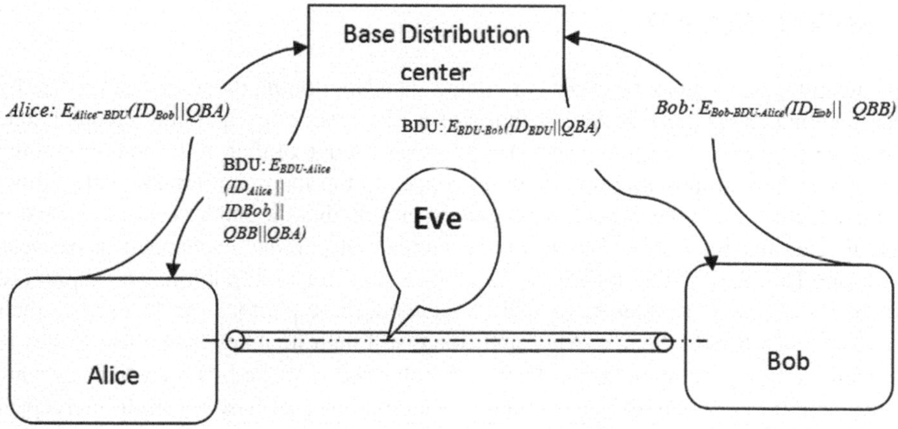

Fig. 5. Quantum base distribution

iii. Bob acknowledge with its base to base distribution center.

$$Bob : E_{Bob-BDU-Alice}(ID_{Bob}||\, QBB)$$

iv. Base distribution center share bobs base with Alice. Alice acknowledges its base via base distribution center to Bob.

$$BDU : E_{BDU-Alice}(IDAlice\,||\, ID_{Bob}||\, QBB||QBA).$$

v. The base distribution center adds connection status in between Alice and Bob and is authenticated to send and receive the message.

vi. Base distribution center starts distributing base in predetermined sequence to encode message to Alice and Bob.

vii. During the communication both Alice and Bob are continuously sharing its bases with base distribution center and during this process if there is increment in number of mismatching bits, then base distribution center can conclude about the intrusion in communication between Alice and Bob.

viii. Base distribution center send the notify signal to Alice and Bob to change their bases.

$$BDU : E_{Bob}(EQB\,(Notify))$$
$$BDU : E_{Alice}(EQB\,(Notify))$$

Another modification to above algorithm is done by calculating the required bits of key by matching the bases at base distribution center and distributing the key in between both entities at particular interval of communication. secure communication [15] is feasible between any nodes in the network by reliable secret key that is controlled by key management server. from user point of view, the quantum key distribution and key management layer can be treated as a black box that supplies them a secure key.

6 Security in QKD

The quantum state cannot be copied or cloned. The No-cloning theorem was verified by Wootters and Zurek [13]. It defines that a user can never copy quantum channel state and never duplicates the qubits without a prior knowledge of their base. so non-cloning theorem make it impossible for an eavesdropper to tap the quantum channel without being detected unless the eavesdropper can exactly predict the sender basis and use this base to measure the qubits. When a large number of photons exchanged in between Alice and Bob they will compare the bases they used for each photon. Those photons in which they used the same basis will be retained, these photons can be used as their "raw key" and discard the rest. Alice and Bob distill their final key from this raw key in consecutive key generation steps. Alice and Bob store only the bits for which they are using same basis. Whereas Eve is making random choice of base for each intercepted bit and for half of bits Eve will have chosen the incorrect basis and photon sent to Bob will be polarized according to mismatched basis [17]. Because of Eve's action, Bobs bit will disagree with Alice's in about quarter cases, even though they used the same base, so Alice and Bob can detect the eavesdropping in their communication

7 Conclusion

In traditional Quantum key distribution protocols like BB84, B92 sender and receiver does not identify the secret key until the last step where they conclude the comparison of their bases as well as they use the public channel to mutually authenticate their keys. These protocols are less efficient and their key distribution are not well managed. in this paper we proposed the third party known as base distribution centre to provide more security and management of keys. the practical quantum key distribution can augment the present cryptographic techniques in order to provide high level of security in future. the free-space quantum key distribution [21, 24] can play an active role in ship-to-ship communication, satellite quantum key distribution, communication between vehicles with fast-beam tracking and filtration methods. The studies suggested that the quantum key distribution at microwave wavelengths [22, 23] is possible and can be used by wireless communications in short range. The mobile quantum key distribution [25] will become an active research area in future.

References

1. Bennett, C.H., Brassard, G.: Quantum cryptography: public key distribution and coin tossing. In: Proceedings of the IEEE International Conference on Computers, Systems and Signal Processing, Bangalore, India, pp. 175–179 (1984)
2. Bennett, C.H.: Quantum cryptography using any two non orthogonal states. Phys. Rev. Lett. **68**, 3121–3124 (1992)
3. Bouwmeester, D., Ekert, A., Zeilinger, A.: The Physics of Quantum Information. Springer, Berlin (2000)
4. Diffie, W., Hellman, M.E.: Privacy and authentication: an introduction to cryptography. In: Proceedings of then IEEE, vol. 67, no. 3, pp. 397–427, March 1979

5. Nielsen, M.A., Chuang, I.L.: Quantum Computation and Quantum Information, p. 13. Cambridge University Press, Cambridge (2010). ISBN 978-1-107-00217-3
6. Gisin, N., Thew, R.: Quantum Communication. Group of Applied Physics, University of Geneva, Geneva (2008)
7. Weinberg, S.: The Quantum Theory of Fields, I edn. Cambridge University Press, Cambridge (2002). ISBN 0-521- 55001-7.2002
8. Heisenberg, W.: Remarks on the origin of the relations of uncertainty. In: Price, W. Chissick, S. (eds.) The Uncertainty Principle and Foundations of Quantum Mechanics. A Fifty Years' Survey, pp. 3–6. Wiley, Hoboken (1977)
9. Bennett, C.H., Brassard, G.: Quantum public key distribution system. IBM Techn. Discl. Bull. **28**, 3153–3163 (1985)
10. Molotkov, S.N.: J. Exp. Theor. Phys. **107**, 28 (2008). https://doi.org/10.1134/S1063776108070030
11. Ma, L., Mink, A., Tang, X.: High speed quantum key distribution over optical fiber network system. J. Res. Natl. Inst. Stand. Technol. **114**(3), 149–177 (2009)
12. Arndt, M., Juffmann, T., Vedral, V.: Quantum physics meets biology. HFSP J. **3**(6), 386–400 (2009)
13. Fan, H., et. al.: Quantum cloning machines and the applications. Beijing National Laboratory for Condensed Matter Physics, Institute of Physics, Chinese Academy of Sciences, Beijing 100190, China (2014)
14. Singh, H., Gupta, D.L., Singh, A.K.: Quantum key distribution protocols: a review. IOSR J. Comput. Eng. **16**(2), 01–09 (2014). Ver. XI
15. Lo, H.K., et al.: Secure quantum key distribution. Nat. Photonics **8**, 595–604 (2014)
16. Diamanti, E., et al.: Practical challenges in quantum key distribution. npj Quantum Inf. **2**, 16025 (2016)
17. Risk, W.P., Bethune, D.S.: Quantum cryptography. Opt. Photonics News **13**(7), 26–32 (2002)
18. Zhang, C.H., et al.: A simple scheme for realizing the passive decoy-state quantum key distribution. J. Lightwave Technol. **36**(14), 2868–2873 (2018)
19. Lucamarini, M., Yuan, Z.L., Dynes, J.F., Shields, A.J.: Overcoming the rate-distance limit of quantum key distribution without quantum repeaters. Nature, May 2018. https://doi.org/10.1038/s41586-018-0066-6
20. Roberts, G.L., Lucamarini, M., Yuan, Z.L., Dynes, J.F., Comandar, L.C., Sharpe, A.W., Shields, A.J., Curty, M., Puthoor, I.V., Andersson, E.: Experimental measurement-device-independent quantum digital signatures. Nat. Commun. **8**, 1098 (2017)
21. Toyoshima, M., Takenaka, H., Shoji, Y., et al.: Polarization-basis tracking scheme in satellite quantum key distribution. Int. J. Opt. **2011**, Article ID 254154, 8 p (2011). https://doi.org/10.1155/2011/254154
22. Ruiz-Alba, A., et al.: Microwave photonics parallel quantum key distribution. IEEE Photonics J. **4**(3), 931–942 (2012)
23. Morozov, O.G., et al.: Universal microwave photonics approach to frequency-coded quantum key distribution. In: Advanced Technologies of Quantum Key Distribution, 20 December 2017 (2017)
24. Gong, Y.H., et al.: Free-space quantum key distribution in urban daylight with the SPGD algorithm control of a deformable mirror. Opt. Express **26**(15), 18897–18905 (2018)
25. Abulkasim, H., et al.: Improved dynamic multi-party quantum private comparison for next-generation mobile network. IEEE Access **7**, 17917–17926 (2019). https://doi.org/10.1109/ACCESS.2019.2894101

Privacy Threat Model for IoT

Shelendra Kumar Jain(✉) and Nishtha Kesswani

Department of Computer Science, Central University of Rajasthan, NH-8, Bandar Sindri, Ajmer 305817, Rajasthan, India
shelendra23@hotmail.com, nishtha@curaj.ac.in

Abstract. In the current era, Internet of Things (IoT) is facilitating ubiquitous and smart services to the users. IoT applications are widely increasing day by day in many fields such as healthcare, military, industries, agricultural and smart city. While, IoT application are facilitating smart services to provide comfortable living, many new issues also have emerged. Among the issues, data and user privacy is a serious issue to be considered on priority. To better understand privacy related parameters; we have presented a methodology of Privacy Threat Modeling along with Identification of the sensitive information in the Internet of Things. Then, An Adversary Model is described to clarify and comprehend the target attacks by the adversary in the architectural layers of the IoT ecosystem. A privacy risk map has been presented that will help to understand and fix priorities of the involved risk in IoT ecosystem. An IoT application called as Patient Data Provider "PDP" has been presented along with the discussion of the data privacy issues and challenges.

Keywords: Sensitive data · Threat modeling · Privacy · Internet of Things · Privacy risk · Information leakage

1 Introduction

Internet of Things (IoT) connects everything to enhance Quality-of-Service, efficiency and save time to make life easy. IoT provides smart solutions in many fields like environmental monitoring, city, home, business, healthcare, security and surveillance, agriculture, transportation, aviation industries, supply chain management, etc. [1–3].

With the evolution of the IoT ecosystem, different types of challenges also arise. Among them privacy protection is of utmost importance. To provide better services to the user, service related data collection is very necessary. Since these huge collected data can include highly sensitive data for which protection should be on the higher priority for the success of the whole system.

In era of the Internet of Things, the amount of the data collection will be vitally higher compared to the past. Sensitive information of users can be collected by the wearable devices like Google Glass, Apple iWatch, Google Fit, Apple Health Kit etc. and this sensitive information can range from user's health information to financial status [4].

N. Nain and S. K. Vipparthi (Eds.): ICIoTCT 2019, AISC 1122, pp. 278–293, 2020.
https://doi.org/10.1007/978-3-030-39875-0_30

The success of the effective privacy preservation mechanism depends on the deep and clear understanding of the root causes of the compromised privacy. In this paper, we aim to enlighten the foundation for the development of the effective privacy protection mechanism. Our contributions and novelties in the paper are as follows:

1. Identification of the sensitive information in the IoT ecosystem.
2. We have proposed a Privacy Threat Model for sensitive information in the Internet of Things.
3. We have proposed a privacy risk map for the IoT that will help to understand priorities of the risk in IoT ecosystem.

In addition to the above contributions, we have determined that which information should be protected, when to protect it and whom access should be granted or restricted in the IoT ecosystem. For this purpose, we have presented an IoT application called as Patient Data Provider "PDP". PDP provides a comprehensive way for the modern and smart healthcare systems to improve their Quality of Services (QoS). Along with this proposed strategy, data privacy related issues, challenges and their solutions are discussed.

The rest of the paper is organized as follows: In Sect. 2, background and related work is given. In Sect. 3, Sensitive information in IoT have been discussed. Section 4 describes privacy threat model in detail. A comprehensive application called "Patient Data Provider" is discussed with the privacy issues in Sect. 5. Finally, Conclusion and future scope are outlined in the Sect. 6.

2 Related Works

Smart things are envisioned to facilitate many application areas like health care, domestic activities and many new applications not identified yet [5]. The Internet of Things has impact on security and privacy of the stakeholders in the system, so measures are required to ensure the architecture's resilience to attacks, data authentication, access control and privacy of client [6].

Accessibility and availability of private health records (like blood pressure, heart rate, glucose level etc. through wearable health monitor) on internet cause serious privacy issues. User information disclosure can lead to tracking, personalization and localizing like privacy threats [7]. Identification of personal information during transmission over the internet is a serious user privacy issue [6–8]. Private data of a user may leak both from cloud and, from and between IoT devices. Many IoT devices send data in the form of plaintext, which results in the disclosure of user information by some IoT devices [9].

Once many sensors, Radio Frequency Identification (RFID) readers, and objects with RFID are connected to a network, it can abruptly lead to likely dramatic threats. Trojan can invade on a household to stake out and disclose information related to the household items [10]. Ziegeldorf et al. [11] analyze privacy issues in internet of things and classify and examine privacy threats. They have defined seven privacy threats namely, Identification, Localization and tracking, Profiling, Interaction and presentation, Lifecycle transitions, Inventory attack, and Linkage. They have also arranged the privacy threats into five different phases of the reference model.

Threat model provides structured approach that can be used to identify attacks and helps in suggesting countermeasures to protect system from exploitation and vulnerabilities [12,13]. Gholami and Laure [13] describes an extension of the Cloud Privacy Threat Modeling (CPTM). For the cloud computing environment, proposed methodology claims that it provides strong methodological support for privacy legislation and regulation.

Li and Xin [14] have given a security threat analysis for perception layer, transport layer and application layer. According to the author, security threat modeling is an important part of security mechanism of the IoT. They have given the threat tree model for IoT that can help to determine under what condition or how the threat can be achieved. Liang and Hongliang et al. [15] analyze the Android's privacy protection mechanism and describe different threats to subject's private data. In this paper, author also discussed the ways to leak sensitive information.

A comprehensive approach for privacy risks and management has been presented in the [16] with the detailed theoretical concept for threat, level of risks and risk management. On the basis of severity and likelihood, feared events are mapped on the risk map and it is helpful to determine order of the treatment for risks.

Huge amount of available literature on the side effects of the smartphone trigger us to understand unknown risks related to effects of the sensor and IoT devices on health. Based on a conclusion by the Working Group of 31 scientists for the assessment of carcinogenic hazard from exposure to radiofrequency electromagnetic fields emission by wireless communication devices, radio, microwaves and other signals, Author [17] observes that, "A risk of hazard exposure from cell phones exists to cause cancer". IoT ecosystem also involves the wireless communication so; it is topic of research that how the IoT devices can affect human health. Among the application areas of IoT, one of the most important application is Remote Healthcare and Monitoring. Adoption of the systematic strategies is necessary requirement for successful remote healthcare monitoring system and it benefits to the patients in many situations i.e. if patient is unable to walk and reduce frequencies to visit hospitals, updates health reports status in the real time etc. Addition of the IoT features into medical devices improves the quality and effectiveness of service and very useful for the patients with chronic conditions, and those requiring constant surveillance [18]. In IoT based healthcare system, real-time data collection is a necessary requirement to detect and diagnose disease with accuracy and on time, whether the patient is in a hospital, clinic, or at home, i.e., as promised by the IoT based service. A systematic and smart strategy for the remote patient's data collection and processing is an important and challenging task and privacy leakage aggravates this challenge.

There has been a lot of research work on the privacy threat modeling in different fields but to the best of our knowledge, only little research in the IoT field and none of them map privacy threats in IoT on a risk map which is important to set priority of a potential threat. In this paper, we have proposed a risk map for potential threats in IoT. We have also tried to identify sensitive information in IoT which form a basis of the proposed risk map for IoT. We have also suggested Patient Data Provider (PDP), an application area that ensure privacy of healthcare information. Our paper provides comprehensive and clear approaches to comprehend privacy threats and move forward to propose privacy protection solutions for the IoT ecosystem.

3 Sensitive and Non-sensitive Information in IoT

The identification of the sensitive data and non-sensitive information is an important step before proceeding to the privacy protection solutions. This section includes detailed overview of sensitive and non-sensitive in the IoT ecosystem. We have categorized data into three parts in a typical IoT ecosystem as shown in Fig. 1.

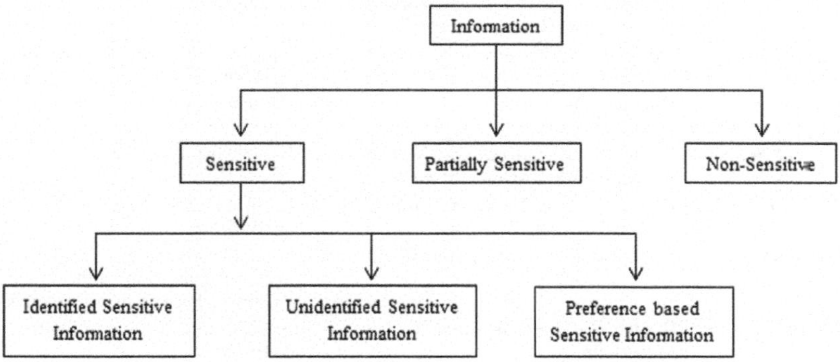

Fig. 1. Data classification in the IoT ecosystem

In the IoT ecosystem, data sensed by the sensor node can be: (1) sensitive information (2) partial sensitive (3) non-sensitive information. Sensitive information is the information which can create fatal consequences if revealed. Partially sensitive is the information which can assist in formation of sensitive information, for example date of birth can be combined with the name and combination of these two can reveal further more private information. Non-sensitive information is the information which may not lead to any serious consequences. Further, Sensitive Information is divided into three sub categories. The identified sensitive information is the information which is already declared as sensitive information, for example location, health information of the patient, financial information etc.

Table 1. Different kind of sensors, their major application area(s) and associated sensitive information.

Sensor name	Main application area(s)	Sensitive
Biometric Sensor	Security	Personally Identifiable Information (PII)
Camera	Surveillance	PII, Security Information, Private Activity
Chemical Sensor	Manufacturing, Agricultural, Healthcare	Presence, Changes in Chemical (Productivity Tracking)
Digital Pill, (Indigestible sensor)	Healthcare	Health Information
Gas Sensor	Manufacturing, Agricultural, Healthcare	Presence, Changes in Gas (Watch Productivity)
GPS Sensor	Location Based Services	Location
Gyroscope Sensor	Automation Industry	Object Rotation/Twist Tracking
Image Sensor	Traffic Monitoring, Automotive, Industry, Military Applications, Healthcare etc.	Private Activity, Health Information
Humidity Sensor	Smart Home, Manufacturing Industry, Smart City, Agriculture, Health Industry, Pharmaceutical Industry	Current and Future Condition of Assets
IR Sensor	Smart Home, Industry, Smart Devices, Smart City, Environment Monitoring, Healthcare etc.	Detect Presence, Health Information
Level Sensor	Oil Manufacturing Plant, Water Treatment, Industry	Amount of liquid, Purchasing Patterns
Motion Detector	Business, Industry	Presence/Absence of People
Optical Sensor	Healthcare, Environment Monitoring, Aerospace and other industries	Health Information
Pressure Sensor	Manufacturing Industry, Controlling System	Fluctuation in Pressure (Productivity Tracking)
Proximity Sensor	Automobile Industry, Retail Industry	Motion Detection, Correlation between Customer and Product, Purchasing Patterns
Smart Dust	Health Care, Environment Monitoring	Health Information, Private Activity
Smoke Sensor	Manufacturing, Accommodation Industry, Home Environment	Habits of Person
Temperature Sensor	Smart Home, Manufacturing Industry, Smart City, Agriculture, Health Industry, Pharmaceutical Industry	Current and Future Condition of Assets
Wearables	Healthcare, Location etc.	Health Information, Location

The unidentified sensitive information is the information which is not identified as sensitive yet but may become sensitive at a particular time or for a specific user. The preference based sensitive information is the information which a specific user wishes to kept private. The level of the sensitivity of sensitive information can be high, or moderate or low. The evaluation of the sensitivity level of the information is a challenging task, especially for the IoT ecosystem due to the resource constraints. There are various kinds of the sensors which are used to provide different kinds of the IoT services. For the sake of brevity and since only sensitive information may be critical while preserving the privacy, we have presented only the sensitive information which are related to sensor. In the Table 1 a brief list of different kind of sensors with its associated sensitive information is identified to provide a comprehensive idea of the consequences of the information leakage.

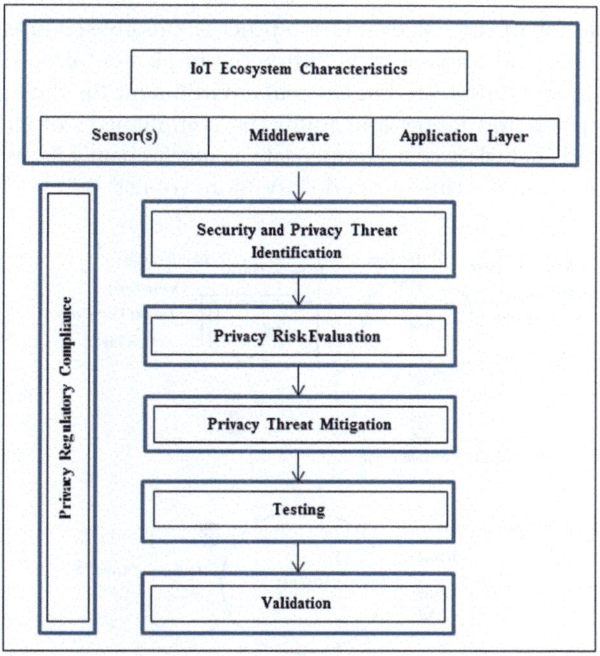

Fig. 2. Overview of IoT_PTM methodology

4 Privacy Threat Model

Privacy preservation for sensitive data has become a challenging issue in Internet of Things. With the motivation from Extended Cloud Privacy Thread Modeling (Extended CPTM) proposed by author [13] and threats in the reference model [11], we have proposed extended privacy threat model and methodology in comprehensive way with the involved risks. Our proposed methodology named as Internet of Things Privacy Thread Modeling (IoT_PTM) and threat model will help for identification and development of the countermeasures against the vulnerabilities in IoT ecosystem.

As shown in Fig. 2, Identification of specific characteristics of the IoT ecosystem is necessary before the identification of involved threats. Some of examples of these characteristics are related to all the components of the system like sensor, middleware, communication channel, software, physical devices, user and service provider requirements and their attitude toward services and benefits etc. after the fixation of these characteristics, all the privacy threats should be identified along with the security threats since security is more close to the privacy and likewise true for threats. In the next step, all the subsequences of the identified threats should be evaluated. A privacy risk map is discussed in the Subsect. 4.2 which helps in determining the level of the risk and likelihood of each threat. After the completion of the risk evaluation process, possible countermeasures can be fixed, developed and implemented. After the implementation of the countermeasures, solutions should tested in the real environment for the final validation. All of these steps should follow and fulfill the compliance's of the privacy regulatory body and inclusion of a compensation mechanism for prejudice due to privacy leakage can make this methodology more robust.

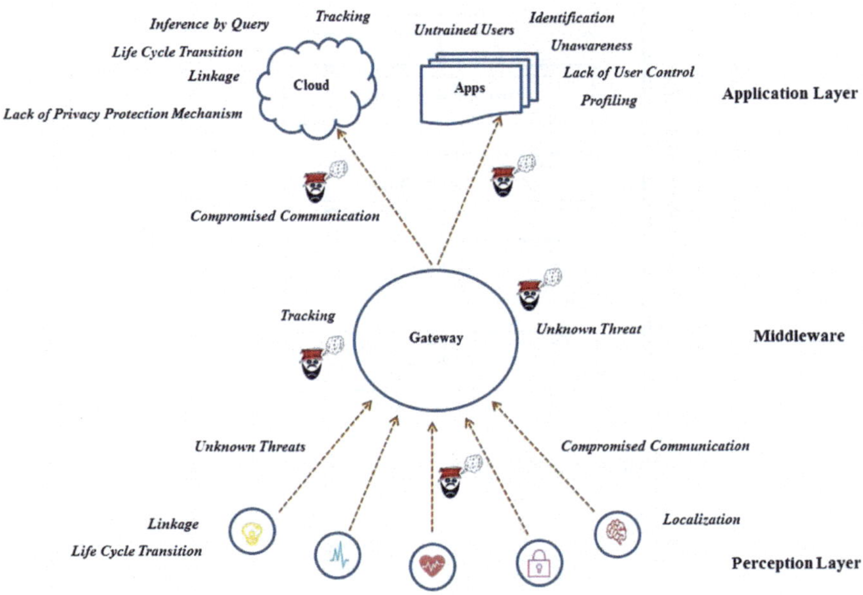

Fig. 3. Extended privacy threat model for IoT

4.1 Threat Model

In this section we propose an extended privacy threat model for sensitive data in IoT. A typical IoT architecture includes three layers namely, Perception Layer, Middleware and Application Layer. We have tried to place different privacy threats according to their proneness with respect to the layer of the typical IoT architecture (Fig. 3). Different privacy threats in addition with the privacy threats as described in [11] have been enlisted briefly in the Table 2.

Table 2. Different threats and brief description.

Threat	Description
Inference by Query (IbQ)	Send query to data store to analyze and make potential inferences
Tracking	Finding different locations of an Individual with the time
Life Cycle Transition	Privacy breaches due to changes in ownership of the products like smart things and devices
Linkage	Combining different IoT information to access or make inferences for private information
Lack of Privacy Protection Mechanism	IoT ecosystem without implementation of proper privacy measures and techniques
Untrained Users	Users in the IoT ecosystem without proper training can lead to the privacy breaches
Identification	Information that can help to identify an individual
Unawareness	Unaware user who don't have knowledge about their sensitive information and consequences of revealing the sensitive information
Lack of User Control	Users in the IoT ecosystem without control over their sensitive information
Profiling	finding interests of an individual
Compromised Communication	Weak communication mechanism lead to compromised user privacy
Unknown Threat	Other unknown threats that lead to privacy breaches with the time
Localization	Finding exact location of an individual

4.2 Privacy Risk Map for IoT

In this subsection, we propose a privacy risk map for IoT that maps the privacy threats which are presented in the privacy threat model. With the motivation and idea from the Risk map presented in the "CNIL, Methodology for Privacy Risk Management" [16]; we have sketched a privacy risk map for IoT that will help to understand priorities of the risk in the context of the IoT ecosystem. In this sketch two dimension namely, risk and its likelihood are used with the range in percentage to quantify the level of the phenomena. We have quantified these two dimensions based on just overview and basic knowledge and it needs further more efforts to determine more accurate and correct quantification of these two phenomena. We have mapped different threats of Threat model (Fig. 3) to the Privacy risk map for IoT (Fig. 4) that will help to set priorities of the risk by the identification of the level of the involved risk and its associated

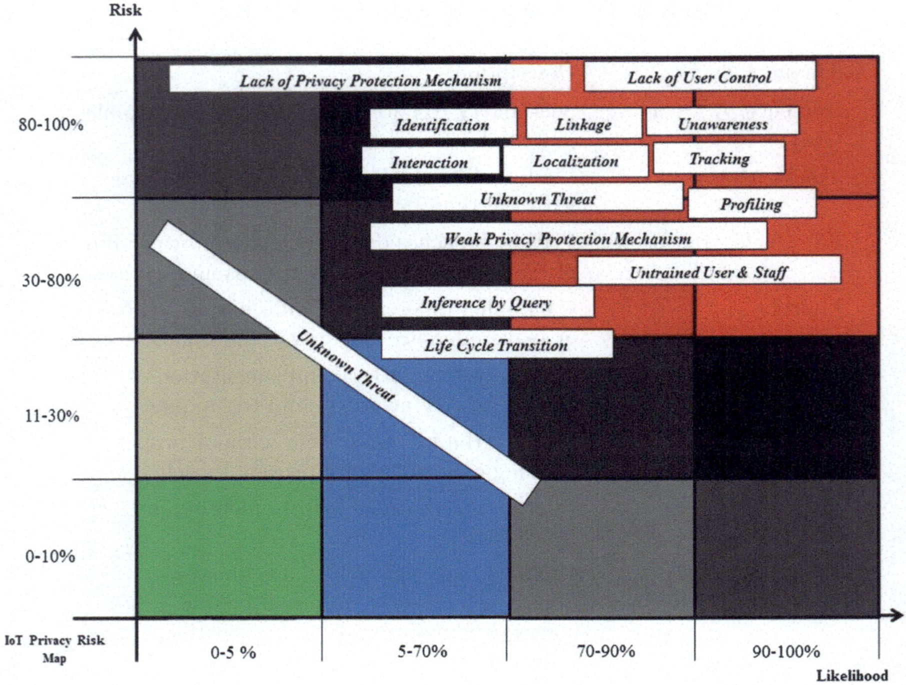

Fig. 4. Privacy risk map for IoT

likelihood; for example, lack of the user control over their private information is more likely and involves high risk that can result in financial and/or physical and/or mental loss(es). We have mapped some major threats according to their definition, characteristics and other features. This mapping provides an accurate envision for the future solutions to be implanted for the privacy preservation in the IoT ecosystem.

5 Patient Data Provider

In the IoT ecosystem, it is important to decide which information should be protected, when to protect it and whom access should be granted or restricted in the IoT ecosystem. We have illustrated this kind of issue with the presentation of a typical application called as Patient Data Provider (PDP). Patient Data Provider is an enhanced emerging application strategy of the IoT based healthcare system which may improve the QoS. Here, we have discussed PDP as an application of the privacy threat model. As presented in the privacy threat model in the Sect. 4, Lack of user control, Inference by Query, Identification, Unawareness, profiling, tracking, untrained user, linkage, lack of proper privacy protection mechanism etc. are privacy threats in the PDP. In the following subsections, detailed PDP strategy is given.

5.1 Basic Elements

An overall PDP application ecosystem for the IoT healthcare consist of the following two basic elements which are necessary for patient data collection, data storage and data processing functionalities.

Sensor(s). Sensor(s) are one basic elements for IoT based remote healthcare monitoring. It is used to collect important and useful data from the remote patient and surrounding environment. A sensor must detect accurate, precise and high quality data for correct prediction about the disease so that correct medication can be given to the patient. Some examples of the useful sensors for the remote healthcare monitoring [19–23] are the following:

1. Accelerometer: It measures body movement of the patient like, steps and sleep patterns.
2. Ambient temperature sensor: it measures body temperature
3. Respiratory devices: It measures respiration rate
4. Pressure sensors: It is used in the blood pressure monitoring equipment
5. Image sensor: It used for mammography, dental imaging, endoscopy etc.
6. Biosensors: It measures blood glucose and cholesterol level
7. Ingestible sensors: It is a digital medicine. It is used for endoscopy applications and it can transmit images and physiological data.
8. Pulse Oximeter: It measure oxygen quantity in the blood.
9. Piezoelectric sensor: It measures breathing rate
10. Phonocardiograph: It measures heart sound
11. Glucose Sensor: It measures blood glucose level
12. Implantable Sensors: It implanted under the skin to measure glucose level and further transmits sensed data to external devices.

Data Provider Company. Second essential element of the PDP is the Data Provider Company (DPC) which owns the data storage/cloud. It is a repository of the patient's medical data. A DPC has a personal account for the each unique patient. All the data sensed by the sensors is transmitted to the data store/cloud within the DPC and medical data is saved in the personal account of the related patient.

5.2 PDP Overview

A PDP consist of the two basic element sensor(s) and DPC. An overview of a PDP is shown in the Fig. 5. Each patient is equipped with the sensor(s) to sense biological, physical and surrounding environment data. In a PDP there exist one or many patients (P_1, P_2, P_3 ... P_n). A DPC has personal account of each of the patient who is enrolled with it. For patient P1 there exists an account P_1P, for Patient P_2, account P_2P and likewise. A record in the each account has attributes and their corresponding values as shown in the Fig. 6. One or more hospitals (H_1, H_2 ... H_n) can approach to the DPC to view and analyze the current status of the patient's medical data which is required for proper medication or to check patient medical history.

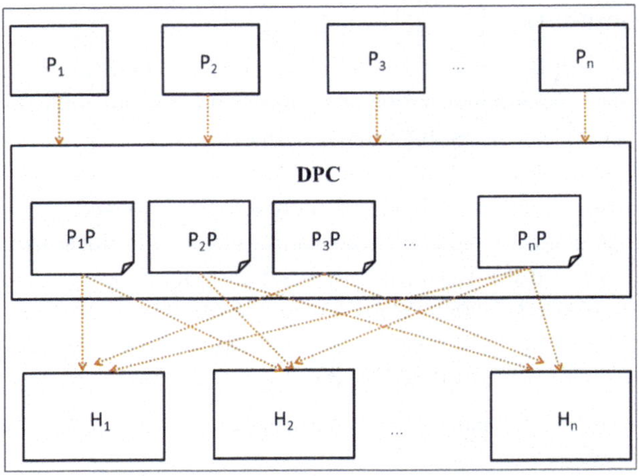

Fig. 5. Overview of a PDP

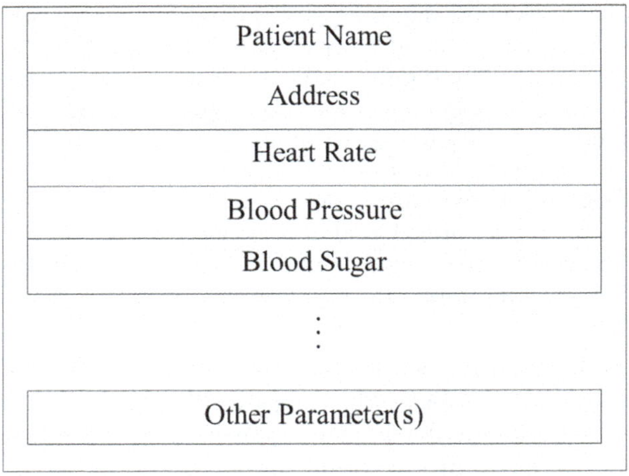

Fig. 6. An example of a record of a patient

5.3 Information Flow in the PDP

A simple and basic information flow in the PDP strategy is given in the Fig. 7. Patient equipped with wearable devices send data to the DPC and it saves those data in the unique account of a patient. A doctor in a hospital may request to DPC for patient's medical data and after the medial data analysis, proper medication is provided to the related patient.

5.4 Privacy in the PDP

Privacy disclosure of user data is a major concern and data protection in the IoT paradigm is a challenging task. The identifications of the security and privacy

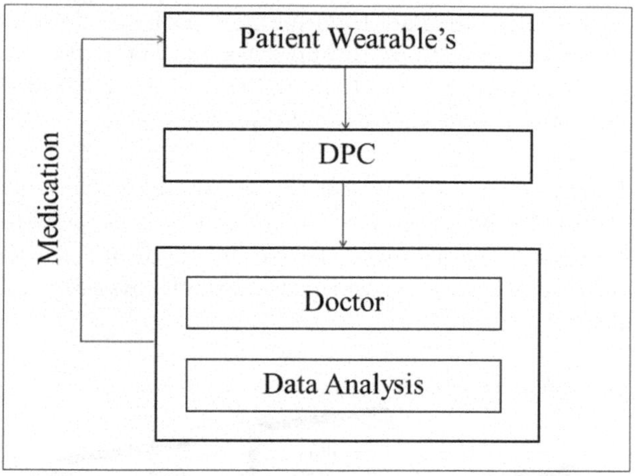

Fig. 7. Information flow in the PDP

loopholes in the IoT based application is also a big challenging job. Data privacy is directly related to the user privacy because if user data is revealed, then user privacy will be affected. A computational system and its related paradigm which provides services to the user must ensure user's data security and privacy, otherwise it will fail to be adopted by the current and future user(s). There are following privacy issues in the PDP system.

Privacy Issues. Privacy issues may arise at the following three levels in the PDP:

At the Remote Patient Level. At this level, unauthorized access of sensor node via physical attack is possible. Privacy can be compromised by eavesdropping in which intruder listen the transmitted information. Inaccurate information may be transmitted by the data modification attack. If camera based surveillance is used for remote patient monitoring then it can reveal extra information which may create dangerous consequences. For example, a person sitting near the remote patient and is using net banking can be under surveillance of the malicious user(s).

At the DPC Level. In the PDP system, A DPC collects patient's personal information. Personal information may be very sensitive in nature, for example Wearables sense patient's heart rate, blood pressure, blood sugar level, body temperature etc. and transmit it to the DPC. A person present in the DPC with nasty intention can continuously watch and analyze this sensitive data and predict much more information of the patient and his/her surrounding environment. The privacy risks increase in the broad scenario (scaling of the IoT paradigm); for example there can be multiple PDPs within a region and two different DPCs of the two different PDPs can share their enrolled patient's data with each other under their mutual collaboration. In this case patient's data sensitivity also increases because after integration of the multiple data attributes, it results in the availability of the more predictable information.

At the Hospital Level. One of the biggest threats at this level is the unauthorized data access. At this level, increased risk to the disclosure of the patient's personal identifiable information. Disclosure of the patient medical sensitive data results in the serious consequence i.e. it can harm the patient physically and/or mentally and/or economically.

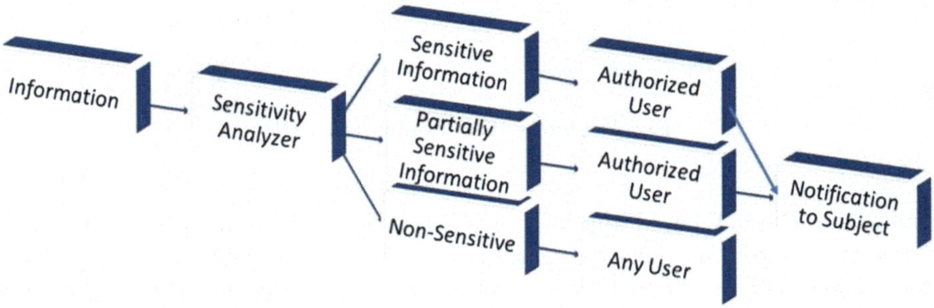

Fig. 8. Role of a sensitivity analyzer

Solutions to Privacy Issues. Above illustration of the PDP application is sufficient to decide that sensitive information must be protected against unauthorized access. For this purpose, identification of the sensitive and non-sensitive data is an important step. An efficient analyzer is necessary to separate sensitive data from non-sensitive data (Fig. 8). The efficiency of the analyzer helps to minimize data privacy risks. After the separation of the information, access to the sensitive information is minimized and regulated according to the purpose. There must be a strong mechanism to grant access to the sensitive data to the authentic user with proper notification to the source of the information.

Currently, the IoT solutions are being proposed in the many application areas. Internet connectivity is an essential part of the IoT based solution to provide service everywhere and in the recent days internet technology is suffering from different types of security and privacy threats and attacks. The success of the proposed PDP application strategy greatly depends on the efficient privacy protection methods, rule and policies. User data privacy must be protected by applying suitable rules and policies at all the three layers where privacy issue arises. Following are some of the solutions for user data privacy preservation in the PDP:

At the Remote Patient Level. A remote patient is equipped with the wearable sensors. Sensor has low computational power and low storage capacity. On this level, there must be some type of the data masking mechanism which can protect original data to disclose to the hackers. If device is hacked there should be a mechanism to send an alarm to the respective authenticated controller. Only limited and purposeful data should be collected with the consent of the patient i.e. data minimization should be followed.

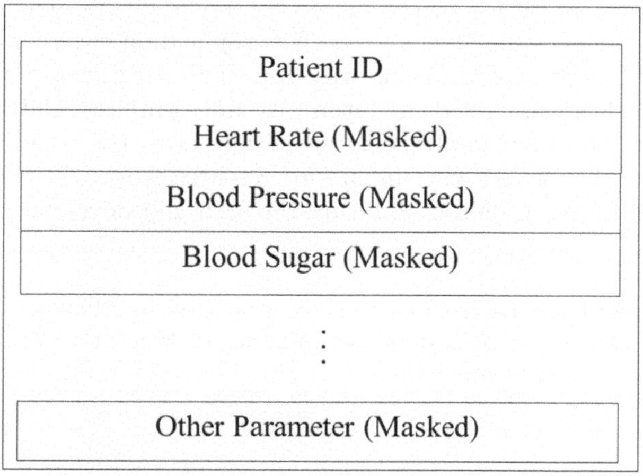

Fig. 9. A record of a patient with privacy

At the DPC Level. Data privacy is at the high risk at this level because data is stored. Proper security mechanism must be ensured at this level so that malicious user(s) are unable to monitor/watch patient personal data. In the Fig. 6, an example of the patient record is shown. This type of the record storage mechanism reveals sensitive data directly, so that data in the record must be saved by using some kind of masking mechanisms. Information which links to the personal identification must not be stored. Instead of the record shown in the Fig. 6, the record shown in the Fig. 9 should be used. Patient ID can be a unique ID assigned by the national authority like Aadhhar ID in the India and SSN in United States. There should be a patient's pre-consent mechanism before sharing the data between two different DPCs and A DPC must not share patient data to other third party without the consent of the real data owner.

At the Hospital Level. On the hospital level only authenticated person (like doctor and nurse who analyze patient data) should be allowed to access. At this level, privacy policy and strong authentication mechanism should be implemented. The hospital administration and authority must not share patient data with other third party without the consent of the patient. Only limited and purposeful data should be displayed to the authentic person.

6 Conclusion and Future Scope

In this paper, we have presented a methodology of Privacy Threat Modeling along with Identification of the sensitive information in the Internet of Things. An Adversary Model is described to clarify and comprehend the target attacks by the adversary in the architectural layers of the IoT ecosystem. We have proposed a privacy risk map for the IoT that will help to understand and fix priorities of the involved risk. Further, we have determined which information should be protected, when to protect it and whom access should be granted or restricted in the IoT ecosystem. For that purpose, an IoT application called as Patient Data Provider "PDP" is presented. Along with the PDP, data privacy related issues, challenges have been discussed. In future, we may propose lightweight privacy protection rules and machine learning algorithms with the edge computing for the resource constraint IoT paradigm. Our research work will help researchers in getting a better idea about the future research and development of the IoT Privacy.

Acknowledgements. This research work is supported by Maulana Azad National Fellowship formulated and funded by the Ministry of Minority Affairs. We wish to thank reviewers for their precious time.

References

1. Bandyopadhyay, D., Sen, J.: Internet of Things: applications and challenges in technology and standardization. Wireless Pers. Commun. **58**(1), 49–69 (2011)
2. Batth, R.S., Nayyar, A., Nagpal, A.: Internet of robotic things: driving intelligent robotics of future-concept, architecture, applications and technologies. In: 2018 4th International Conference on Computing Sciences (ICCS), pp. 151–160. IEEE (2018)
3. Nayyar, A., Puri, V., Nguyen, N.G.: BioSenHealth 1.0: a novel internet of medical things (IoMT)-based patient health monitoring system. In: International Conference on Innovative Computing and Communications, pp. 155–164. Springer (2019)
4. Perera, C., Ranjan, R., Wang, L., Khan, S.U., Zomaya, A.Y.: Big data privacy in the Internet of Things era. IT Professional **17**(3), 32–39 (2015)
5. European Commission: Iot privacy, data protection, information security. http://ec.europa.eu/information_society/newsroom/cf/dae/document.cfm?doc_id=1753
6. Weber, R.H.: Internet of Things-new security and privacy challenges. Comput. Law Secur. Rev. **26**(1), 23–30 (2010)
7. Porambage, P., Ylianttila, M., Schmitt, C., Kumar, P., Gurtov, A., Vasilakos, A.V.: The quest for privacy in the Internet of Things. IEEE Cloud Comput. **3**(2), 36–45 (2016)
8. Roman, R., Zhou, J., Lopez, J.: On the features and challenges of security and privacy in distributed Internet of Things. Comput. Netw. **57**(10), 2266–2279 (2013)
9. Baker, F., et al.: Internet of Things (IoT) security and privacy recommendations, November 2016. https://www.bitag.org/documents/BITAG_Report_-_Internet_of_Things_(IoT)_Security_and_Privacy_Recommendations.pdf. Accessed 10 Oct 2018. 11:22 AM

10. Radomirovic, S.: Towards a model for security and privacy in the Internet of Things. In: Proceedings of First International Workshop on Security of the Internet of Things (2010)
11. Ziegeldorf, J.H., Morchon, O.G., Wehrle, K.: Privacy in the Internet of Things: threats and challenges. Secur. Commun. Netw. **7**(12), 2728–2742 (2014)
12. Swiderski, F., Snyder, W.: Threat Modeling. Microsoft Press, Redmond (2003)
13. Gholami, A., Laure, E.: Advanced cloud privacy threat modeling. arXiv preprint arXiv:1601.01500 (2016)
14. Li, Z., Xin, T.: Threat modeling and countermeasures study for the Internet of Things. J. Convergence Inf. Technol. **8**(5), 1163–1171 (2013)
15. Liang, H., Wu, D., Xu, J., Ma, H.: Survey on privacy protection of android devices. In: 2015 IEEE 2nd International Conference on Cyber Security and Cloud Computing, pp. 241–246. IEEE (2015)
16. CNIL: Methodology for privacy risk management (2012). https://www.google.com/url?sa=t&rct=j&q=&esrc=s&source=web&cd=1&ved=2ahUKEwjYv9_ho_ndAhUJurwKHdgOAn8QFjAAegQICBAC&url=https%3A%2F%2Fwww.cnil.fr%2Fsites%2Fdefault%2Ffiles%2Ftypo%2Fdocument%2FCNIL-ManagingPrivacyRisks-Methodology.pdf&usg=AOvVaw1ZpBEi2P0NGSe0mIJpvduH. Accessed 09 Oct 2018. 5:22 PM
17. Miakotko, L.: The impact of smartphones and mobile devices on human health and life. New York University (2017). (Online Journal), Tersedia dari. https://www.nyu.edu/classes/keefer/waoe/miakotkol.pdf. Accessed 2 Oct 2018
18. Canino, G., Scarpino, M., Cristiano, F., Mirarchi, D., Tradigo, G., Guzzi, P.H., Cuda, G., Veltri, P.: Geoblood: a web based tool for geo-analysis of biological data. Procedia Comput. Sci. **98**, 473–478 (2016)
19. Bhelkar, V., Shedge, D.: Different types of wearable sensors and health monitoring systems: a survey. In: 2016 2nd International Conference on Applied and Theoretical Computing and Communication Technology (iCATccT), pp. 43–48. IEEE (2016)
20. Sharma, S., Tripathi, M., Mishra, V.: Survey paper on sensors for body area network in health care. In: 2017 International Conference on Emerging Trends in Computing and Communication Technologies (ICETCCT), pp. 1–6. IEEE (2017)
21. Kiourti, A., Shubair, R.M.: Implantable and ingestible sensors for wireless physiological monitoring: a review. In: 2017 IEEE International Symposium on Antennas and Propagation & USNC/URSI National Radio Science Meeting, pp. 1677–1678. IEEE (2017)
22. Yuce, M.R., Dissanayake, T.: Easy-to-swallow antenna and propagation. IEEE Microw. Mag. **14**(4), 74–82 (2013)
23. Luis, J., Roa Romero, L., Gómez-Galán, J., Hernández, D., Estudillo-Valderrama, M., Barbarov-Rostán, G., Rubia-Marcos, C.: Design and implementation of a smart sensor for respiratory rate monitoring. Sensors **14**(2), 3019–3032 (2014)

IoT Based Smart Luggage Monitor Alarm System

Sowmya Valluripally[1], Deepak Sukheja[1(✉)], Kriti Ohri[1(✉)], and Suyash K. Singh[2(✉)]

[1] VNR VJIET, Hyderabad, India
sowmya.valluripally29@gmail.com, deepak_s@vnrvjiet.in, kriti.ohri@gmail.com

[2] Accenture Solution Ltd., Bengaluru, India
Singhsuyash007@gmail.com

Abstract. The Internet of Things is the integrated global network, which allows connecting everything to everyone. Internet of things spreads from smart Cities to smart connected homes to wearables to health-care. Based on IoT there are various security and tracking devices, which are used for tracking and monitoring personal things. But not many of those devices serve flexibility for the users. "IOT smart luggage monitor alarm" is the device based on IoT, it is a portable device connected to the user's smart phone through Bluetooth connectivity. The device is composed of Arduino, BLE(HC-05) and Piezo speaker (Buzzer), GSM Module(SIM-800L), GPS Module. When the Bluetooth connection between the user mobile and the device is established, then it means that the luggage is safe and it is with the user. when the Bluetooth connection between the device and the user's mobile is lost then the user is given an audio alert from the buzzer of the device and also by an SMS sent to the user mobile along with GPS location of the luggage which is connected to the device, and also provides a continuous tracking of the luggage and sends the location in some time intervals. The location is tracked till the device is connected to the user's mobile. Comparatively based on the existing solutions it is portable, feasible and easy to install.

Keywords: IoT · GSM · Bluetooth

1 Introduction

In Internet of Things applications can be found in every sector of society and industry with a diversity of application for travel and transportation like vehicle tracking system, tracking kitchen inventory, asset tracking systems [1]. Based on the above tracking applications or systems we also need to develop a tracking system for luggage which should also monitor the luggage continuously. Taking a step ahead we use IOT for securing our luggage, based on the above scenarios there are devices which provides the tracking of luggage only when the user founds the bag is stolen or whenever the user is searching for the missing luggage. We need to develop a system or device which alerts the user when the luggage is away from them because in many cases bags are stolen from

N. Nain and S. K. Vipparthi (Eds.): ICIoTCT 2019, AISC 1122, pp. 294–302, 2020.
https://doi.org/10.1007/978-3-030-39875-0_31

hotel lobbies, even slipped from beneath a traveler's chair while they sip their drinks or when they sleep [2]. So, the tracking system should be developed in a way that it should alert the user when he/she is leaving the luggage behind or when the distance between the bag and the user is more than 15 m. Because if we discover the theft quickly or earlier, we might have some recourse but the longer it takes us to notice that the bag is missing the smaller our chances of recovering our baggage or luggage [3]. The paper is organized in Sect. 7 in which the existing systems and their demerits are explained in Sect. 2. Proposed system is detailed in the Sect. 3. System Design and Analysis is mentioned in the Sect. 4. The implementation of the proposed system is detailed in Sect. 5. Results and testcases are shown in Sect. 6. Finally, the paper is concluded in Sect. 7.

2 Existing System and Its Demerits

Now-a-days people are facing a lot of mess in handling their baggage, where they are misplacing their baggage by forgetting where they have placed their bags and also from thieving. The existing solutions or methods are some products like TILE, TRACR which tracks our bag whenever we identify that our bag is missing and open the application to find the location using crowd sourcing GPS which is not supported in all places [4]. There are also some travel bags like BLUESMART [5] bags which have provided with a GPS tracking feature, but the bag is very costlier where a common man cannot afford and also the system is stable and fixed to the bag i.e. it is not portable There are also some papers which provided solution to this problem i.e. [6]. An accurate real time GPS tracking system, which only keep track of the location of the item to which it is attached. On the basis of this GPS tracking system some other similar systems like biometrics based security system with GPS and GSM technology was developed [7]. In this system the luggage tracking starts only when the user fails to match his fingerprint while opening the luggage, but all these GPS tracking systems are useful whenever the user identifies that the bag is missing they don't alert when someone is trying to steal the bag or whenever the user forgets the luggage by placing it somewhere [8]. RFID based luggage tracking system is also one of the existing solution but the problem with the system is whenever the RFID tag is dirty or damaged the tracking of luggage is not possible. So the existing solutions are providing systems where we can track our luggage or baggage when or after we mislay it [9]. In the above cases regaining our luggage is possible only sometimes. So we need to develop a system where we can discover the theft swiftly and increase the chances of regaining our luggage as soon as possible.

Objectives:

- To propose a system or a device which alerts the user from misplacing his bag and mobile.
- To enhance the proposed system with add on feature i.e. Live tracking of the misplaced bag.

3 Proposed System

To deal with the problems of the existing products and solutions this paper provides a solution in the name of IOT smart luggage monitor alarm, which is a portable device and can be installed in any bag easily. The device consists of components like a micro-controller, Bluetooth module and GSM and GPS modules. The proposed solution works in three phases they are: Bluetooth connection, Alert and Tracking.

In this proposed system we use HC-05 Bluetooth module to establish Bluetooth connection between the device and the user's mobile. Whenever the connection between the device and user's mobile is lost then the device alerts the user with an audio indication using the buzzer connected to the device, and also sends an SMS to the user mobile using the GSM module that the luggage is away from the user. The tracking feature is implemented by using the GPS module to track the location of the luggage when it is away from the user and sends the location of the luggage to the user via GSM module. The device is kept in the user's bag which consists of any important valuables and the user is connected to the device via Bluetooth connection through mobile. When the user omits his/her bag and tries to leave the place or whenever a thief steals the user's bag and runs after a certain distance Bluetooth connectivity of the device and the mobile is lost. As the connection is lost the device gives buzzer to alert the user and also sends SMS to the user mobile that he/she is away from the bag or any of his valuables. After alerting the user with an SMS the device tracks the location of the bag and sends the location of the bag to the user's mobile for tracking the where the bag is located. Power to the device is provided by rechargeable battery. The baggage will not produce any alarm as long as it is in the Bluetooth range. As Bluetooth range is too far it produce alarm with in short distance away from the user. The user can also disable the alarm option if at home. This device provides us a two-way protection i.e. The user mobile and the device are connected via Bluetooth connection so whenever the user misplaces his mobile it alerts the user with buzzer from the device as the Bluetooth connection is lost. The proposed solution works on motive that "prevention is always better than cure", as all the existing systems provided solution for tracking the luggage when it is lost, but the proposed solution in the paper doesn't allow the user to misplace his/her bag. If at all if the bag is missed it also provided a tracking feature for the device.

Though if the bag is lost then it must send the location of the bag to the users mobile and it should also send the location to the user whenever they want to know the location of the luggage. The device sends the location till the user gets back the luggage. This paper proposed a GPS based tracking system with composed open source software like Google Earth and Google Map.

4 System Design and Analysis

Consider Designing goes through logical and physical stages of development, logical design reviews the present physical systems, prepare input and output specification, details of implementation plan and prepare a logical design walkthrough.

4.1 Bluetooth Connection Module

Initially after setting up the device, we need to turn on the device whenever we are travelling or whenever we want to protect our luggage. In this module user will establish a Bluetooth based connection between the device present in the luggage and user's mobile through an Arduino Bluetooth control application. In this device we use HC-05 Bluetooth for establishing connection with the user mobile. We can also setup a password for the HC-05 Bluetooth module so, whenever the user enters the correct password only then they are allowed to connect with the device. After establishing the connection between the user's mobile and device whenever, the user omits the luggage or when someone is trying to steal the luggage the Bluetooth connection between the mobile and the device gets lost and the device gives an audio indication with buzzer which is connected to the device. So, that the user can alert as soon as possible and can protect the luggage and the valuables in it. Here we have an add on feature that is when someone is trying to steal your mobile then also it alerts through audio indication because the Bluetooth connectivity is lost.

4.2 Tracking Module

As mentioned above the device connected to the user's mobile and when the connection is lost it alerts the user with buzzer sound. If in a bad case if the user is not aware of the buzzer alert and the bag is lost then we cannot protect our luggage. So, in order to avoid those situations, the device has GPS module to track the location of the luggage. The GPS module connected to the device continuously tracks the location of the luggage but it will provide the location initially when the Bluetooth connection is lost between the users mobile and the device. After that it provides the location of the luggage continuously in some intervals of time till we connect the device to Bluetooth again. The location is sent to the user in the form of a link through which we can find the location of the luggage on the google maps. It sends the location to us till again the device is connected to the users mobile.

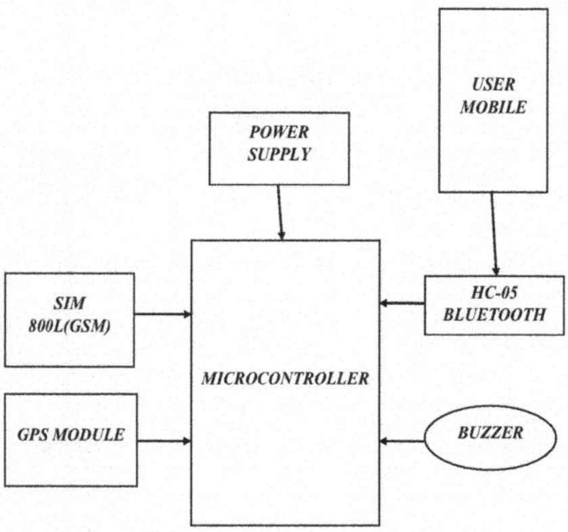

Fig. 1. System architecture.

4.3 Notification Module

In the notification module we use GSM module to send notification or SMS to the user mobile whenever the Bluetooth connection is lost between the user mobile and the device. Whenever the connection is lost it sends an SMS that the luggage is away please take care of it. Because, sometimes we may not hear the buzzer sound in crowded areas and noisy areas so that we alert the user with an SMS which is sent to their mobile using the GSM module. The location link which is been tracked by the GPS module is also sent to the user via GSM module as an SMS to the mobile of the user.

4.4 Block Diagram

The above Fig. 1 is the block diagram of IOT smart luggage monitor alarm in which the Bluetooth module HC-05, buzzer is connected to the Arduino microcontroller. To alert the user with an SMS we have connected GSM module to the microcontroller, GPS module is connected to the microcontroller to track the location of the luggage. We need to provide the power supply to power on the device and whenever we want to power off the device, we can stop the power supply. User's mobile is connected with the device via Bluetooth.

4.5 Data Flow Diagram

See Fig. 2.

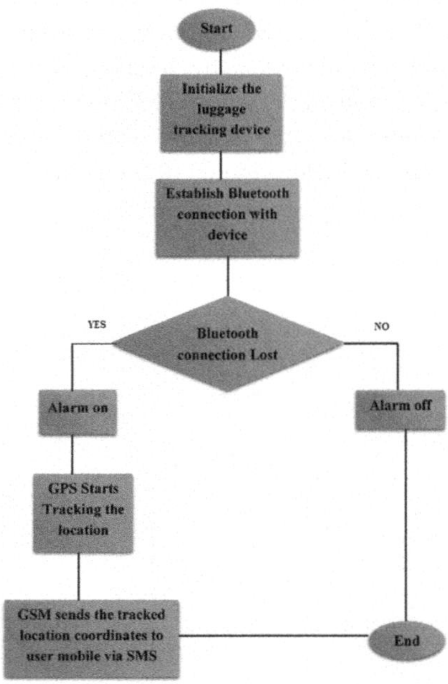

Fig. 2. Data flow diagram of luggage monitor device

5 Implementation

The most crucial stage is achieving a new successful system and in giving confidence on the new system for the users that it will work efficiently and effectively. It involves careful planning, investigation of the current system and its constraints on the implementation, design of methods to achieve the changeover and an evaluation of change over methods apart from planning. The more complex the system being implemented, the more involved will be the systems analysis and design effort required just for implementation.

5.1 Connections

(a) Arduino to Bluetooth connections

ARDUINO	HC-05 BLUETOOTH
A0	STATE
VCC	5 V
GND	GND

(b) Arduino to buzzer connections

ARDUINO	BUZZER
8	Phase wire
GND	Ground wire

(c) Arduino to GSM connections

ARDUINO	GSM
0(RX)	TX
1(TX)	RX
GND	GND

(d) Arduino to GPS connections

ARDUINO	GPS
12(RX)	TX
11(TX)	RX
GND	GND

5.2 Device Setup

Now install Arduino IDE software to upload the source code or sketch. Configure the Serial Port (also known as the COM Port) to dump the code in to the device. After setting the com port we need to upload the code in to the device which we have setup after then, open Arduino Serial Monitor, set the baud rate to “9600” and command line ending drop down (the one next to baud rate) to “BOTH NL & CR” Now after opening the serial monitor we need to set the baud rate. At the right bottom of the serial monitor baud rate is set. Now we need to download an application Bluetooth Terminal HC-05 from playstore to connect the mobile to the device via Bluetooth connection in our mobile now turn on the Bluetooth and open the application and connect to the device HC-05. Whenever our mobile is connected to the device we get a message from the device that we are connected. Whenever the luggage is away from the user then the device gets disconnected and an alert SMS is sent to the mobile along with location link. Now for every constant time interval it sends the location as an SMS till we get back our luggage i.e. till the mobile and device are connected via Bluetooth. We can track the location using the link which pins the location of luggage in Google Maps (Figs. 3 and 4).

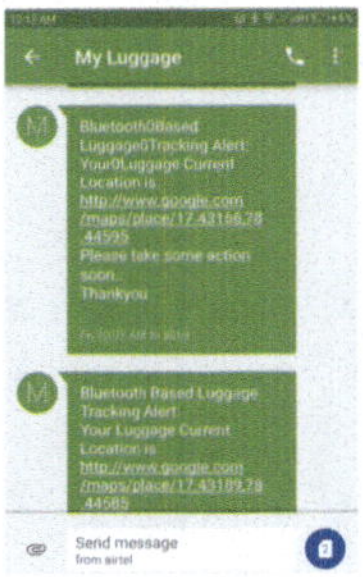

Fig. 3. Message alert when device is disconnected

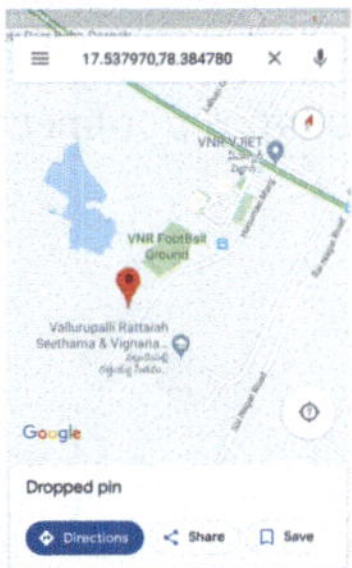

Fig. 4. Location pin in Google Map

6 Result Analysis and Test Cases

In Test case is a set of input data and expected results that exercises a component with the purpose of causing failure and detecting faults. The test case is an explicit set of instructions designed to detect a particular class of defect in a software system, by bringing about a failure.

Test Cases: The test cases are described on the basis of expected output and actual outputs. Test cases are categorized into two different categories mentioned below:

- Positive test cases
- Negative test cases

Positive test cases are conducted with the developer’s intention is to get the output shown in Table 1. In negative test cases are conducted by the developer intention is to don’t get the output shown in Table 2.

Table 1. Positive test cases

S. No	Test case Description	Actual value	Expected value	Result
1	Connection to the device Bluetooth	It will ask you the passcode when you are connecting for the first time	When the passcode is entered correct it will allow you to connect to the device	Connected successfully
2	Checking the alert from buzzer	It alerts as soon as the connection is lost	Buzzer provides alert successfully	Alert provided successfully
3	Checking the GSM module whether the messages are sent	Messages are sent to the user mobile if the Sim is inserted properly	Messages are received successfully	Messages received successfully
4	Checking the GPS module whether the location sensed was correct	Location is sent to the user whenever the connection is lost	Accurate location is sent to the user	Location sent successfully

Table 2. Negative test cases

S. No	Test case Description	Actual value	Expected value	Result
1	Check the difference between the audio alert and message alert	They both differ in time. Audio alert is immediate and whereas the message takes a while	They both gives the alert at the same time to the user	There is 15–30 s delay in receiving message after buzzer alert

7 Conclusion

This Bluetooth based luggage tracking system is very much essential in public transport like buses and trains, and in all crowded places. People mostly use this device when they are carrying any important valuables with them in their luggage. This project is a simple very powerful project. It continuously monitors the passenger luggage and if needed it protects the luggage from theft through an audio indication and an SMS alert with location tracking of luggage. The design and implementation of microcontroller-based security system (using mobile phone) has been proven to be a reasonable advancement in security system technology and access control. Then this is a major breakthrough in digital design and technological advancement. The security concerns raised by the typical chain-and-lock arrangement have been completely overcome by using our Luggage monitor Alarm.

Security is fundamental criteria for all kind of applications. So, this device works on the motive that "prevention is better than cure". Securing the luggage as soon as possible when it is nearby rather than tracking the luggage when it is lost when the user recalls about it. Though if the bag is lost then it must send the location of the bag to the users mobile and it should also send the location to the user whenever they want to know the location of the luggage. The device sends location till the user gets back the luggage. This paper proposed a GPS based tracking system with composed open source software like GoogleEarth and Google Map. Our system offers a real-time tracking system using a client-server model

References

1. Lakshmi, C., Solomon, S.S., Sandhya, V.A.: IOT and GSM integrated multi purpose security system. In: 2018 Fourth International Conference on Advances in Electrical, Electronics, Information, Communication and Bio-informatics (AEEICB), Chennai, pp. 1–4 (2018)
2. Quadri, S.A.I., Sathish, P.: IoT based home automation and surveillance system. In: 2017 International Conference on Intelligent Computing and Control Systems (ICICCS), Madurai, pp. 861–866 (2017)
3. Shah, T., Venkatesan, S.: Authentication of IoT device and IoT server using secure vaults. In: 2018 17th IEEE International Conference on Trust, Security and Privacy in Computing and Communications/12th IEEE International Conference on Big Data Science and Engineering (TrustCom/BigDataSE), New York, NY, pp. 819–824 (2018)
4. Waleed, A., Kharchenko, V., Uzun, D., Solovyov, O.: IoT-based physical security systems: structures and PSMECA analysis. In: 2017 9th IEEE International Conference on Intelligent Data Acquisition and Advanced Computing Systems: Technology and Applications (IDAACS), Bucharest, pp. 870–873 (2017)
5. Mrinal, M., Priyanka, L., Saniya, M., Poonam, K., Gavali, A.B.: Smart home automation and security system based on sensing mechanism. In: 2017 Second International Conference on Electrical, Computer and Communication Technologies (ICECCT), Coimbatore, pp. 1–3 (2017)
6. Dafallah, H.A.A.: Design and implementation of an accurate real time GPS tracking system. In: The Third International Conference on e-Technologies and Networks for Development (ICeND2014), Beirut, pp. 183–188 (2014)
7. Chadil, N., Russameesawang, A., Keeratiwintakorn, P.: Real-time tracking management system using GPS, GPRS and Google earth. In: 2008 5th International Conference on Electrical Engineering/Electronics, Computer, Telecommunications and Information Technology, Krabi, pp. 393–396 (2008)
8. Gupta, P., Chhabra, J.: IoT based Smart Home design using power and security management. In: 2016 International Conference on Innovation and Challenges in Cyber Security (ICICCS-INBUSH), Noida, pp. 6–10 (2016)
9. Prarthana, R.J., Dhanzil, A.M., Mahesh, N.I., Raghul, S.: An automated garage door and security management system (a dual control system with VPN IoT & biometric database. In: 2018 Second International Conference on Electronics, Communication and Aerospace Technology (ICECA), Coimbatore, pp. 1468–1472 (2018)

Text Steganography Based on Parallel Encryption Using Cover Text (PECT)

Subhash Panwar[1](✉), Mukesh Kumar[2], and Sakshi Sharma[2]

[1] Department of IT, Government Engineering College, Bikaner 334004, India
Panwar.Subhash@gmail.com
[2] Department of CSE, Government Engineering College, Bikaner 334004, India

Abstract. This paper presents a self-checking text steganography which could be used efficiently where large data is involved in various IoT applications. The space complexity of its cover text is one-fourth of its message size which is very less as compared to other techniques. The cover text of this method is generated from the private message itself. Later it is used to check the correctness of the data on receiver side. This helps to reduce the time overhead in case of corruption of any part of the data. Thus, the security and trust issues of IoT are easily handled even on high traffic channels.

Keywords: Internet of Things · Complexity · Hiding · Text Steganography · Plain text · Cover text · Encryption · Parallelism · Self-checking

1 Introduction

IoT security and trust are the major area of research and study nowadays. The objective of this research is to secure our information from unauthorized users therefore hacker, cracker in IoT environment. Our main aim is to hide our plain text so that it cannot be figured out with intrusive eyes with some encryption method. The two encryption techniques that are currently used are Cryptography and Steganography. Both the approaches are different from each other. Steganography is hiding our whole secret message under a message known as cover message whereas Cryptography is hiding of our contents of the secret message using either symmetric or asymmetric encryption method. Therefore in Steganography, the presence of our secret message is hidden i.e. only the sender and receiver are aware of its presence and can retrieve it whereas in Cryptography it is not, third party is aware of the presence of secret message.

This is the main advantage of steganography. Message in steganography can be transmitted through communication channel by using different kinds of cover messages such as text, image or media in order to effectively conceal the existence of the message, which is a major limitation in cryptography technique. One more reason is there for using steganography because some of our countries like U.S.A and British government have banned encryption as third party become aware of some important communication being done which is not at all secure. So, steganography techniques are using much

N. Nain and S. K. Vipparthi (Eds.): ICIoTCT 2019, AISC 1122, pp. 303–313, 2020.
https://doi.org/10.1007/978-3-030-39875-0_32

more powerful way for hiding plain text into cover text. Steganography algorithms is used to hide the data in a cover media such as text, image, video etc. as the cover medium and changes the structures and features of text message so that it does not get identified by human eyes. However, using text medium as a cover text is relatively difficult as compared to other methods of covering information because of lack of available redundant information in text data. In this paper, we are presenting an overview of existing text steganography approaches and our purposed approaches of steganography methods. The drawbacks of many existing methods of Text Steganography is using large cover text for encryption and decryption for hiding small plain text message and also it takes too much time in encryption and decryption. We introduce new approaches for text Steganography which is easy to hide the cover text and also taking very less time for encryption.

The shortcoming of many existing methods of Text Steganography is that there is no cross checking of the data at the receiver end i.e., if there is a third-party interruption during data transfer then it would be difficult for the currently existing techniques to detect the interruption and also it takes too much time in the process of encryption and decryption. A new method is introduced based on Text Steganography which takes very less time for encryption and also easily hides plain text original data in the cover text. This paper presents a self-checking text steganography which could be used efficiently where large data is involved. The space complexity of its cover text is 1/4th of its message size which is very less as compared to other techniques. The cover text of this method is generated from the private message itself. Later it is used to check the correctness of the data on receiver side. This helps to reduce the time overhead in case of corruption of any part of the data. Thus, it can be used even on high traffic channels also. So now let us study some existing approaches of text steganography before we discuss our method. We will illustrate an example to describe our approach. Finally, we will compare our approach with other approaches and discuss its merits and demerits. Future work will be related to removing the demerits of our method.

2 Related Work

Steganography hides our plain data while passing through the communication channel. It needs some cover media to hide the content. The most popularly used cover media are image, audio, text or a movie clip/video file. Among these different cover files, we preferred text data because it uses lesser memory storage as compared to other cover media files mentioned above [1]. To encrypt more bytes of plaintext in less amount of storage and to communicate more information with less execution time text steganography is preferable and commonly used. Text steganography is trickiest among all as it don't have any repeatable information like image, video or an audio file. Randomness of cover file is more in text steganography. it's also faster to execute as well as easy to read makes it more user-friendly technique as compared to other of Steganography methods. It can be broadly classified into three types: Format editing methods, Random sequence and Statistical methods and lastly Language based methods.

A. Format Based Methods

Format-based methods use formatting of the cover-text to hide data. They do not change any word or sentence, so it does not change the original meaning of the cover-text. Use of white space, changing colour, size or type of fonts, etc. are some example of format based text steganography method [3]. A format-based text steganography are open space method, feature coding method. Use of White spaces and deliberate misspelling to hide secret data can be recognized by computer while it is taken as a normal text by human users. In the same way human can also recognize these following changes of changing type, colour or size of text which are considered as normal text by computer. Reformatting or Retyping of text lost our secret data.

B. Random and Statistical Generation

Random and statistical generation methods are used to generate cover-text according to the statistical properties of language. Random generation method used to avoid known plain text attack [3]. A probabilistic context-free grammar is a commonly used language model to generate random cover text. The random cover sequence must appear random after hiding secret information to everyone, except the ones who are communicating. Another approach to this type of method is to generate random words by using statistical properties of word-length and letter frequency. Because of randomness both random character sequence and random word sequence are meaningless, that is why it is suspicious.

C. Linguistic Steganography

The linguistic method considers the linguistic properties of the text to modify it. This method uses linguistic structure of the message to hide secret information [6]. Syntactic method is a linguistic steganography method where some punctuation signs like comma (,) and full-stop (.) are placed in proper places in the document to embed a data. This method needs proper identification of places where the signs can be inserted. Another linguistic steganography method is semantic method. Existing Approaches:

In this sub-section, we discuss some of the common existing approaches of text steganography.

A. Open Space Method

In an Open space method whitespace are used to hide secret data. This method does not change the meaning of original text and casual reader considers additional white spaces as a normal text. Basically, there are three methods used white spaces to hide secret data. They are Inter-Sentence Space method, End-of-Line Space method and Inter-Word Space method.

B. Line Shift Method

In this method, the lines of text are shifted vertically to some degree and information are hidden by creating a unique shape of the text. This method is used for printed text. This method uses up and down shifting method to hide "0" and "1" bit respectively. Line is shifted to some degree like 1/400 inch, so distance between two lines will not be noticeable. This method cannot use for normal typing as shifting to this extent can't be possible. This method also has drawbacks that retyping of text would destroy our secret information. Even lines are used to hide secret bits and odd lines are considered as control lines and used while decoding. So even lines are moved as per hidden information and odd lines are left unmoved.

C. Word Shift Method

This method is almost same as line shifting. But instead of shifting line, words are shifted horizontally depend on secret bits in this method. A word is move left or right, while the words immediately adjacent before or after are left unmoved. If the text is to be decoded, original document is required in this method. The main problem with this document is that if anyone gets the original document and s/he is aware about this method, then s/he able to get secret data by comparing original document with modified document. Another disadvantage of this method is, retyping the document or character reorganization programs can destroy secret data.

D. Syntactic Method

By placing punctuation like comma (,) and full stop (.) in proper places to hide secret information. Punctuation could be used to hide secret binary data. This method requires identifying proper places for putting punctuation. This method hides very less information as hides data in binary format and also reformatting or retyping could lose our secret data. The amount of hidden information is trivial using this technique. Sometime it can change meaning of text by placing punctuation on wrong places so care must be taken for applying these techniques.

E.g., the phrases "bread, butter, and milk" and "bread, butter and milk" are both considered correct usage of commas in a list. To hide "1" bit, first phrase can be used and to hide "0" bit, second phrase can be used.

E. Semantic Method

This method has a great advantage over others as it protects the information in case of retyping. In this method, synonyms of actual words of our cover-text are used to hide the secret information. Bit '1' is being hidden by using actual word whereas bit '0' is being hidden by using synonyms of that word. This method is used in OCR programs.

F. Feature Coding Method

This method is based on the features of characters of the cover text. The features of the characters are been modified according to the secret data in form of bits. Some of the common features used are text size, colour or style. Example: By altering the vertical line length of words likh b, d, h, etc., by extending or shortening it depending on secret bits. The advantage of this method is that it can hide a large amount of data as compare to previously described techniques. And its disadvantage is that if retyping or reformatting of the text in done then our secret information will be lost. And we need the original document to decode our secret data; if original document comes in wrong hand then it will cause a major loss to us.

G. Capital Alphabets Shape Encoding (CASE) Method

This method is based on random character generation and hiding method. In this method English letters are divided into different groups as per their format like their vertical position, circle, curve, horizontal line. In this method one-character use to hide one character of secret information. Encoding techniques defines 8-bit encoding techniques for digits, alphabets or symbols [2]. As each group will have maximum 8 character which requires three digits to select one of them. This method is better than above existing methods but still time overhead are too high and also for hiding n bytes it requires at least n bytes of cover text.

H. Parallel Encryption with Digit Arithmetic of Cover Text (PDAC) Method
This method [7] uses three operations additions, subtraction and x-or for encryption of plain text. The digits of cover text are added and subtracted to get the new cover text, and then this generated new cover text is x-ored parallelly with plain text for encryption process. In this method, four characters of plain text forms one character of cover text. The overhead time and space requirement for storage of cover text both are very less as x-ored is performed parallelly. This is a good method but in it, at least two characters of cover text are required to hide plain text of more than four characters. Our approach is to minimise number of bytes of cover text so that the space occupied by it can be utilized at somewhere else. So, this is how we came with our new approach.

In all of the above approaches, there is no method of checking that whether there is a third-party intrusion present or not. This makes our approach above all these approaches. Also, both the memory requirement for storage of cover text and the time required for execution of this approach are very less. Therefore, the proposed approach reduces bytes of cover text that are required to encrypt (hide) message or information. In our approach, characters of cover text are used to check the correctness of data on receiver side. The very next section describes our approach along with its implementation.

3 Proposed Approach

The proposed technique (PECT) is based on simple mathematical operations between ASCII code of each character of cover text and the characters of secret message. To check the data parallelly on receiver side we introduce a self-checking method on the basis of vowels and special characters. We generate a different key for each character of secret message and because of such different key values, it makes our method complex and hard to decipher by intruders.

PDAC [2] approach use parallel method to reduce the time consumption of CPU and small cover text helped to cover larger amount of private data in small storage. But there were not any checking methods in PDAC. Thus, in case of any packet of data getting harmed then complete data have to be sent again. To remove this situation, we divide the PDAC in several steps and all the steps are processed parallelly for reducing time overhead.

In starting steps, we introduce the self-checking method on the basis of vowels and special characters. We check the place of vowels and characters in the data and make a bit arrangement according to it. Then, on the basis of the obtained bit pattern we divide the bits in group of four to obtain the decimal number. The obtained decimal number is then converted into the ASCII code after adding 60. This ASCII code forms the cover text for the private data. Thus, we don't need any extra space for checking our data on receiver side. Message that is sent has only secret message and the cover text. Thus, instead of having a random cover text as in PDAC, in PECT, we have an automated generated cover text. The reason for this is that, it would not increase the space. Thus, it uses the same amount of space as PDAC. After obtaining the cover text, PDAC approach is followed to obtain the encrypted data. The decimal number are obtained from the alpha-numeric cover text. First of all, the cover text is obtained from the encrypted data by every fifth character of the data. After that its decimal form is obtained. Then, 10 is added and subtracted from the four-digit obtained. The decimal numbers of cover text are themselves added and subtracted. Now, the obtained four digits have the exclusive OR operation with the private data. The obtained result is the enciphered text which is then encrypted with the key. the result is sent to the receiver by adding cover text to every fifth position. The process starts from generating random cover text of size 1/4th of our secret message. For less time overhead and hiding large amount of data within cover text we are using parallelism. After generating encipher message, it hides into same cover text message in such forms that one character of cover text with four characters of enciphered message which have encrypted using same cover text. This method is tricky in comparison to existing method of steganography so it would hard to calculate original text from enciphered text for casual reader. This approach using self-checking based on vowels, consonants and special characters with simple addition, subtraction and X-OR Operation for encryption of plain text and hiding into our random cover text so it's fast to compute. We required maximum n/4 bytes of cover text to hide n bytes of secret message.

3.1 Methodologies Used (PECT Method)

Figure 1 shows the sender side flow chart of the message. First of all, the secret message is imported and out of which we generate the 0–1 string according to the alphabets and characters, then cover text is obtained from string and 8-bit random key. After that, we obtain the stego key from the cover text by the mathematical calculation. Further the message is encrypted with stego key. Finally, the message is delivered to the receiver in encrypted form.

Figure 2 shows the receiver side flow chart. At the receiver end the message is decrypted and if there is any change in the pattern of the vowels and consonants then the message alert is send to the sender about the corrupted message.

Here is the detailed example of PECT approach using sample plain text "ARMOURER" which is to be encrypted and decrypted again.

A	R	M	O	U	R	E	R

Figure 3, describes the example of the sender side of the proposed approach (PECT). Figure 3(a) gives the cover text from the private data. Figure 3(b) is the encryption process of the private data. Figure 3(c) performs the X-ORing of the private data and the digits obtained from the encryption process. In Fig. 3(d), the values obtained from the previous step are converted into their respective ASCII values after this, we insert the cover text at every fifth position. The finally obtained message is transmitted to the receiver.

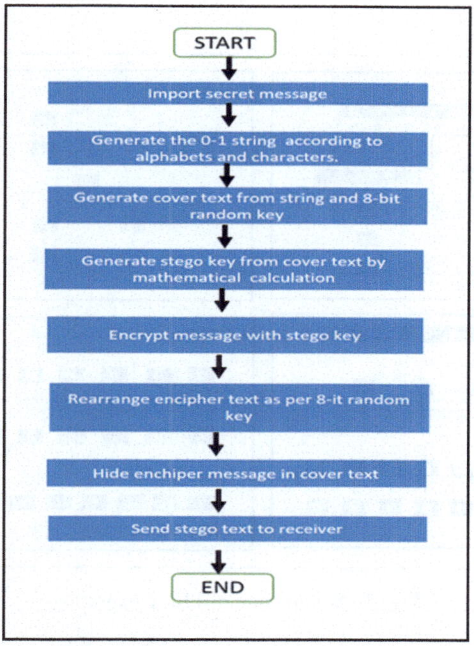

Fig. 1. Sender side flow chart of PECT approach

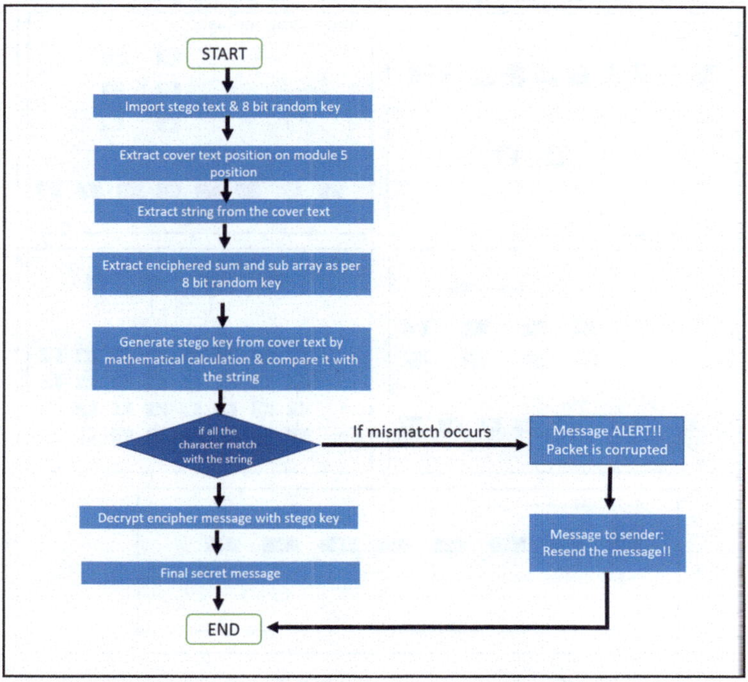

Fig. 2. Receiver side flow chart of PECT approach

Sender Side:

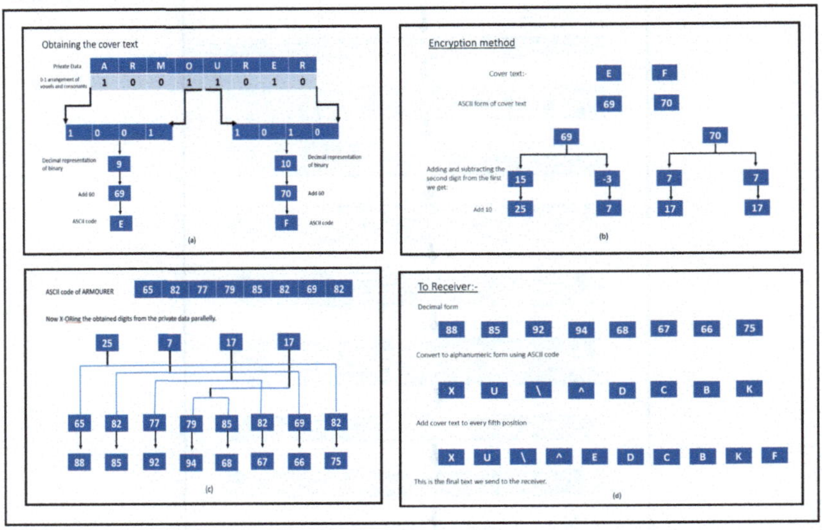

Fig. 3. Sender side of PECT example

Receiver Side:

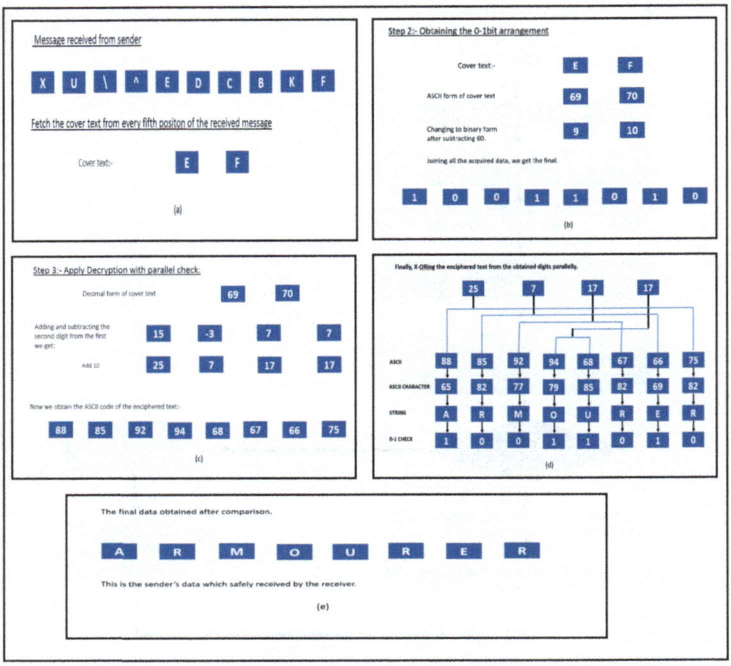

Fig. 4. Receiver side of PECT example

Figure 4, in continuation gives the decryption process of the private message received from the sender. In Fig. 4(a) cover text is fetched from the received message. Based on the cover text the 0-1bit pattern is obtained (Fig. 4(b)). In Fig. 4(c), we apply the decryption process to obtain the ASCII code of the enciphered text. Finally, in Fig. 4(d) we perform X-ORing of the enciphered text with the obtained digits. Parallelly we also carry out the 0–1 check in order to identify whether the received message is safe or not.

4 Results

The memory consumption as space complexity of our approach has not increased at all because we are using same size of message text and cover text similar to previous approaches, but time overhead has been reduced drastically. This can be used to keep a check on third party interventions as well as network security. We have compared our approach with the existing approaches and have found that memory consumption and time usage of others is more than our method. By simple addition, subtraction and comparison, we have been able to encrypt the data dynamically. Time complexity overhead of comparison is ½ of the total number of bytes in secret message. Due to parallelism as we compare first and last ASCII code of encrypted data together, the increased overhead time is n/2. But we can see that in case of large message this method is useful to detect any alteration in the message before completion of whole decryption. We can also ask the sender to send the data only for only the corrupted part. This will reduce the time and space complexity of message that is sent again.

Experiments show that our approach having following cases with different time overheads which was calculated through compilation.

Case 1- The overhead time is 6–8 ms, when the whole packet gets corrupted or the fault is found in start or end of the message.

Case 2- the overhead time is 12–15 ms, when there is a fault in middle of the message.

Case 3- the overhead time is 16–20 ms, when the message has no faults.

Hence the average overhead time in proposed PECT approach is 12–16 ms (Table 1).

Table 1. Comparison of various text steganography techniques

Text steganography techniques	Message text size (in bytes)	Cover text size (in bytes)	No. of bytes can hide	Time overhead (ms)
Feature coding [1]	600	1980	52	20,289
Inter word space [13]	600	1980	45	22,604
CASE [4]	600	1980	200	1,666
PDAC [2]	600	1980	600	15–20
PECT (Proposed approach)	600	1980	600	12–16

The Graph. 1 shows the comparison of various text steganography on the basis of cover text required to hide 200 bytes and 1000 bytes. Graph. 2 shows the time overhead of different text steganography approaches.

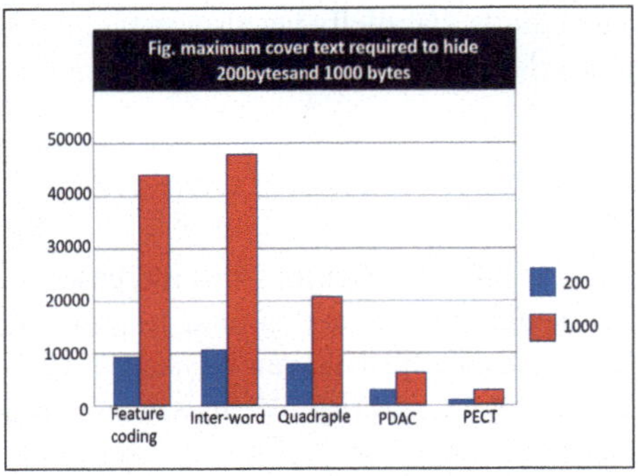

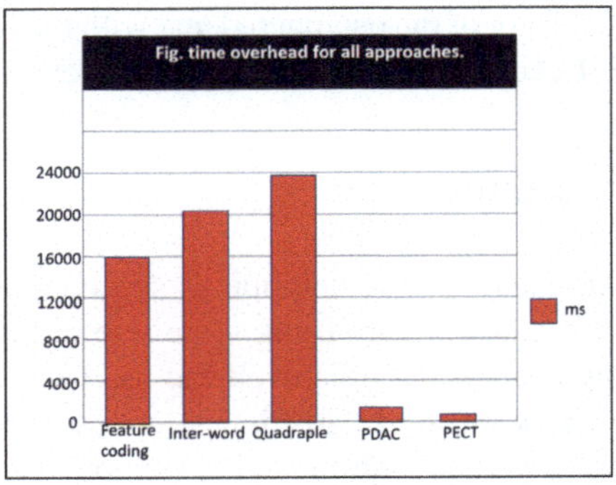

5 Conclusion

We conclude that PECT approach is highly secure as it consequently checks the characters of secret message obtained in the receiver side. The binary pattern used for the vowels and consonants is cross examined to obtain the same pattern during the decryption process. Also, after applying PECT, we have made it efficient for big data considerations. Now it can be used even on high traffic channels. As only the part that is defective is needed to be sent again by sender instead of full message. With the increase in technology and more use of digitalisation we would need steganography that would work on large data and that can be used on high traffic channels. In addition to this, storage requirement and time complexity of PECT is very less as compared to other text steganography methods.

References

1. Gaur, M., Sharma, M.: A new PDAC (Parallel Encryption with Digit Arithmetic of Cover Text) based text steganography approach for cloud data security. Int. J. Recent Innov. Trends Comput. Commun. **3**(3), 1344–1352 (2015)
2. Kataria, S., Singh, B., Kumar, T., Shekhawat, H.S.: PDAC (Parallel Encryption with Digit Arithmetic of Cover Text) Based Text Steganography (2013)
3. Chaudhary, S., Mathur, P., Kumar, T., Sharma, R.: A capital shape alphabet encoding (CASE) based text steganography. In: Conference on Advances in Communication and Control Systems, (CAC2S 2013), India (2013)
4. Kataria, S., Singh, B., Kumar, T., Shekhawat, H.S.: An efficient text steganography using digit arithmetic. In: Fourth International Joint Conference of Advances in Engineering and Technology in Elsevier Science and Technology, NCR India, December 2013, vol. 6, pp. 155–163 (2013)
5. Agarwal, M.: Text steganographic approaches: a comparison. Int. J. Netw. Secur. Appl. (IJNSA) **5**(1), 91–106 (2013)

6. Kataria, S., Singh, K., Kumar, T., Nehra, M.S.: ECR (encryption with cover text and reordering) based text steganography. In: 2013 IEEE Second International Conference on Image Information Processing (ICIIP), Shimla, December 2013, pp. 612–616 (2013)
7. Xiang, L., Sun, X., Luo, G., Gan, C.: Research on steganalysis for text steganography based on font format. In: Third International Symposium on Information Assurance and Security, IAS 2007, August 2007, pp 490–495 (2007)
8. Shirali-Shahreza, M., Shirali-Shahreza, M.H.: Text steganography in SMS. In: 2007 International Conference on Convergence Information Technology, November 2007, pp. 2260–2265 (2007)
9. Shirali-Shahreza, M.: Text steganography by changing words spelling. In: 2008 10th International Conference on Advanced Communication Technology, ICACT 2008, February 2008, vol. 3, pp. 1912–1913 (2008)
10. Shirali-Shahreza, M.H., Shirali-Shahreza, M.: A new synonym text steganography. In: 2008 International Conference on Intelligent Information Hiding and Multimedia Signal Processing, IIHMSP 2008, August 2008, pp. 1524–1526 (2008)
11. Khairullah, M.: A novel text steganography system using font color of the invisible characters in microsoft word documents. In: 2009 Second International Conference on Computer and Electrical Engineering, ICCEE 2009, December 2009, vol. 1, pp. 482–484 (2009)
12. Changder, S., Ghosh, D., Debnath, N.C.: Linguistic approach for text steganography through Indian text. In: 2010 Computer Technology and Development (ICCTD) (2010)

Intelligence in Station-of-Things

Soumya Saha[1(✉)], Tejashree Bhave[1], and Manisha Nene[2]

[1] Department of Applied Physics, Defence Institute of Advanced Technology, Pune 411 025, Maharashtra, India
soumyasaha.eie@gmail.com

[2] Department of Computer Science and Engineering, Defence Institute of Advanced Technology, Pune 411 025, Maharashtra, India

Abstract. The study in this paper proposes a notion to represent a *Station* containing a *Thing* of interest termed as *Station-of-Things*. The proliferation in communication, computation and sensors technology has enabled to build intelligent *station-of-things* environments. A *station-of-things* is pertinent to many scenarios related to occupancy of chairs in a classroom, seats in a theatre or auditorium, cars in parking-lot, books in a bookshelf, etc. The use of occupancy of *station* and size by a *thing* leads to extract the information useful to derive intelligence for the *station-of-things*. In this paper, the proposed communication model, network architecture and experimental results demonstrate the merit of the work which lays a foundation to estimate and evaluate the occupancy along with the data-types or parameters suitable to enrich machine learning models to derive artificial intelligence.

Keywords: Smart parking system · Intelligent parking system · Internet-of-Things · Infrared sensors · Ultrasound sensors · Artificial intelligence · Station-of-Things

1 Introduction: Station-of-Things

In real world, the humans always thrive to quantify the available resources around them. These resources are specific to the human needs; and are stationed for the use. The study in this paper proposes a notion to represent the state of a *Station* containing an object or a *Thing* of interest termed as *Station-of-Things*. A *station-of-things* is pertinent to many scenarios related to, occupancy of chairs in a classroom, seats in a theatre or auditorium, cars in parking-lot, books in a bookshelf, etc. Further, the use of occupancy of *station* and *size* by a *thing* leads to extract the information useful to derive intelligence for the *station-of-things*. The proliferation in communication, computation and sensors technology enable the ability to build such intelligent *station-of-things* environments.

1.1 Smart Environments and Station-of-Things

With the advent of technology, there is perpetual need to build smart environments around us. Smart parking systems being one such environment typically obtain information

N. Nain and S. K. Vipparthi (Eds.): ICIoTCT 2019, AISC 1122, pp. 314–320, 2020.
https://doi.org/10.1007/978-3-030-39875-0_33

about available parking spaces in an area and processes it in real-time to facilitate parking with the required information to ultimately optimize on utilization of space. Sensors are devices that detect the presence of an object and respond accordingly. Sensors play a critical role in the design and implementation of smart parking systems. Finding a parking space has always been challenging for the driver.

The study in this paper considers a representation of *Station-of-Things* where a parking-lot represents a *Station* and the cars represent the *Things*.

1.2 Related Work

Last two decades have been vigilant of the stupendous investigation to resolve the parking mess. Parking spaces are governed by the organization. Every organization needs parking and the rules and design of smart parking are based on the functionality of the organization and the infrastructure available. Authors [1–4] have introduced and deployed parking nodes at each parking space to conscious the presence or absence of a vehicle, while authors [5–7] have also proposed the monitoring of parking spaces collectively without the requirement of installation of parking node at each parking space. Sensors facilitate this consciousness which differs from one another in the mechanism to achieve consciousness, also each sensor has its own peculiarities and challenges, the proposed systems fail to address the issue of false positive detection of vehicles. The reporting and conversion of this conscious data into meaningful information is another area of contrast stated by the authors in [2, 5, 7–12].

1.3 Objective and Outline of the Paper

The objective of the study is to develop intelligence in *station-of-things* here; the scope of application is restricted to a parking-lot.

The paper is organized as follows: Sect. 2 describes the proposed *station-of-things* concept along with the methodology; Sect. 3 describes the implementation, platform, results and challenges; Sect. 4 concludes the work.

2 Proposed Station-of-Things

2.1 Assumptions and Network Architecture

In this paper, the *Station-of-Things* is assumed to be a parking space. The parking space for implementation is semi closed type; and has roof to provide shelter to the car with the following dimensions: 450 cm × 250 cm × 300 cm. It is the effective area to station the vehicle. Each parking space is Wi-Fi enabled and has a 220 V power supply to energize the nodes.

Ref. Fig. 1. the four-tier network architecture is setup, described as follows: tier 1 consists of sensors along with their communication protocol, tier 2 the sensor nodes along with the data processing aspects, tier 3 the base-stations with the data storage, gateway and tier-4 consists of the server , accessible globally.

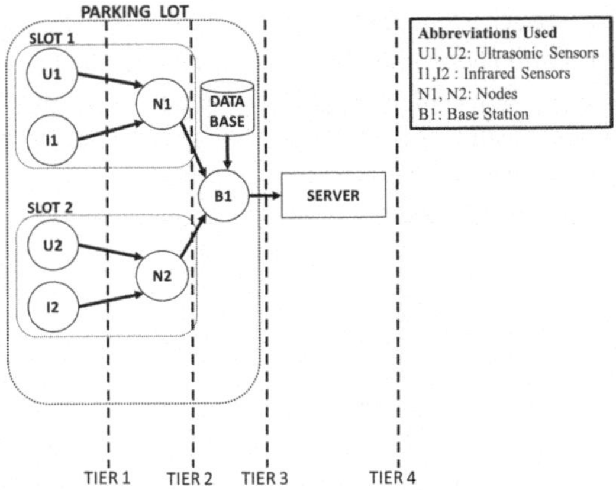

Fig. 1. Station-of-Things: proposed network architecture.

2.2 Data Flow and Algorithm

See (Fig. 2).

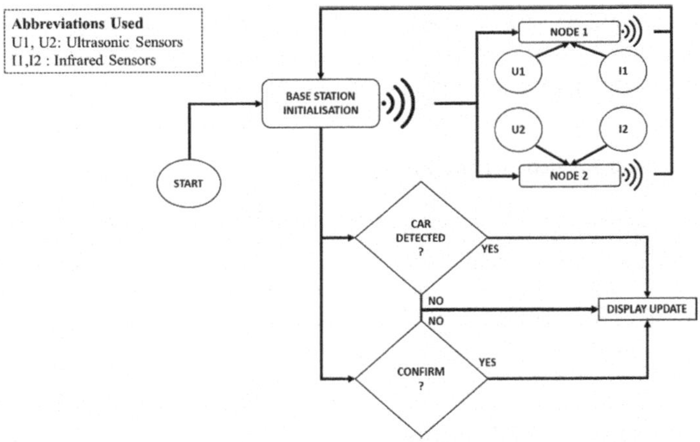

Fig. 2. Station-of-Things: data flow.

Ref. Fig. 3. the algorithmic steps are employed, with the following representations:

`IR_Sensor_Value` is the output of infrared sensor
`US_Sensor_Value` is the output of ultrasonic sensor
`IR_THRESHOLD` is the threshold set for the output of infrared sensor
`US_UP_THRESHOLD` and `US_DN_THRESHOLD` are the upper and lower threshold set for the output of ultrasonic sensor
`UP_Hatchback_Sedan` and `DN_Hatchback_Sedan` are the upper and lower threshold set for the prediction of Hatchback/Sedan type of car
`UP_Crossover` and `DN_Crossover` are the upper and lower threshold set for the prediction of Crossover type of car
`UP_SUV` and `DN_SUV` are the upper and lower threshold set for the prediction of SUV type of car

```
1.  Read IR_Sensor_Value
2.  Is IR_Sensor_Value lesser than IR_THRESHOLD?
3.  Yes, Read US_Sensor_Value.
4.  {
5.  Is US_Sensor_Value lesser than US_UP_THRESHOLD and
    greater than US_DN_THRESHOLD?
6.  Yes, Output "It's A Car"
7.  {
8.  Case 1: IR_Sensor_Value lesser than UP_Hatchback_Sedan
    and greater than DN_Hatchback_Sedan?
9.  Output "Hatchback/Sedan Type Car Detected"
10. Case 2: IR_Sensor_Value lesser than UP_Crossover and
    greater than DN_Crossover?
11. Output "Crossover Type Car Detected"
12. Case 3: IR_Sensor_Value lesser than UP_SUV and greater
    than DN_SUV?
13. Output "SUV Type Car Detected"
14. Default: Output "Unknown Car Type Detected"
15. }
16. No, Output "It's Not A Car"
17. }
```

Fig. 3. Station-of-Things: algorithmic steps.

3 Experiments and Results

This section describes the firmware used, the results and the challenges associated to the realization of *station-of-things*.

3.1 Firmware: Hardware and Software

Node MCU (Microcontroller Unit) [13] is a low cost ESP8266 Wi-Fi based single board microcontroller. It is represented as N1 and N2 in Fig. 1.

The HC-SR04 is an US (Ultrasonic) Sensor operates on the theory of SONAR (Sound Navigation and Ranging) and RADAR (Radio Detection and Ranging) to determine the distance to an object. It is represented as U1 and U2 in Fig. 1.

The SHARP GP2Y0A02YK [14] is an active IR (Infrared) Sensor operates on the theory of optical triangulation to determine the distance to an object. It is represented as I1 and I2 in Fig. 1.

The latest edition of the Raspberry Pi 3, the Model B+ is powered by a 1.4 Ghz quad-core processor and features dual band wireless LAN, an HDMI output and various USB inputs. It is represented as B1 in Fig. 1.

The 5 V/1.5 A power adapter is an AC to DC convertor, is used to energize the sensors and microcontroller board via 5 V/3.3 V dual supply module.

3.2 Logic Conversion

The HC-SR04 and SHARP GP2Y0A02YK requires 5 V supply to operate, however, the Node MCU cannot handle voltages greater than 3.3 V, and a potential divider circuit is built to safeguard the microcontroller board.

3.3 Sensor Calibration and Deployment

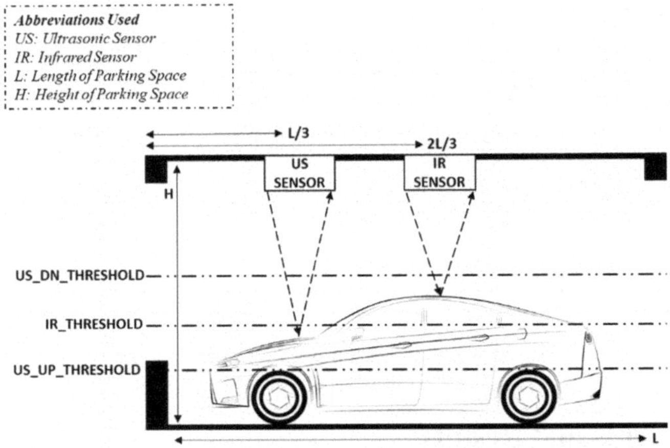

Fig. 4. Station-of-Things: deployment.

Table 1. Height prediction of the vehicles using *station-of-things*.

Car Type	Model	Height (in cm)
Hatchback/Sedan	Maruti Suzuki Alto 800	147.5
	Renault Kwid	147.8
	Maruti Suzuki Ciaz	148.5
	Hyundai Elantra	146.5
SUV	Honda CR-V	168.9
	Renault Duster	169.5
Crossover	Hyundai i20 Active	155.5
	Maruti Suzuki S-Cross	159.5

3.4 Results and Challenges

The proposed deployment model for *Station-of-Things*, Ref. Fig. 4 along with the occupancy detection of the *Station*, also enables the prediction, Ref. Table 1. The ability to estimate the dimension parameter demonstrates the merit of the work which lays a foundation to estimate the size-parameters of a *Thing* in a *Station*, Ref. Table 1.

Here, the remote measurement of the height of the car enabled to predict the car type. Eventually, such data parameters derived using sensor nodes at right position in the *Station* demonstrate the potential to predict the car type and car model. These sensor-node generated data feeds provide the basic data-types to enrich machine learning models and derive intelligence. The challenges associated here is the placement of sensors to measure the exact parameters, sensitivity factor of a sensor-node and environmental factors that affect the correctness in the readings.

4 Conclusion and Future Work

With the advent of Internet of Things (IoT) and Wireless Sensor Networks (WSN), the research on building the smart parking systems has evolved rapidly. Variety of smart parking systems exist. A smart parking solution also contributes to bring efficiency in human-efforts.

The study in this paper proposes the notion of *Station-of-Things*, here a parking-lot represents a *Station* and a vehicle represents a *Thing*. In this paper, the technique utilizes ultrasonic and infrared sensors. The proposed deployment model for *Station-of-Things*, Ref. Fig. 4 along occupancy of the *Station*, also enables the prediction, Ref. Table 1. The ability to estimate the dimension parameter demonstrates the merit of the work which lays a foundation to estimate the size-parameters of a *Things* in a *Station*.

The two sensors bring reliability in the process to identify the occupancy by a *thing* and detection accuracy. Further, the communication method, circuit/setup complexity and ability to prediction on the type of *things* enable data gathering to suit the generation of useful information. The experimental results demonstrate the merit of the work which lays a foundation to measure, estimate and evaluate the data-types or parameters that are suitable to enrich machine learning models to derive artificial intelligence.

References

1. Wang, H., He, W.: A reservation-based smart parking system. In: 2011 IEEE Conference on Computer Communications Workshops (INFOCOM WKSHPS), Shanghai, pp. 690–695 (2011)
2. Banerjee, S., Choudekar, P., Muju, M.K.: Real time car parking system using image processing. In: 2011 3rd International Conference on Electronics Computer Technology, Kanyakumari, pp. 99–103 (2011)
3. Khanna, A., Anand, R.: IoT based smart parking system. In: 2016 International Conference on Internet of Things and Applications (IOTA), Pune, pp. 266–270 (2016)
4. Lookmuang, R., Nambut, K., Usanavasin, S.: Smart parking using IoT technology. In: 2018 5th International Conference on Business and Industrial Research (ICBIR), Bangkok, pp. 1–6 (2018)
5. Al-Kharusi, H., Al-Bahadly, I.: Intelligent parking management system based on image processing. World J. Eng. Technol. **02**, 55–67 (2014)
6. Kumar, L., Khan, M.H., Umar, M.S.: Smart parking system using RFID and GSM technology. In: 2017 International Conference on Multimedia, Signal Processing and Communication Technologies (IMPACT), Aligarh, pp. 180–184 (2017)

7. Rane, S., Dubey, A., Parida, T.: Design of IoT based intelligent parking system using image processing algorithms. In: 2017 International Conference on Computing Methodologies and Communication (ICCMC), Erode, pp. 1049–1053 (2017)
8. ur Rehman, M., Shah, M.A.: A smart parking system to minimize searching time, fuel consumption and CO2 emission. In: 2017 23rd International Conference on Automation and Computing (ICAC), Huddersfield, pp. 1–6 (2017)
9. Abdulkader, O., Bamhdi, A.M., Thayananthan, V., Jambi, K., Alrasheedi, M.: A novel and secure smart parking management system (SPMS) based on integration of WSN, RFID, and IoT. In: 2018 15th Learning and Technology Conference (L&T), Jeddah, pp. 102–106 (2018)
10. Na, K., Kim, Y., Cha, H.: Acoustic sensor network-based parking lot surveillance system. In: Proceedings of the 6th European Conference on Wireless Sensor Networks, EWSN 2009, pp. 247–262. Springer, Heidelberg (2009)
11. Orrie, O., Silva, B., Hancke, G.P.: A wireless smart parking system. In: IECON 2015 – 41st Annual Conference of the IEEE Industrial Electronics Society, Yokohama, pp. 004110–004114 (2015)
12. Lee, C., Han, Y., Jeon, S., Seo, D., Jung, I.: Smart parking system for Internet of Things. In: 2016 IEEE International Conference on Consumer Electronics (ICCE), Las Vegas, NV, pp. 263–264 (2016)
13. NodeMcu – An open-source firmware based on ESP8266 wifi-soc. http://www.nodemcu.com/index_en.html
14. SHARP GP2Y0A02YK IR Sensor Datasheet

Mathematical Modeling and Numerical Simulation of a Double Touch-Mode Pressure Sensor with Graphene as the Sensing Element

Smiti Tripathy, Shiyona Dash, and Sumit Kumar Jindal(✉)

School of Electronics Engineering, Vellore Institute of Technology, Vellore, Tamil Nadu, India
sumitjindal08@gmail.com

Abstract. Touch mode capacitive pressure sensors have been a vital utilisation of Micro Electro Mechanical Systems (MEMS) offering better exhibitions on account of their efficiency, precision and utility. Graphene has been viewed as a promising material for MEMS because of its high elasticity, high tensile strength and tuneable elastic modulus. Graphene MEMS outperforms Silicon as Graphene can increase the sensitivity of the MEMS up to a great extent. This paper reports a Graphene and Aluminium nitride (AlN) based Double Touch Mode Capacitive Pressure Sensor (DTMCPS). A set of mathematical calculations has been presented for the sensor under study, so as to achieve high sensitivity. We used MATLAB in our paper to interpret theoretical evaluations in the form of graphical plots. This research aims at enhancing the performance of MEMS based DTMCPS so that they can be used for numerous industrial applications operating in a harsh environment.

Keywords: Double Touch Mode Capacitive Pressure Sensor · Graphene · Sensitivity

1 Introduction

MEMS based capacitive pressure sensors have a wide scope of uses because of their higher pressure sensitivity, lower power consumption, dependence on temperature is low and has robust structure. Basically, they consist of series of parallel plate capacitors with electrodes on top of the diaphragm made of various materials such as Graphene, Silicon, Silicon nitride and the bottom electrode forms on the substrate which is stationary and is of Graphene or silicon or other insulating material. The shape of the diaphragm is distorted in response to the pressure applied on it, thus affecting the capacitance of the sensor.

Touch mode capacitive pressure sensors have been overcoming various challenges, in recent times. They were designed to work in a specified range of pressure values in which the diaphragm makes contact with the substrate which is a thin insulating layer present between conductive portion of membrane and the conductive substrate. These MEMS based sensors find widespread application in automobile, industrial and commercial application fields. MEMS based capacitive pressure sensors have gained

N. Nain and S. K. Vipparthi (Eds.): ICIoTCT 2019, AISC 1122, pp. 321–329, 2020.
https://doi.org/10.1007/978-3-030-39875-0_34

wide importance in healthcare and biomedical field, as in case of blood pressure sensors and biomedical implants. Capacitive Blood pressure sensors are used to obtain accurate pressure of the blood flowing through artery and veins.

Graphene is a developing material in the field of MEMS. It's a two dimensional material, which is made up of thin layers of atoms of carbon, provides special properties such as high mechanical strength and thickness (in nanometres). As Graphene permits extensive elastic deformation, it has enormous potential in various mechanical structures such as cantilever beam resonator, micro-springs and suspended diaphragm [6]. In view of these alluring properties, Graphene has been constantly utilised as mechanical structure material in designing various pressure sensors. Graphene membrane structure has overcome the reliability problems that are faced in SiC and other Touch mode capacitive sensors made of various polymers. The deposition of SiC needs extremely high temperature that restricts the usage of on-chip integrated circuitry. In addition to it, the usage of sacrificial layers in the processing of Silicon carbide puts couple of confinements on the architecture and the material design of the device; materials are supposed to be resistant against aggressive etchants and should be steady to overcome the capillary force [2, 3].

Aluminium nitride (AlN) is used as an insulation layer. It was proposed due its compatibility, good thermal shock resistance, higher electrical resistivity, higher mechanical strength and non-toxicity. It has been found that AlN can remain stable at high temperatures and has a high melting point of about 2800 °C. Additionally, it is impervious against the attack from acids, strong soluble bases, liquid salts [1].

In research works, it has been proven that Double Touch Mode Capacitive Pressure sensor (DTMCPS) has more accuracy and efficiency [1, 4].

This paper mainly focuses on the basic structure of the DTMCPS of Graphene, mathematical calculation for evaluation of capacitance and sensitivity, interpretation of the mathematical calculations in the form of graphical plots.

2 Elementary Structure of DTMCPS

The proposed device has a diaphragm which is flexible and is in fact one of the electrodes. The second electrode is the underlying substance. The electrodes are coated with Aluminum Nitride which acts as an insulator in order to prevent the chance of short circuiting between the electrodes. The device utilizes Graphene as the material for the diaphragm. The advantages of using Graphene are: it has higher elasticity, high tensile strength and tunable elastic modulus. These properties lead to an improvement in the properties of capacitive sensors by increasing the pressure range and an increase in pressure-sensitivity.

The working principle of DTMCPS is that the pressure difference between the surrounding environment and the underlying cavity increases when an external pressure is applied on the diaphragm [1].

3 Working of DTMCPS

The proposed device works when there is an application of external pressure on the diaphragm which leads to the decrease in distance between the underlying substrate and the diaphragm. The two primary regions in which there are observable changes in the sensitivity of the capacitive pressure sensors are: Normal mode and Touch mode. Beyond this, the sensor enters into the saturation mode and the sensitivity then lowers to zero and the membrane is now fully compressed into the underlying cavity [5].

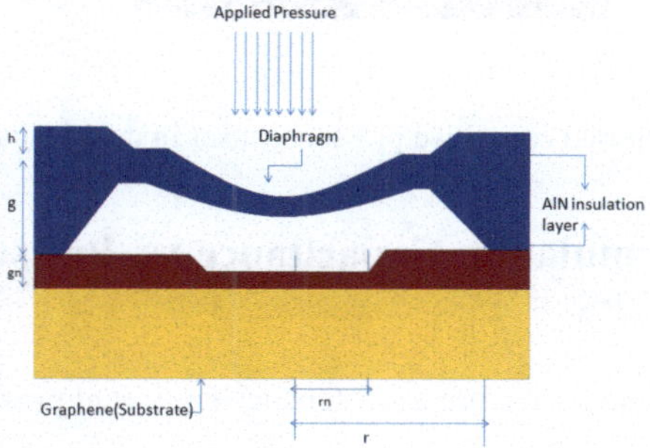

Fig. 1. Proposed DTMCPS (Graphene) device in normal mode.

Touch mode capacitive pressure sensor develops a closer linear capacitance to pressure (C-P) relationship [7]. The schematic figure of touch mode capacitive pressure sensor for normal and touch mode is as shown in Figs. 1 and 2 respectively. In Fig. 1, in the Normal mode, there is no contact between the diaphragm and the cavity. However as it can be seen in Fig. 2 after the applied pressure is increased till a certain value Touch Point Pressure 1 (TP1) the diaphragm touches the base of the substrate and the capacitance increases with the external pressure in a non linear fashion. This, in turns, puts the sensor transit into the touch mode. Above the transit-mode pressure range, the diaphragm makes a direct contact with the insulating layer. This region is typically the preferred region of operation in MEMS.

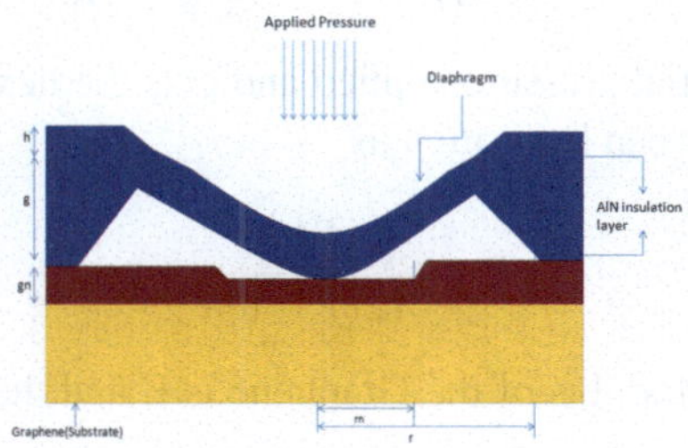

Fig. 2. Proposed DTMCPS (Graphene) device in touch mode at pressure TP1.

It can be observed in Fig. 3 that on increasing the pressure till the value Touch Point Pressure 2 (TP2), the diaphragm makes a contact with the first notch point of the substrate. In the DTMCPS, the pressure of the notch leads to a significant increase in the properties of the capacitive pressure sensor. It should be known that sensitivity of the sensor is inversely proportional to touch point pressure, thus the first touch point ensures high sensitivity at TP1 pressure while second touch point ensures high linearization of touch mode region in the proposed design [4].

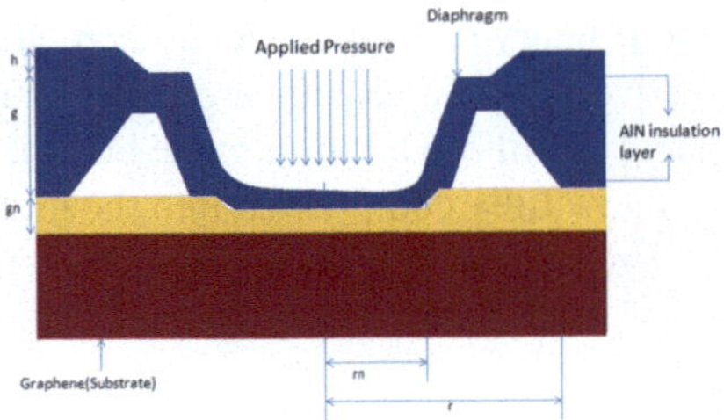

Fig. 3. Proposed capacitive pressure sensor in double touch mode.

4 Theoretical Formulation Capacitance vs. Pressure for Touch Mode in DTMCPS

The capacitance of the capacitive pressure sensor, if external pressure is applied, can be summarized as:

$$C = \int_0^{2\Pi} \int_0^{a} \frac{\varepsilon_o \varepsilon_i \varepsilon_a r\theta dr d\theta}{\varepsilon_a t + \varepsilon_i (g - \omega(r, \theta))} \tag{1}$$

Here, the permittivity of the vacuum is taken as ε_o, the dielectric constant of air is ε_a, the dielectric constant of the insulation layer is ε_i, the thickness of the insulation layer is considered to be t, the depth of the gap is g and the deflection of the diaphragm is taken as $\omega(r, \theta)$.

The deflection of the diaphragm (**ω**) due to the external pressure applied (P) at a distance r from the centre is

$$\omega(r) = \frac{P(R^2 - r^2)}{64D} = \omega_o \left[1 - \left(\frac{r}{R}\right)^2\right]^2 \tag{2}$$

Here, P is the differential pressure applied and D is the flexural rigidity.

The flexural rigidity D can be written as

$$D = \frac{Eh^3}{12(1 - v^2)} \tag{3}$$

Where, the Young's Modulus of the Graphene is E and the Poisson's ratio is taken as v.

For the proposed DTMCPS design, the total capacitance can be calculated as:

$$\mathrm{C} = \mathrm{C}_1 + \mathrm{C}_2 + \mathrm{C}_3 + \mathrm{C}_4 \tag{4}$$

The individual capacitances can be formulated as:

$$C_1 = \frac{\Pi \varepsilon_o \varepsilon_i r_{n1}^2}{t} \tag{5}$$

$$C_2 = \left(\frac{2\Pi \varepsilon_o \varepsilon_a \varepsilon_i}{(g + g_n)\varepsilon_i + t\varepsilon_a}\right)\left[(r_{n1})^2 k_1 + r_{n1} r_{n2} k_2\right] \tag{6}$$

Here $k_1 = \frac{1}{2}\frac{\arctan h(\sqrt{s})}{\sqrt{s}}$

$$k_2 = \frac{\arctan\left(\frac{\sqrt{s}}{\sqrt{\sqrt{s}-s}}\right)}{2\sqrt{\sqrt{s}-s}} + \frac{\arctan(\frac{\sqrt{s}}{\sqrt{\sqrt{s}+s}})}{2\sqrt{\sqrt{s}+s}} \quad (7)$$

$$C_3 = \frac{\Pi\varepsilon_o\varepsilon_i(z^2 - r_n^2)}{t} \quad (8)$$

$$C_4 = \left(\frac{2\Pi\varepsilon_o\varepsilon_a\varepsilon_i}{\varepsilon_i(g) + \varepsilon_a t}\right)\left[(z_2)^2 k_1' + z_1 z_2 k_2'\right] \quad (9)$$

Here $k_1' = \frac{1}{2}\frac{\arctan h\left(\sqrt{s'}\right)}{\sqrt{s'}}$

$$k_2' = \frac{\arctan\left(\frac{\sqrt{s'}}{\sqrt{\sqrt{s'}-s'}}\right)}{2\sqrt{\sqrt{s'}-s'}} + \frac{\arctan\left(\frac{\sqrt{s'}}{\sqrt{\sqrt{s'}+s'}}\right)}{2\sqrt{\sqrt{s'}+s'}} \quad (10)$$

5 Evaluation of Capacitive Sensitivity

The change in capacitance for a given change in the pressure applied to the diaphragm is termed as capacitance sensitivity.

$$\mathrm{S} = \frac{dC}{dP} \quad (11)$$

$$S = 2\left(\frac{\Pi\varepsilon_o\varepsilon_i r_{n1}^2}{t}\right)\left[1 + 2\left(\frac{k_1 - k_2}{(g + g_n)t + 1}\right)\frac{1}{r_n^2}\right](r_{n1})^2\frac{dr_{n2}}{dP} +$$
$$\left[2\left(\frac{k_2}{\left(\frac{(g+g_n)\varepsilon_i}{t}\right) + 1} - 1\right)\right]\frac{dr_n^2}{dP}$$
$$+ 2\left[\frac{\Pi\varepsilon_o\varepsilon_i R^2}{t}\right]\left[1 + 2\left(\frac{k_1' - k_2'}{gt + 1}\right)\frac{1}{R^2}\right]z_2\frac{dz_2}{dP}$$
$$+ \left[2\left(\frac{k_2'}{(\frac{g\varepsilon_i}{t}) + 1} - 1\right)\right]\frac{dz_2}{dP} \quad (12)$$

6 Structural Parameters and Design Specification of the Proposed Sensor

Parameters	Design values
Graphene's Young's modulus (E)	5×10^9 N/m^2
Thickness of Diaphragm/substrate (h)	5×10^{-6} m
Graphene's Poisson's ratio (υ)	0.149
Radius of the substrate/diaphragm (R)	180×10^{-6} m
Distance taken between the diaphragm at the 1st notch (g)	2×10^{-6} m
The 2nd notch height (g_n)	1×10^{-6} m
Permittivity of vacuum (ε_0)	8.854×10^{-12} F/m
Dielectric constant of Aluminium nitride (AlN)	9.14
Dielectric constant of cavity (ε_a)	1
Insulation layer thickness (t)	0.2×10^{-6} m
Range of pressure in normal mode	0.0–0.18 $\times 10^6$ Pa
Range of pressure in touch mode	0.2–2 $\times 10^6$ Pa

7 Results and Discussions

The paper focuses on the mathematical modeling of the proposed DTMCPS device which is made of Graphene. The numerical equations were derived in the Sect. 4 [1]. The improvement in the results obtained can be attributed to the various physical parameters of the substrate (Graphene) such as Young's Modulus, Poisson's ratio, and the structural parameters of the sensor such as: radius, thickness of the insulation layer, gap depth and radius of the second notch etc.

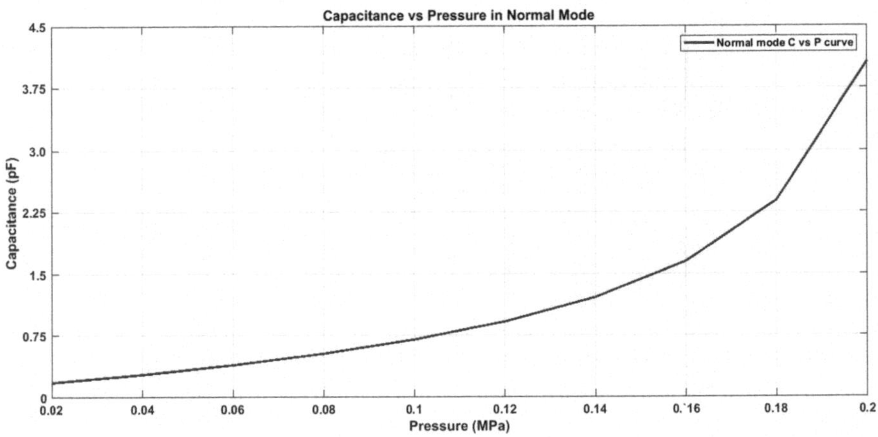

Fig. 4. Capacitance versus Pressure curve in normal mode in DTMCPS (Graphene) device

Figure 4 shows the graph of capacitance vs pressure in normal/non-touch mode, in the pressure range of 0–0.2 MPa and the maximum capacitance at 0.2 MPa is 3.75 pF. However after the pressure is substantially increased the device enters into touch mode. However there is transition mode when this change occurs the mode is usually ignored because it is non-linear in nature and creates noise [4].

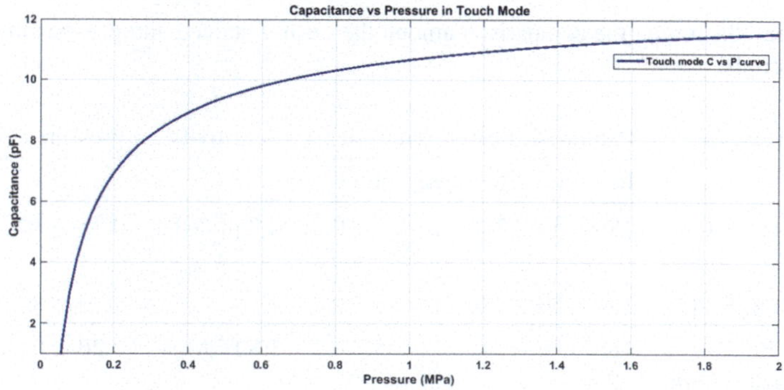

Fig. 5. Capacitance versus Pressure curve in the touch mode of DTMCPS (Graphene) mode

Figure 5 depicts the variation of capacitance with respect to the external applied pressure for the proposed DTMCPS Graphene device in touch mode. It can be observed that in the linear region of the touch mode, capacitance varies linearly with the pressure. However after additional pressure, the device enters into the saturation region of touch mode, there is minimal variation of capacitance and it ultimately reaches the peak capacitance level of 11 pF for an applied pressure of 2 MPa.

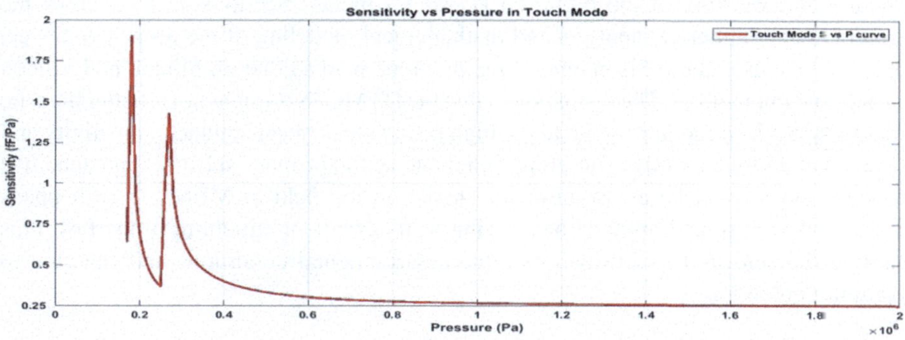

Fig. 6. Sensitivity versus Pressure curve in the touch mode of DTMCPS (Graphene) model

Figure 6 shows the variation of sensitivity with respect to the external applied pressure for the proposed DTMCPS Graphene device in touch mode. In the applied pressure range of 0–2 MPa, the peak value of sensitivity that is obtained is 1.8 fF/MPa. The sensitivity value obtained for the device is significantly better than the values obtained for other models.

The comparative values obtained for the two touch modes of operation and for the different substrates show that the notch present in the device has no significance in the normal mode of operation. However after the diaphragm membrane touches the notch and the membrane is fully compressed into the cavity; it is observed that, in the case of Graphene, high sensitivity and high linearity of touch mode operation are realized.

Table 1. Table illustrating the comparison among the result obtained and pre-existing works on different materials.

Type of sensor	Parametric design values (μm)						Sensitivity	Pressure (MPa)
	R	r_n	g	g_n	h	t		
DTMCPS (SiC) [1]	180	50	2.0	1	5	0.2	4.0 pF/MPa	0.25–1.80
DTMCPS (Si) [4]	180	50	2.0	1	5	0.2	–	0.30–1.80
DTMCPS (Graphene) Our research	180	50	2.0	1	5	0.2	1.8 fF/Mpa	0.20–2.0

8 Conclusion

The paper aimed to analyze a theoretical and mathematical model for DTMCPS having a Graphene-AlN-Graphene structure using MATLAB. Section 1 lists the advantages of using Graphene as a substrate for MEMs over the existing models made of Silicon and Silicon Carbide. Sections 2 and 3 provide a brief introduction to the elementary structure and working of the proposed DTMCPS model. Sections 4 and 5 show the theoretical and numerical analysis and mathematical modeling of the proposed device. Table 1 provides a comparison among the previous works done on Silicon and Silicon Carbide and our method. The results show that the DTMCPS (Graphene) is better than the previously realized models owing to the high peak values of capacitance, sensitivity and linear range in touch mode. Therefore Graphene as the primary substrate in capacitive pressure sensors will be a revolutionary change in the field of MEMs since it opens an arena of widespread applications, owing to its excellent structural properties, thus enabling the sensors to withstand extreme environmental conditions and function to maximum potential.

References

1. Jindal, S.K., Varma, M.A., Thukral, D.: Comprehensive assessment of MEMS double touch mode capacitive pressure sensor on utilisation of SiC film as primary sensing element: mathematical modelling and numerical simulation. Microelectron. J. **73**, 30–36 (2018)
2. Dai, C.-L., Lu, P.-W., Chang, C., Liu, C.-Y.: Sensors **9**, 10158–10170 (2009)
3. van Spengen, W.M.: Microelectron. Reliab. **43**, 1049–1060 (2003)

4. Jindal, S.K., Raghuwanshi, S.K.: Capacitance and sensitivity calculation of double touch mode capacitive pressure sensor: theoretical modeling and simulation. Microsyst. Technol. **23**(1), 135–142 (2016)
5. Berger, C., Phillips, R., Pasternak, I., Sobieski, J., Strupinski, W., Vijayaraghavan, A.: Touch-mode capacitive pressure sensor with Graphene-polymer heterostructure membrane 2D Mater. **5**, 015025 (2018)
6. Zhang, Y., Gui, Y., Meng, F., Gao, C., Hao, Y.: Design of a Graphene capacitive pressure sensor for ultra-low-pressure detection. In: 2016 IEEE 11th Annual International Conference on Nano/Micro Engineered and Molecular Systems (NEMS) (2016)
7. Eswaran, P., Malarvizhi, S.: MEMS capacitive pressure sensors: a review on recent development and prospective. Int. J. Eng. Technol. **5**, 2734–2746 (2013)

Quality Evaluation Model for Multimedia Internet of Things (MIoT) Applications: Challenges and Research Directions

Malaram Kumhar(✉), Gaurang Raval, and Vishal Parikh

Institute of Technology, Nirma University, Ahmedabad, India
{malaram.kumhar,gaurang.raval,vishalparikh}@nirmauni.ac.in

Abstract. Multimedia Internet of Things (MIoT) is uniquely addressable broad network of interconnected objects, it covers many types of different fields of utilization. There are several heterogeneous requirements related applications exists for the same purpose. The MIoT revolution is reforming modern healthcare system with promising technological, economic and social prospects. The extensive distribution of MIoT devices has made the evaluation for performance of MIoT applications very important aspect as relevant placements are present in every daily life activity. Managing the quality of MIoT applications would be natural consequence of the trends. However, evaluating the quality of MIoT applications is substantially diverse from calculating the quality of software systems. This paper attempts to review and understand the quality evaluation parameter to evaluate the overall quality of a MIoT application and challenges in developing good quality evaluation model.

Keywords: MIoT · Quality of Service(QoS) · Quality of Experience(QoE) · QoD · QoI

1 Introduction

Internet of Things(IoT) is uniquely addressable very broad network which comprises of inter-connected objects and with all types of objects, within standard communication protocols basis it covers many types of different fields of utilization [1]. There are several heterogeneous requirements related applications exists for the same purpose. The extensive distribution of IoT devices has made the quality evaluation for performance of IoT applications very important aspect as relevant placements are present in every daily life activity.

As a result of fast growth in communication technologies, an enormous type of networking technologies with different scales are used in our day to day life. The communication enabling technologies of IoT heavily depends on rapid technical innovation. The key enabling technologies are, RFID, Sensor Technology (nanotechnology), Wireless Communication, Near Field Communication (NFC) and Cloud Computing.

Due to deployment of wide range of networks, the idea of IoT can be broadly categorized in terms of network size as shown in Fig. 1.

N. Nain and S. K. Vipparthi (Eds.): ICIoTCT 2019, AISC 1122, pp. 330–336, 2020.
https://doi.org/10.1007/978-3-030-39875-0_35

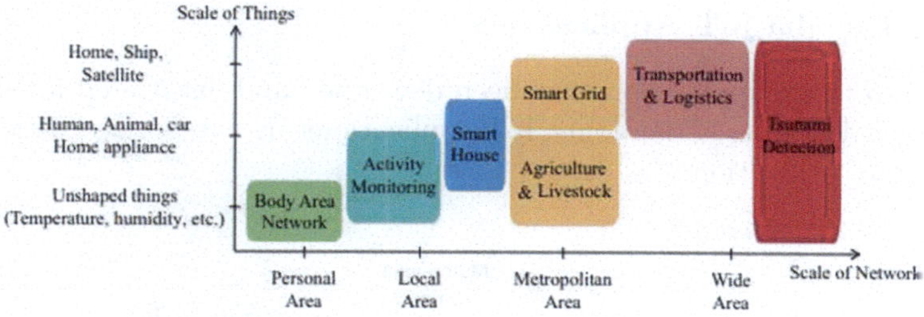

Fig. 1. IoT systems examples and classification [2]

2 Multimedia Internet of Things (MIoT)

IoT is expanding more and more into market with newest inventions of next generation smartphones, wearable devices, smart cars, different types of sensing devices, which introduces so many multi-media contents into the facility of everyday activities. The Internet of Things is mainly defined as very broad network which comprises of inter-connected objects where information can be gathered from physical environment and make it available on wide spread Internet medium. Multi-media IoT is defined as a collection of inter-connected objects in network which are capable to get multi-media data from the physical world. The multimedia devices are capable to obtain multimedia data from the surroundings, equipped with cameras and microphones. However, all of the layers in IoT are referred as MIoT, every layer has its own purpose where multimedia data are analyzed, processed, merged and stored into their respective layers. Presentation layer is also included because multimedia content is used for respective application development and control functions.

Following can be three different scenarios with respect to applications of the multi-media content in MIoT [3]:

- Multi-media as IoT input: acquired multi-media content inputs provided by multi-media devices (objects) and is in use for providing proper service which serves as an IoT application;
- Multi-media as IoT output: acquired signals and data with information such as scalar data that are represented as multi-media content (such as by animations using and generating graphs, etc.) by an IoT application.
- Multi-media as IoT input and output: acquired multi-media content by multi-media devices (objects), which are represented as multi-media fashion by an IoT design algorithm;

Multi-media content such as video, audio, images etc., require more memory resources, higher processing power and more bandwidth to process the multi-media information acquired by them. Bring together multi-media objects adopts a widespread numbers of usage in both military, commercial applications. However, some extra functionalities and the revision of existing ones is essential for increasing IoT systems with multimedia devices, as content is not simple which also involves taking to a specific subset of IoT, which we state as MIoT [4].

2.1 Multimedia IoT Applications

There are so many applications has been developed and discovered for different goals and service areas. Commonly, IoT applications are categorized with respect to the area of utilization.

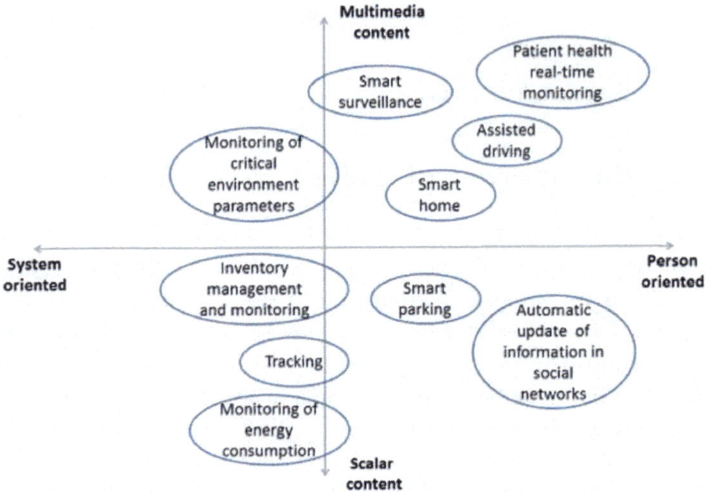

Fig. 2. Classification of IoT applications [3]

Figure 2, shows IoT Applications with respect to each IoT domain. The applications are categorized based on multimedia or scalar content (acquired content) they manage and the type of end-user environment which can be system or person oriented. To create specific tasks and make particular decisions, system-oriented IoT application automatically acquire, manage and control data and human interaction and evaluation are not essential for these applications. Since person-oriented IoT applications are the native-end-users who benefit from the information and content provided, by IoT application, so many people have a crucial role in the quality evaluation of IoT applications [3].

3 Quality Evaluation of MIoT Applications

Managing the quality of IoT applications would be natural consequence of the trends. However, evaluating the quality (QoS and QoE) of IoT applications is substantially diverse from calculating the quality of software systems. IoT applications are very mix blend of a variety in terms of technologies such as embedded, wireless networking communication, connectivity and various sensors. However, the numbers of IoT applications are quite enormous, so how to perform the quality assessment is very tedious in a fast growing situations. The measurement of the performance of IoT (in terms of usage) applications is becoming very essentials as number of related deployments are all over the place in our regular daily life activities.

Quality indicators for MIoT applications

- Quality of Service(QoS): QoS is the metrics measured usually to verify the quality of service provided by each IoT device. The QoS metrics measured are delay, packet loss, throughput and jitter. Being technology-centric, the QoS measurement is more suitable for Machine to Machine (M2M) communication [5].
- Quality of Experience(QoE): For multimedia systems and service, the QoE is well-defined as the degree of delight or annoyance of the user of an application or service [6]. QoE goal is to assess, how end-users personally perceive the quality of the application. Being user-oriented, the QoE offers a complete understanding of the systems impact parameters with respect to QoS approach.

There are very less works that have started addressing quality evaluation of IoT applications. Very limited studies have been conducted with reference to the QoE, which targets at evaluating how end-users subjectively perceive the quality of an application or a service. Factors influencing the QoE are grouped in three types, namely Human, System, and Context. Human factors consider the property and characteristic of a human user. System factors refer to characteristics and properties that determine the technically produced quality of an application or service (including the QoS). With respect to the user perception, the QoE provides better indications to what degree of the applications will be used by the users which can make real impact on the quality of human life.

Li et al. in [7] and Duan et al. in [8] focused on QoS evaluation defining an IoT system with three layers namely Sensing, Networking and Application layer. For every layer, resource allocation is done by monitoring module as a function of used QoS metrics. Information precision, sensing accuracy and energy utilization at the Sensing layer; bandwidth, throughput and delay at the Network layer; load, performance time, service performance cost and reliability at the Application layer.

Floris et al. in [3], a layered QoE management framework is proposed for measuring and monitoring the QoE of MIoT applications, where people are involved as the end-users of data. The layered approach was chosen since the goal was to evaluate different types of QoE influence factors in all the layers and then to combine these actions in order to maximize the final QoE perceived by the end-user. Each layer emphases on a particular QoE domain, so that the overall quality can be assessed as a combination of all domains.

A QoE layered framework is also proposed by Ikeda et al. in [9], where the large numbers of quality metrics are defined as physical metrics and organized into four layers: User interface, Device, Network and Computing, Metaphysical metrics are also introduced, which are defined as the quality metrics that users of IoT applications request. QoE of applications is modeled as a function of physical and metaphysical quality metrics in the IoT architecture.

Floris et al. in [10], a different approach for QoE prediction of IoT applications is proposed, which consists in the analysis of the data and information collected by the IoT application. The information and data that can be used to understand users QoE such as usage of data, user action and unsatisfactory service.

New Quality Indicators for the MIoT applications

The Multimedia IoT applications strongly depend on the data acquired by sensors and objects (which can also be multimedia objects) as well as on the huge amount of information acquired and processed to provide specific services. Therefore, two further quality indicators should be considered, i.e., the Quality of Data (QoD) and the Quality of Information (QoI).

The QoD consists of data accuracy, data completeness, data truthfulness, and up-to-date data [11]. The data accuracy reflects the correctness of data collected. The data completeness corresponds to the ratio of collected data amount to the amount of all required data. The truthfulness of data specifies the degree of reliability in data resource. The up-to-date data reflects the validity of data for decision-making, i.e., delay in data to assist decision making, it is useless. These four aspects jointly manage the overall QoD.

The QoI is defined as the body of tangible data available that can be used to make decisions about the fitness-of-use and utility of information products [12]. Floris et al. in [13], the QoI was introduced for measuring the obtained information from sensor networks, motivated by the significance that quality information plays in enhanced situation awareness, effective decision making and action taking.

Floris et al. in [3], Li et al. in [7] and Duan et al. in [8] have focused on QoS and QoE metrics to enhance the end users satisfaction for MIoT applications. Furthermore, in M2M communications QoE is not suitable as we need to connect the many objects without human involvement. Traditionally in QoS measurement only parameters such as packet loss, delay and jitter are taken into consideration, and ignore other factors such as device and environmental factors. Here, the idea of Quality of Things (QoT) came into picture for M2M communications in MIoT applications.

Karaadi et al. in [14], proposed the QoT model based on cross layered architecture. It emphases on the quality of data needs to be collected, processed and delivered between two or more devices. The purpose of the QoT is to meet the required quality for an IoT application. Thus meeting the minimum requirements of an IoT object on an IoT application, the system will avoid over provisioning of quality and hence, it reduce the bandwidth requirement.

4 Challenges and Research Directions

This section provides open issues and future research directions related to MIoT applications.

- Multimedia data play an important role in many of the IoT applications. In MIoT especially for M2M communication, QoE models are not appropriate, there is a need to develop good quality model for the relationships among new quality metrics and conventional QoS/QoE to evaluate the overall quality of MIoT applications.
- The processing power for MIoT multimedia devices is limited,therefore the operation of existing multimedia encoding schemes on these devices is infeasible. So, computationally less complex encoding techniques are required to minimize the network bandwidth.
- Multimedia sensing demands high processing and continuous data acquisition, which results in higher energy consumption. Thus,efficient energy harvesting procedures need to be invented to energize sensors and increase the network lifetime.
- MIoT devices are expected to be low-cost and tiny, possessing limited memory resources. Thus, efficient data acquisition methodologies should be developed to efficiently utilize memory resources.
- Conventionally, the multimedia devices are considered dumb and all the intelligence is kept on the control centers. However, embedding low level local intelligence on small multimedia devices can enable enormous application in automation and robust system operation.

5 Conclusion

In this paper, we have surveyed the need for evaluating the quality of MIoT applications. Present research on multimedia IoT to evaluate the quality of application is mainly focused on QoS and QoE metrics. However, in MIoT especially for M2M communication, QoE models are not appropriate. In this case other quality indicators are to be taken care of, such as QoD, QoI and QoT. In future MIoT systems, there is a need to develop good model for the relationships among new quality metrics and conventional QoS and QoE to evaluate the overall quality of MIoT applications.

References

1. Atzori, L., Iera, A., Morabito, G.: The internet of things: a survey. Comput. Netw. **54**(15), 2787–2805 (2010)
2. Kawamoto, Y., Nishiyama, H., Kato, N., Yoshimura, N., Yamamoto, S.: Internet of things (iot): present state and future prospects. IEICE TRANS. Inf. Syst. **97**(10), 2568–2575 (2014)
3. Floris, A., Atzori, L.: Managing the quality of experience in the multimedia internet of things: a layered-based approach. Sensors **16**(12), 2057 (2016)

4. Alvi, S.A., Afzal, B., Shah, G.A., Atzori, L., Mahmood, W.: Internet of multimedia things: vision and challenges. Ad Hoc Netw. **33**, 87–111 (2015)
5. Simiscuka, A.A., Bezbradica, M., Muntean, G.-M.: Performance analysis of the quality of service-aware networking scheme for smart internet of things gateways. In: 2017 13th International Wireless Communications and Mobile Computing Conference (IWCMC), pp. 1370–1374. IEEE (2017)
6. Brunnström, K., Beker, S.A., De Moor, K., Dooms, A., Egger, S., Garcia, M.-N., Hossfeld, T., Jumisko-Pyykkö, S., Keimel, C., Larabi, M.-C., et al.: Qualinet white paper on definitions of quality of experience (2013)
7. Li, L., Li, S., Zhao, S.: Qos-aware scheduling of services-oriented internet of things. IEEE Trans. Ind. Inf. **10**(2), 1497–1505 (2014)
8. Duan, R., Chen, X., Xing, T.: A QoS architecture for iot. In: 2011 International Conference on Internet of Things and 4th International Conference on Cyber, Physical and Social Computing, pp. 717–720. IEEE (2011)
9. Ikeda, Y., Kouno, S., Shiozu, A., Noritake, K.: A framework of scalable QoE modeling for application explosion in the internet of things. In: 2016 IEEE 3rd World Forum on Internet of Things (WF-IoT), pp. 425–429. IEEE (2016)
10. Floris, A., Atzori, L.: Towards the evaluation of quality of experience of internet of things applications. IEEE Internet Things J. (2017)
11. Wu, Q., Ding, G., Xu, Y., Feng, S., Du, Z., Wang, J., Long, K.: Cognitive internet of things: a new paradigm beyond connection. IEEE Internet Things J. **1**(2), 129–143 (2014)
12. Bisdikian, C., Kaplan, L.M., Srivastava, M.B., Thornley, D.J., Verma, D., Young, R.I.: Building principles for a quality of information specification for sensor information. In: 2009 12th International Conference on Information Fusion, pp. 1370–1377. IEEE (2009)
13. Floris, A., Atzori, L.: Quality of experience in the multimedia internet of things: definition and practical use-cases. In: 2015 IEEE International Conference on Communication Workshop (ICCW), pp. 1747–1752. IEEE (2015)
14. Karaadi, A., Sun, L., Mkwawa, I.-H.: Multimedia communications in internet of things QoT or QoE? In: 2017 IEEE International Conference on Internet of Things (iThings) and IEEE Green Computing and Communications (GreenCom), pp. 23–29. IEEE (2017)

A Study of Intrusion Detection System in Wireless Sensor Network

Atul Agarwal(✉) and Narottam Chand Kaushal

National Institute of Technology, Hamirpur, Hamirpur 177005, H.P., India
atul2841996@gmail.com, nar@nith.ac.in

Abstract. Wireless Sensor Networks are most widely used in networking because of their large application domain. These networks due to their nature and appealing features, like multihop wireless communication, deployment in a hostile unprotected environment, low installation cost, auto-configurable etc, are very prone to security. Sensors are small devices deployed in the unprotected region and are vulnerable to attacks. To collect information from the surroundings sensor node senses information and follows multi-hop communication and the data reaches to sink. In these networks, there is no monitoring of information flow, hence security is a big concern. To provide security in Wireless Sensor Network operations, all kinds of intrusions should be detected and appropriate action must be taken against them in order to ensure that there is no harm done to the sensor network. This paper presents a review of Intrusion Detection Systems in Wireless Sensor Networks. Out of several detection techniques, this paper focuses on signature-based, anomaly-based and hybrid-based techniques. Various detection models are examined based on certain parameters. This paper summarizes various Intrusion Detection algorithms with their features which are used in Wireless Sensor Networks.

Keywords: Wireless sensor network · Intrusion detection system · Misuse-based · Anomaly-based · Hybrid-based

1 Introduction

Wireless Sensor Network (WSN) has a number of sensor nodes. These sensor nodes are having the property of auto-configuration and self-organization. Network architecture follows decentralization and is distributed in nature. WSN has a number of application in the areas of environmental monitoring like air, water and soil, structural monitoring, human activity, and behavioral monitoring, military surveillance, asset tracking and so on [1]. The advantage of WSN nodes is their small size and these can be placed at some place where surveillance by a wired network or human is not possible. These small nodes are deployed in thousands in number to monitor temperature, pressure etc. These nodes are always prone to physical damage because these are deployed in the open and unprotected environment. Self-organizing and auto-configuration in nature, limited

N. Nain and S. K. Vipparthi (Eds.): ICIoTCT 2019, AISC 1122, pp. 337–343, 2020.
https://doi.org/10.1007/978-3-030-39875-0_36

battery, and bandwidth, distributed and decentralization, multihop communication, are some characteristics which may lead to expose this network to many security attacks in all layers of OSI model. In order to provide protection to wireless sensor networks, many solutions such as cryptography and secure routing, key exchange and authentication are proposed. These methods are used to provide security from outside attack up to some level but these cannot eliminate all security attacks [2]. To detect an inside attack Intrusion Detection System (IDS) is introduced which can deal with wide range attacks in WSN.

Security attack in wireless sensor network is either active or passive. In passive attack attacker generally, hide and collects data by tapping communication link but does not modify it. Traffic analysis, malfunctioning of a node, eavesdropping comes under passive attack. The active attack affects the operation of the network. An active attack can lead to termination or degradation in networking services. Denial-of-service (DoS, DDos), Network jamming, warmhole, blackhole, sinkhole attacks are grouped in active attacks [3]. Solutions to these attacks (active or passive) in any network involves Prevention, Detection, and Mitigation [4]. In prevention step, the technique provides defense against attack i.e. it prevents the attack. Detection step comes into action when an attacker has found a way to pass prevention technique. This involves to detect or identify the node which has been attacked (being aware of the presence of attack). Mitigation aims to remove the attack detected in the detection step by taking action against the compromised node.

Intrusion is any activity in a network which is not authorized which affects network either passively or actively. If prevention of such activity is not provided in the first line of defense in WSN security then Intrusion Detection System comes into play as the second line of defense. Intrusion detection is performed by the network members nodes by detecting any suspicious behavior. After detection, it cannot take action on it but it raises an alarm to the controller. IDS provides information to the controller about intruder information (identification of node or region, location, time and date of intrusion, an activity of intrusion, type of intrusion etc). This information is used in the third line of defense i.e. in mitigating the attack.

IDS functionality is defined by the detection method which are mainly of two types- Anomaly-based, Misuse-based. Anomaly-based technique tries to finds the deviation from normal behavior. To flag operation as an anomaly, the regular observation of system must be there to accommodate system changes. Misuse-based technique tries to detect previously known attack with high detection rate by comparing the new attack signature with known signatures.

This paper presents a survey of existing IDS in WSN. The paper is organized as follows: In Sect. 2, our main focus is to examine existing intrusion detection techniques in the wireless sensor network. Section 3, specifies the main characteristics and features of different IDS with their comparison. Section 4, concludes the paper.

2 Literature Review

Mainly there are two types of IDS- Rule-based IDS and Anomaly-Based IDS [5]. Rule-based IDS can detect previously known attacks with high precision because it takes help of built-in signatures. A rule-based technique was developed [6] which compares the incoming information with known information. But the problem was that signature-based IDS cannot detect a new attack because its signature is not present. Anomaly-based IDS detects intrusion by monitoring statistical behavioral, can detect the new attack as well as common attacks having more false positive rates i.e. normal packet declared as abnormal. One particular method of threshold was developed to detect the new attacks [7]. Also, there can be multiple anomaly attacks or some attacks which are consisting of both the attacks, can be detected using Hybrid IDS (HIDS) [8]. In HIDS, anomaly-based IDS is used as a filter and another one is used as the second level of IDS.

In [9], knowledge-based Intrusion Detection System is (KB-IDS) applied on cluster-based WSN in order to secure the network. Their method keeps a record of networks internal nodes behavior. They placed knowledge base at the base station and inference engines were used by cluster heads to access the knowledge base. Because of the continuous monitoring of nodes they were able to sense any potential threat and generate events against them to tackle the attack. This information goes to the base station through the inference engine. The cluster head concludes and reports back to the cluster head for the suspicious node.

In [10], HIDS on Cluster-based WSN (CWSN) is discussed. Cluster head collects the data from other sensor nodes in that cluster and aggregates it because CH has the higher capability. In this paper it they used 3 models - anomaly detection, misuse detection with decision-making model. First two models were able to detect intrusion from a large number of packets and then results from these two models were combined by decision-making model i.e. if an intrusion has occurred then this model will classify its type. This model then will inform to the administrator about the attack with details.

In [11], three different individual intrusion detection systems were proposed for the heterogeneous wireless sensor network. An IDS was designed each for the sink, cluster head (CH) and sensor node (SN). The capability of IDS was depending upon the attack on a particular node. For sink node, IDS has the learning capabilities which helps network to deal with unknown attacks. For cluster head node, a host-based IDS was proposed without learning capability. This helps the network to detect known attacks and avoid resource wasting, efficiently. To detect and update the class of attack it uses a feedback mechanism. A simple and fast misuse IDS was proposed for SNs.

In [12], an intrusion detection system is presented which is very effective low-power WSN. This IDS mechanism uses a Random Neural Network (RNN). The IDS detects any performance degradation anomaly attack. It does not require any dedicated hardware and its attack detection accuracy was found to be 97.23%. This model detects any anomaly by identifying any deviation of an event from previously learned normal network operations. This mechanism can also detect previously unknown attacks.

In [13], an intrusion detection system is discussed which uses Energy trust in WSN. This method specifically focuses on detecting hybrid DoS attack using effective node energy analysis. This method predicts energy consumption and correlates it with the security of nodes. The method assumes that a sensor node can monitor their own energy level and consumption. It also faces a problem when a node is being under attack by flooding then enemy may try to control the node which can show fake energy information.

In [14], KMP Pattern matching technique was used to detect an intruder. In this, after receiving a pattern, a pattern matching linear time algorithm is used. When features are matched, accordingly a rule is applied to the data packet. Then it calls a plug-in, which is used to find the intrusion. If there is no problem and received packet found to be with the correct pattern then it is forwarded to its neighbors. If a suspicious pattern is detected then it performs functions accordingly, and trace instruction to trace out the intruder.

In [15], IDS based on improved AdaBoost- Radial basis function in Support vector machine is discussed. This system can detect and resist against DoS attack effectively. This system uses RBF- SVM as AdaBoost classifier by training. The IABRBFSVM algorithm is proposed by using the influence of parameter "σ" and model training error "e" on AdaBoost weights. Significantly improves network performance and lifetime, with short computation time and higher detection rate.

In [16], the intrusion detection method is discussed for cluster WSN, which uses trust values and multi-agent framework functioning. Trust value calculation and accuracy are calculated by Mahalanobis distance. Reduction in false positive rate is done by calculating tolerance factor in trust value calculation. This system was scalable as it uses a multi-agent framework. The proposed system was fault tolerant and can detect multiple attacks at the same time with a high detection rate.

In [17], An anomaly-based IDS was used to detect new attacks. Then, in order to understand routing in WSN for intrusion detection, they found a set of features. These features can be applied to all protocols. This IDS was able to detect main attacks in WSN but active sinkhole attacks can be detected effectively. Also, this method consumes less power as it does not require communication between nodes.

In [18], an IDS for cluster-based WSN is discussed. This IDS has different detection frameworks for different levels. The first framework runs at IDS agent at a low level, which is a specification based protocol. The second framework runs at the head node of the cluster (CH, medium level), which is a binary classification protocol. At Base station (higher level) a voting mechanism is applied which is used by cluster head to monitor its cluster head neighbors. This system was able to detect blackhole, warmhole, flooding, and selective forwarding attacks. The detection rate was almost 100%. Time, energy consumed to detect was very low.

3 IDS - Features and Limitations

This section compares features & limitations of different IDS based on different parameters. Table 1 shows various intrusion detection techniques with their detection methods listed. Features and limitations of each IDS detection method is also given in table to compare and determine that some methods were best suited for specific kinds of attacks.

Table 1. Study of various intrusion detection systems

S.N	Proposed system	Method	Features	Limitations
1	Mehmood et al. [9]	Cluster-based IDS, uses knowledge base for storing patterns, inference engine	Traffic is monitored and any suspicious event generated by an attacker node is blocked by the CH	KB-IDS puts a load on a single node inside the cluster, faster battery drainage for cluster head
2	Saeed et al. [12]	Uses Random Neural Network, without any dedicated hardware	Very effective in low-power WSN, detects any performance degradation anomaly attack, can also detect previously unknown attacks	Computation time, energy consumption was high as compared to others at the cost of accuracy
3	Jianjian et al. [15]	Improved AdaBoost-Radial basis function in Support Vector Machine	Detects DoS attack efficiently, improves network performance, short computation, high detection rate	Only Focused on DoS attack, can;t detect multiple attacks
4	Jin et al. [16]	Uses trust values and multi-agent framework functioning, uses Mahalanobis distance	Reduction in false positive rate, scalable system, fault tolerant, can detect multiple attacks at the same time	Trust value calculation and accuracy are calculated by Mahalanobis distance
5	Meng et al. [19]	K-nearest neighbor (KNN) algorithm with filtering. Built with three components	Resolves issues of network packet overload, false alarm rate. Provides efficient signature matching	Expensive in terms of time for computation
6	Gowrisona et al. [20]	Uses Neural Network with KDD Cup99 data	An adaptive method with higher detection accuracy. Detects DoS attacks with enhanced rules by learning. Efficient computational complexity of O(n)	No knowledge management system, implementation has to be in cloud-based environment

4 Conclusion

Applications of wireless sensor networks are so wide that WSN is reaching in all areas from military to medical. In all applications, data plays an important role and hence it must be kept secure. Various possible attacks try to modify data or analyze it. An outsider attack can be detected or prevented using Cryptography and other security mechanisms. When the attacker resides inside the network, in this situation the node which is compromised and under attack can lead to various security problems. Using this node any data/ information passing through this node can be manipulated. So, it is important to detect these inside attacks and prevent any data loss. Intrusion detection systems are designed for this purpose.

In this paper, a detailed literature survey of the IDS is presented. Some of the IDS system able to detect known attack while some were able to detect unknown attacks too. Hybrid systems have also been designed in order to detect all kinds of security attacks in the wireless sensor network. Accuracy and high detection rate of some of these detection mechanisms were about 100% and hence low false positive rate. Some IDS can detect all the attacks, while on the other hand some of those were designed to detect specific attacks efficiently. Furthermore, features and applications of different detection methodologies have been compared. Conclusively, this paper concludes various types of IDS. In the process of analyzing different detection techniques, it is observed that anomaly-based detection methods normally use statistical algorithms whereas signature-based detection methods generally uses knowledge-based algorithms. Also, it can be inferred that hybrid systems are more efficient in terms of detection rate with low false positive rate as compared to the anomaly and signature-based techniques.

References

1. Akyildiz, I.F., Su, W., Sankarasubramaniam, Y., Cayirci, E.: Wireless sensor networks: a survey. Comput. Netw. **38**(4), 393–422 (2002)
2. Ping, Y., Xinghao, J., Yue, W., Ning, L.: Distributed intrusion detection for mobile ad hoc networks. J. Syst. Eng. Electron. **19**(4), 851–859 (2008)
3. Padmavathi, G., Shanmugapriya, M., et al.: A survey of attacks, security mechanisms and challenges in wireless sensor networks. arXiv preprint arXiv:0909.0576 (2009)
4. Fuchsberger, A.: Intrusion detection systems and intrusion prevention systems. Inf. Secur. Tech. Rep. **10**(3), 134–139 (2005)
5. Khan, S., Loo, J., Ud Din, Z.: Framework for intrusion detection in IEEE 802.11 wireless mesh networks. Int. Arab J. Inf. Technol. **7**(4), 435–440 (2010)
6. Jha, S., Hassan, M.: Building agents for rule-based intrusion detection system. Comput. Commun. **25**(15), 1366–1373 (2002)
7. Xie, M., Han, S., Tian, B., Parvin, S.: Anomaly detection in wireless sensor networks: a survey. J. Netw. Comput. Appl. **34**(4), 1302–1325 (2011)
8. Sedjelmaci, H., Feham, M.: Novel hybrid intrusion detection system for clustered wireless sensor network. arXiv preprint arXiv:1108.2656 (2011)

9. Mehmood, A., Khanan, A., Umar, M.M., Abdullah, S., Ariffin, K.A.Z., Song, H.: Secure knowledge and cluster-based intrusion detection mechanism for smart wireless sensor networks. IEEE Access **6**, 5688–5694 (2018)
10. Yan, K.Q., Wang, S.C., Liu, C.W.: A hybrid intrusion detection system of cluster-based wireless sensor networks. In: Proceedings of the International MultiConference of Engineers and Computer Scientists, vol. 1, pp. 18–20 (2009)
11. Wang, S.-S., Yan, K.-Q., Wang, S.-C., Liu, C.-W.: An integrated intrusion detection system for cluster-based wireless sensor networks. Exp. Syst. Appl. **38**(12), 15234–15243 (2011)
12. Saeed, A., Ahmadinia, A., Javed, A., Larijani, H.: Random neural network based intelligent intrusion detection for wireless sensor networks. Procedia Comput. Sci. **80**, 2372–2376 (2016)
13. Jinhui, X., Yang, T., Feiyue, Y., Leina, P., Juan, X., Yao, H.: Intrusion detection system for hybrid dos attacks using energy trust in wireless sensor networks. Procedia Comput. Sci. **131**, 1188–1195 (2018)
14. Kalnoor, G., Agarkhed, J.: Detection of intruder using kmp pattern matching technique in wireless sensor networks. Procedia Comput. Sci. **125**, 187–193 (2018)
15. Jianjian, D., Yang, T., Feiyue, Y.: A novel intrusion detection system based on IABRBFSVM for wireless sensor networks. Procedia Comput. Sci. **131**, 1113–1121 (2018)
16. Jin, X., Liang, J., Tong, W., Lei, L., Li, Z.: Multi-agent trust-based intrusion detection scheme for wireless sensor networks. Comput. Electr. Eng. **59**, 262–273 (2017)
17. Eik Loo, C., Yong Ng, M., Leckie, C., Palaniswami, M.: Intrusion detection for routing attacks in sensor networks. Int. J. Distrib. Sens. Netw. **2**(4), 313–332 (2006)
18. Sedjelmaci, H., Senouci, S.M., Feham, M.: An efficient intrusion detection framework in cluster-based wireless sensor networks. Secur. Commun. Netw. **6**(10), 1211–1224 (2013)
19. Meng, W., Li, W., Kwok, L.-F.: EFM: enhancing the performance of signature-based network intrusion detection systems using enhanced filter mechanism. Comput. Secur. **43**, 189–204 (2014)
20. Gowrison, G., Ramar, K., Muneeswaran, K., Revathi, T.: Minimal complexity attack classification intrusion detection system. Appl. Soft Comput. **13**(2), 921–927 (2013)

Energy Oriented Routing in Wireless Sensor Networks: Hardware Implementation

Dipanshu(✉) and A. Nagaraju(✉)

Department of Computer Science, Central University of Rajasthan, Ajmer 305801, Rajasthan, India
dipanshu.acad@gmail.com, nagaraju@curaj.ac.in

Abstract. In the recent Internet Technology world, sensor-oriented network domain has significant importance for real-world application to perform user's day to day activities. Energy efficiency in a sensor field is an extremely important characteristic for various on demand needs. The related literature on energy efficient path selection and message scheduling in a sensor-enabled wireless network is a predominant factor in the Internet of Things domain. The proposed work implements an energy-efficient routing scheme and provides design details for sensor nodes. The sensor nodes are divided into three types of nodes – the base station node, the cluster head node, and the normal sensor node. The base station node is responsible for setting up the network and the normal sensor node performs the various operations designed for the network while the cluster head node acts as a cluster head. A new simple application layer protocol is also presented. This application layer protocol provides a simple and powerful enough interface to the user. This scheme and architecture when employed produces a reliable and efficient network. The reliability and efficiency are measured and illustrated using real-time implementation. The real-time implementation is performed on custom-designed sensor nodes based on AVR Atmega8a and nRF24L01 transceiver chip. The Sensor Nodes run on a custom-designed firmware which provides basic and enough security. The custom-designed sensor nodes provide a reliable Wireless Sensor Network and can be deployed in real-world applications.

1 Introduction

The ubiquitous presence of Wireless Sensor Networks (WSN) demand new techniques for routing data reliably and efficiently, it is highly required for the ongoing demand of IoT applications. IoT and Wireless Sensor Networks (WSN) are among the most useful and essential technology in the evolution of smart city development. The implementation of these technologies is a very challenging task because of the discrepancies between real life hardware and implementation details on paper. The IoT infrastructure and WSNs are typically built on

N. Nain and S. K. Vipparthi (Eds.): ICIoTCT 2019, AISC 1122, pp. 344–357, 2020.
https://doi.org/10.1007/978-3-030-39875-0_37

numerous homogeneous small form factor computers. In case of IoT infrastructure, the individual computing devices are generally fully fledged computers running a typical operating system while in WSNs the individual computing unit is relatively a low power computationally limited device called a Wireless Sensor Node (SN). A Wireless Sensor Node (SN) is usually built using an 8-bit or 32-bit micro-controller, communications chip, power source, some sensors, and some external circuitry to provide extra features such as switching and actuation. Recent trends in miniaturization of chips and development and a healthy market of SoCs (System on Chips) will eventually lead to more sophisticated SNs running more sophisticated operating systems such as Linux.

The problem of providing power to SNs is one of the most researched topics in the research area of WSNs. The most straight forward way would be to provide power through wires but that would defeat the very purpose of WSNs. So the other way to provide power would be through batteries and this is what is usually done in the current trend. So the methods to provide power to SNs constitutes a very import design constraint. Instead of focusing on power sources it is relatively easy to focus on and improve and reduce power usage. An SN should usually consume around less than tens of mA of current.

The best way through which we can minimize the consumption of power is through improving the software and specifically improving the routing support. Good routing support in a particular deployment area would result in a better lifetime of the network. A routing protocol generally and in almost all cases uses the shortest path available to transmit the packets and doesn't care about the position, packet size and scheduling all in general. A routing algorithm proposed by Latif et al. [6] gives a good scenario and a routing protocol that takes into account the size of packets, position of the nodes and use a scheduling approach. The protocol consists of two algorithms. The first algorithm is named Shortest Path and Less number of Links (SPLL) and the second one is named Longer Hop (LH).

The first algorithm uses the criteria of the shortest distance in a hop and uses it to forward or route the packets in a network. The second algorithm uses the position of the nodes and schedules or sorts the packets according to their priority which in turn is based on the energy loss equation. The sensor nodes are generally divided into clusters using some specific techniques, like using the k-means algorithm on the location information of the sensor nodes. A cluster head (CH) is appointed in each cluster to collect the data from non-CH nodes. This ensures that the path on which the data travels is short as possible since only CH can communicate with the base station (BS). Both algorithms are combined to create a complete system for routing in wireless sensor networks.

To implement any theoretical work, the designer needs to have a sensor node which can be easily designed and modified according to the needs. Our aim is to ultimately propose a hardware design of sensor node and methods which can be used by researchers to test their ideas and further improve them.

1.1 Motivation

The motivation for the proposed work is from smart city development point of view in which the sensors are communicating to control the environment. So an efficient data transmission technique is very essential for better service delivery. The proposed work made an attempt to emulate the SPLL and LH routing mechanism using a custom-designed sensor node powered by AVR ATmega8a Microcontroller and Nordic semiconductor's nRF24L01+ transceiver. The hardware is chosen such that to maintain the system with high reliability, high efficiency, and high battery life which are the most obligatory and indispensable features for any network system.

1.2 Contributions of the Proposed Work

The main contributions of the proposed paper as follows.

- The proposed work is a real-time implementation of sensors for use in commercial and critical conditions.
- New open hardware design for wireless sensor nodes.
- Implementation of the routing algorithm in a small form factor sensor devices such as our custom designed sensor node.

Further, in this paper, we will showcase a new simple and straight forward application layer protocol for wireless sensor nodes.

2 Literature Survey

A **Wireless Sensor Network** is a network of devices called **sensor nodes** that are connected through some wireless means. The wireless connection is typically accomplished by using radio waves. Various proprietary and open radio communication protocols are present in the market. The networks typically work in the Industrial Scientific and Medical (ISM) bands and various lower frequency bands. Our implementation works in the 2.4–2.5 GHz range. A Sensor Node replies to requests and transmits, from and to a controlling device, called a **Base Station**. A Base Station is responsible for controlling the network, managing the data, management of Sensor Nodes and data requests. Base Station is typically a powerful computing device.

Routing Protocols are techniques, methods, algorithms that are employed in a network that guarantees and helps in transferring a data packet to its destination. They have acquired a prime interest in researches related to WSNs. Routing Protocols are generally categorized into two types, Network structure and Protocol Operation based. Network structure based protocols are employed in the lower level of the network while the Operation based protocols are designed for a higher level of the network. Routing protocols is a heavily researched topic but the implementation is never talked about.

Routing Table is a data structure that stores information about the address of destination and address of next hop. A routing protocol implementation uses a routing table to figure the next hop if the destination address is given. To populate a routing table is to generate data of addresses according to a certain algorithm. In generating the data is the only step in the process where any routing algorithm will be used. After the generation of address data, the program just needs to query the routing table.

A **Printed Circuit Board** (PCB) is a thin board onto which electrical circuits are put on and various components are soldered on. They are typically manufactured in a sophisticated environment in a computer controlled machinery. The computer reads a PCB describing file called a **Schematic**. The Schematic describes all the features, components position and dimensions, drill locations and various other things about the board. They are typically generated and designed using computer-aided design and drafting (CAD) software.

Farhan et al. [6] has proposed an energy efficient path selection and message scheduling mechanism in wireless sensor networks to improve the energy consumption in transmitting the data packets from source node to destination node. In the proposed work authors has suggested the shortest path with less number of links algorithm. The location of the message sender and the number of utilized processors are decided to improve the energy efficiency in the IoT. A system model is developed by authors who use the M/M/1 queuing analysis to showcase the effective cooperation of Shortest Path and Less Number of Links (SPLL) and LH towards energy efficiency.

Chang et al. [5] has proposed a novel tree-based power saving scheme is proposed to reduce the energy consumption and to extend the lifetime of the network, throughput, and transmission overhead. In the proposed work authors considered a mobile sink and some fixed sensor nodes. In the proposed model nodes are considered as non-CH Nodes, hop nodes, and CH nodes. Based on the location of the sensor nodes and distance between the sensor nodes and residual energy the structure will be defined and proposed scheme define routing structure based on the above information. Accounting for the sensing area, the number of sensor nodes and the movement of the sink node, a tree structure will be defined and multi-hop concept is used to create energy efficient architecture.

Mukherjee et al. [9] has proposed a scheme for integrating MANETs with WSN to improve the connectivity between ongoing number of objects which are going to share the information through the internet to perform various Internet of Things (IoT) applications. The most challenging part in the integration is various kind of devices, platforms and heterogeneous nodes grouped together to share information through wireless devices.

The description of hardware modules is necessary to move forward and discuss the proposed implementation.

A microcontroller is a small computation device that houses RAM, ROM and various communication protocols on the chip. The microcontroller market has various leading manufacturers which are Atmel, Microchip and STMicroelectronics. The choice of a microcontroller is entirely application dependent.

A microcontroller typically has an 8-bit or 32-bit architecture. The 8-bit architecture is well suited for various applications while the 32-bit architecture is suitable for tasks which require high computation power such as cryptography, data storage, data processing, and implementation of high-level network layers.

Atmel developed the AVR ATmega8, a cheap and powerful microcontroller that has been part of various projects throughout its lifetime. It supports free and open source programming tools and compilers. It is a low power device and can work on up to 16 MHz clock speed. The ATmega8 microcontroller is an Advanced RISC Architecture device with instruction executed in a single cycle. It has a decent memory profile. It has 8 KB of In-System Self-programmable Flash Program, 512 Bytes of ROM (EEPROM) and 1 KB internal SRAM. Its data retention is of about 100 years at room temperature. Various communication protocols such as USART, SPI, I2C. Watchdog timer and two 8-bit timers and one 16-bit timer are present. It has 23 programmable I/O pins out of which 8 are 10-bit ADC channels. It operates in the range 2.7–5.5 V and consumes about 3.6 mA and 0.5 A in active mode and power down mode respectively.

The radio transceiver chip, the nRF24L01+ by Nordic Semiconductor is a popular, powerful, easy to use and programmable radio wave communications chip. It is an ultra-low-power chip that works in 2.4 GHz ISM band. The device can be programmed through its SPI interface. The 29 programmable registers control the entire functioning of the chip. It supports automatic packet acknowledgment by using its Enhanced ShockBurst technology. It can receive data on six different address that makes it easy to implement the broadcasting feature of any Wireless Sensor Network scheme since we don't need to change the receiver address every time after and before broadcasting. It supports 250 kbps, 1 Mbps, and 2 Mbps on-air data rates. The operation of the device is typically done at 3.3V.

Lewis [8] and Al-Karaki et al. [3] gave an overview of wireless sensor networks in general and routing in wireless sensor networks. These two sources and the literature detailed here and in reference [1,2,7,10] guided us to design a good entry level implementation.

3 Proposed System Model and Architecture

Sensor nodes used for implementation are custom designed, housing Atmel AVR ATmega8a microcontroller [4] as the main processing unit and Nordic Semiconductor's nRF24L01+ [11] as the communication unit. Both devices are powerful enough for the implementation and use very low power and even lower on certain modes. The main firmware running on the microcontroller uses very minimal features and have very straightforward capabilities such as sending, receiving and routing the data, and sensing, actuating and processing is also included.

The sensor node can be run in three modes normal, cluster head and base station. In normal mode the sensor node behaves normally in the sense that it can receive, send, route, sense, actuate and process. In cluster head mode the sensor node behaves normally and also acts as a sensor node where each node of its cluster sends the data to. In base station mode the sensor node acts as a communication unit for a much powerful system. For example, in base station mode the sensor node together with a general purpose desktop computer can generate routing table and distribute it, ask for data, send commands to cluster heads or directly to other non-cluster head sensor nodes. In base station mode the sensor node requires a more powerful computation device because alone it cannot work.

The transceiver chip used can at maximum transmit at 0 dBm. The range of the device is environment and antenna depended. The custom designed hardware uses the on-chip meander line antenna. External antenna such dipole antenna can also be used. With the on-chip antenna, at 0 dBm, the indoor environment range advertised is around 100 m, the outdoor range is around 250 m. The outdoor range is significantly increased, up to 1 KM, when PA+LNA version (with dipole antenna) is used.

3.1 Hardware Specifications

The sensor node is custom designed and contains two main components, Atmel AVR ATmega8a, and Nordic nRF24L01+ RF Transceiver.

The modules are of two types, Type-1, and Type -2, and different on the basis of the nRF24L01+ module type. Type- 1 has SMD (Surface Mount Device) nRF24l01+ module while the Type-2 has generic through hole module. Their schematics are shown in Figs. 1 and 2.

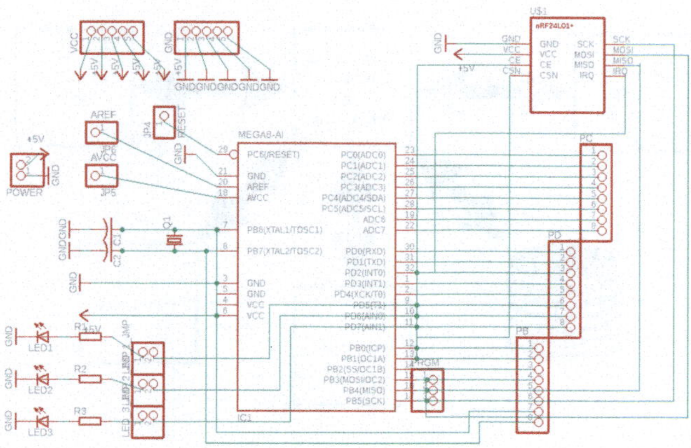

Fig. 1. Type 1 schematic.

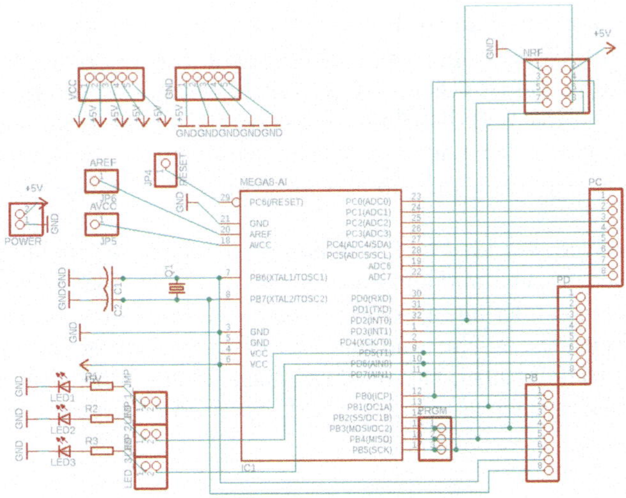

Fig. 2. Type 2 schematic.

3.2 Software Modules Developed

The software for the sensor node and base station are different. The sensor node software is written in C and the basic proposed architecture of the proposed model is given in Fig. 3.

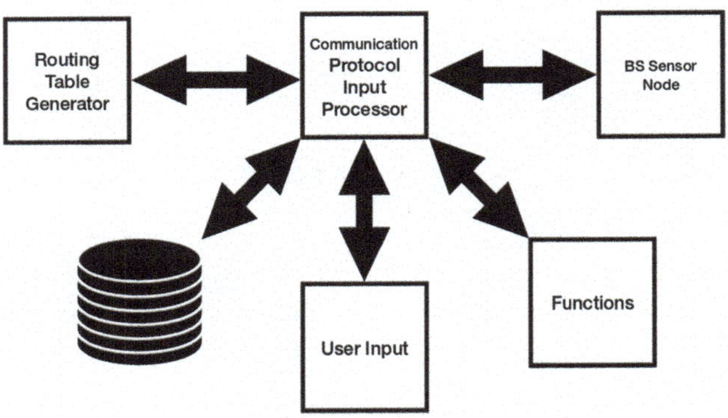

Fig. 3. Software architecture.

3.2.1 Routing Table

The routing table stores routing information related only to that particular node. Thus only require three columns and it is visualized in the Routing Table Generator section below.

3.2.2 Protocol Processor

The Protocol Processor (PP) queries the Routing Table (RT). It can also update, insert and delete table entries. By default, the RT can accommodate a maximum of 25 entries. The maximum entries size can be changed before compiling the firmware.

The PP is responsible for decoding the data packets and accordingly generating the reply packets. The PP can be modified by adding more packet types i.e. user can add more packet types that the Sensor Node can understand.

3.2.3 Communication Processor

The Communication Processor (CP) directly interacts with the communication chip (nRF24L01+). The CP is responsible for sending and receiving data packets. The CP can reject data packets if the checksum does not match. The CP, after successfully receiving data packets, transfers the data packet to the PP.

In sensor node software the incoming data is handled by the communication processor. The communication process further calls the protocol processor. The protocol processor identifies and decodes the packet and accordingly generates a reply packet using the routing table and calls CP with the reply packet as the argument. The communication processor then sends the packet to its destination.

3.2.4 Communication/Protocol/Input (CPI) Processor

In base station software either the data can come in through Base Station Sensor Node or User Input and then the Communication/Protocol/Input (CPI) Processor decodes the data and takes action accordingly. It can use Functions for various algorithms, Database for database functionality and Routing Table Generator for routing table generation. Each of these modules replies with appropriate data when they are called. The CPI processor then outputs appropriate information to the user or uses the Base Station Sensor Node to reply accordingly to the node.

3.2.5 Routing Table Generator

The Routing Table Generator or RTG is, as of now, capable of generating the routing table using the SPLL and LH routing scheme described in the Literature Survey Section of the paper. The routing table is visualized in the Fig. 4.

source	next_hop	destination
0xA000000006	0xA000000004	0xA000000003
0xA000000006	0xA000000004	0xA000000004
0xA000000007	0xA000000002	0xA000000005
0xA000000007	0xA000000004	0x0000000008

Fig. 4. An example of routing table generated by the Routing Table Generator after the application of the routing scheme.

3.2.6 Database

The database will store all the data coming from the network. The database of choice is SQLite and the choice is flexible and is user dependent provided that a proper plugin is developed so that the database can work with Base Station Software seamlessly.

3.2.7 User Input

This module interacts directly with the user and is responsible for data or command input/output. The way the user interacts with UIO, as of now, is through text commands i.e. it has Command Line Interface (CLI).

3.2.8 Functions

This module contains various predefined or user-defined program functions that can be used by the CPI P. The functions such as table traversal, lookup, etc. reside here.

3.2.9 Base Station Sensor Node

The Sensor Node connected to the High Computation Device (HCD) acts as a gateway between the sensor network and the HCD. This Sensor Node is not specially designed or has different firmware. In this implementation, it is connected to the HCD through RS-232 Serial Communication.

3.3 Application Layer Protocol

The application layer protocol is simple and straight forward. The packet is 32 bytes long. The example in the next paragraph explains the packet given in the Fig. 5.

5 Bytes source	5 Bytes destination	1 Bytes packet type	2 Bytes command
2 Bytes argument	15 Bytes data		2 Bytes checksum

Fig. 5. Application layer packet format.

ABCDEVWXYZRSN01000000000000000+0x73A1 'ABCDE' is *source*, 'VWXYZ' is *destination*, *R* is packet type, *SN* is command, *01* is argument, *000...000* is data and *0x73A1* is the checksum.

We found the packet size and packet parts to be sufficient for complete data transmission in WSN.

3.4 Network

As of now, the network supports only one request at a time in the network. The reason for this is that the transport layer is not implemented in this version of the implementation. Certain TDMA techniques can be used to get data from all the sensor nodes.

Phases of Operation

1. Initial Setup
   If the position of the nodes is static and known the data can be directly used as an input to the Base Station Program (BSP) and the Routing Table Generator (RTG) will generate the routing table and call the Communication/Protocol/Input Processor (CPI P) to populate the routing table in network.
2. Normal operation
   The user can get the data from any position or node by interacting with the User Input/Output (UIO).

The entire software suite is written in C and Python, and its architecture is given in the Fig. 3. The entire software suite can be found on https://github.com/c3rl/snfw (Fig. 6).

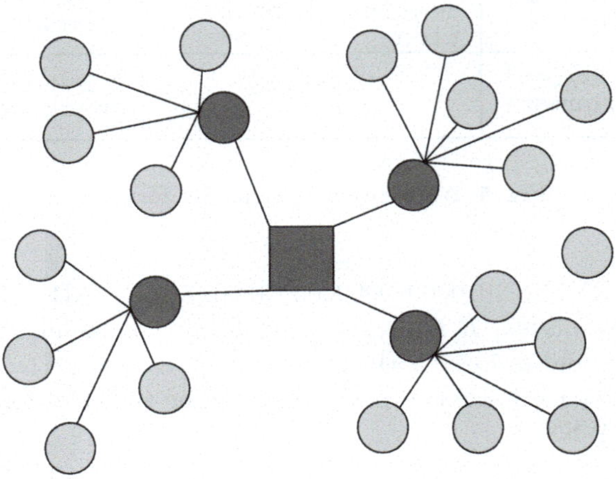

Fig. 6. An example cluster head based wireless sensor network.

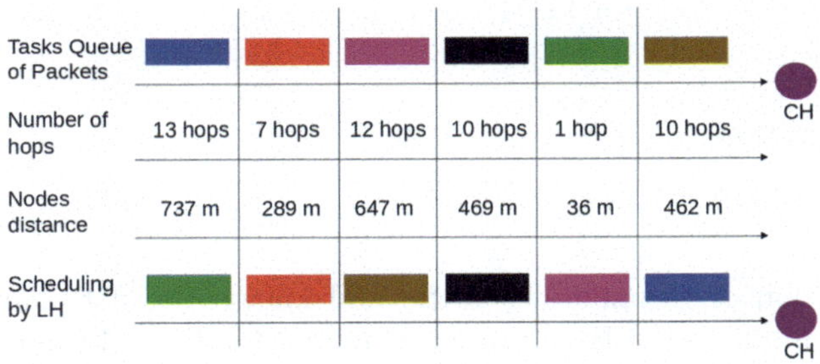

Fig. 7. LH scheduling.

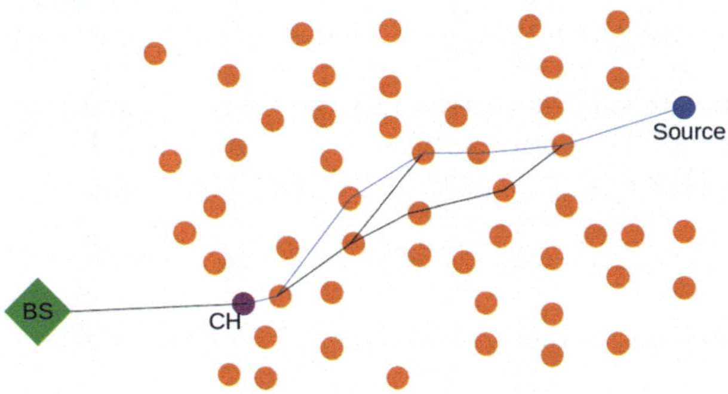

Fig. 8. SPLL scenario.

4 Methodology

Figure 1 depicts the typical working of the SPLL algorithm. This algorithm first calculates the shortest path available. If there are multiple shortest path available then it selects the one which has nodes which have the least number of links to other nodes. This idea enables the network to have the least number of paths which have common nodes thus reducing congestion and packet delay.

The LH algorithm has to do with the nodes situated at the border of the network. The LH algorithm is a scheduling algorithm. It schedules according to the hop distance from the Cluster Head. If two packets have the same distance then it takes into account the physical distance. If the physical distance is also same then, in our implementation, we just use the first in first out method. The packet which came first will be transmitted first. This algorithm is depicted nicely by the figure in Fig. 2.

5 Experimental Results

In the proposed work we have considered the network composed of our custom designed sensor devices, which can be of type 1 and type 2 and both of them can act either in normal mode or base station mode. The software modules are integrated with the sensor devices. The range of the data transmitted, speed and quality of received data and power consumption between type 1 and type 2 were the parameters chosen to be tested.

The reliability checks of the implemented hardware were performed using two nodes. Using only two nodes provides us with the range of data transmission, speed, quality of received data and power usage between the two nodes. The routing scheme was tested on a five nodes network.

Figures 7, 8 and 9 shows the packet drop in percentage with increasing distance between the two nodes at three different speeds - 250 Kbps, 1 Mbps, and

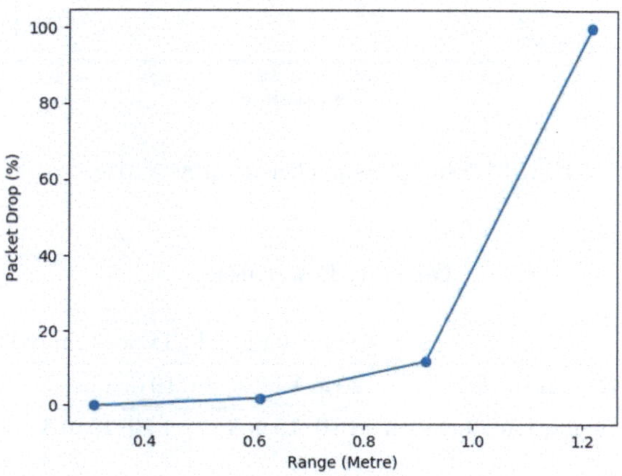

Fig. 9. Range vs packet drop at 250 Kbps

2 Mbps. Table 1 shows the power usage for model Type 1 and Type 2. While the power usage depends on the model the packet loss does not (Figs. 10 and 11).

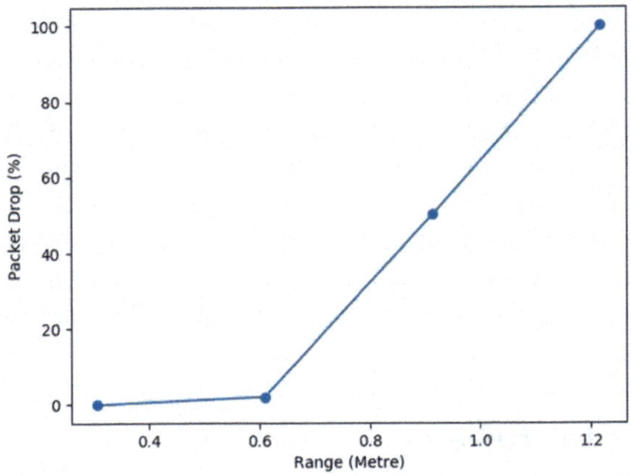

Fig. 10. Range vs packet drop at 1 Mbps

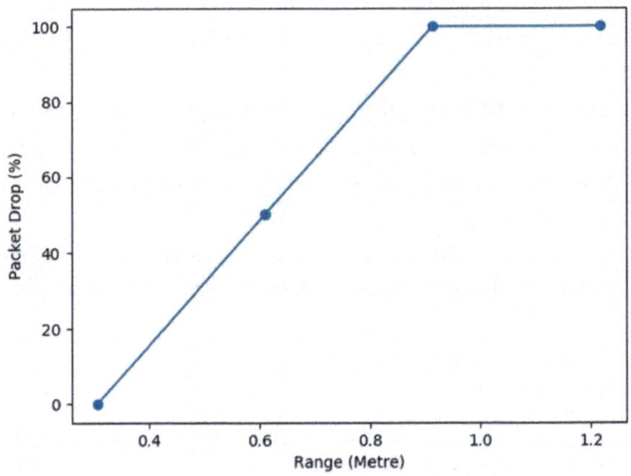

Fig. 11. Range vs packet drop at 2 Mbps

Table 1. Power usage

	Type 1 model	Type 2 model
Normal mode	20 mA	10 mA
During transmission	+10–15 mA	+10–15 mA

6 Conclusions and Future Scope of the Proposed Work

The proposed research work is a preliminary base framework for a practical scenario and tested on a five nodes network. In the future, there will be the scope of implementation for hardware and software point of view.

Due to the nature of the used radio transceiver, we got a very low and unusable range. This issue can be easily resolved using better radio chip.

The next version will have the following things implemented

- Hardware
  - Embedded Sensors
  - More power and efficient micro-controller and Transceiver
  - Onboard non-volatile memory
- Software
  - Transport Layer
  - More Routing Protocols
  - Automated Data Collection
  - More Energy Efficient Code
  - More Modularity

References

1. Akkaya, K., Younis, M.: A survey on routing protocols for wireless sensor networks. Ad Hoc Netw. **3**(3), 325–349 (2005)
2. Akyildiz, I.F., Su, W., Sankarasubramaniam, Y., Cayirci, E.: Wireless sensor networks: a survey. Comput. Netw. **38**(4), 393–422 (2002)
3. Al-Karaki, J.N., Kamal, A.E.: Routing techniques in wireless sensor networks: a survey. IEEE Wirel. Commun. **11**(6), 6–28 (2004)
4. Atmel. 8-bit avr microcontroller atmega8a datasheet complete
5. Chang, J.-Y., Shen, T.-H.: An efficient tree-based power saving scheme for wireless sensor networks with mobile sink. IEEE Sens. J. **16**(20), 7545–7557 (2016)
6. Farhan, L., Kharel, R., Kaiwartya, O., Hammoudeh, M., Adebisi, B.: Towards green computing for internet of things: energy oriented path and message scheduling approach. Sustain. Cities Soc. **38**, 195–204 (2018)
7. Karlof, C., Wagner, D.: Secure routing in wireless sensor networks: attacks and countermeasures. In: Proceedings of the First IEEE International Workshop on Sensor Network Protocols and Applications, 2003, pp. 113–127. IEEE (2003)
8. Lewis, F.L.: Wireless sensor networks. Smart environments: technologies, protocols, and applications. pp. 11–46 (2004)
9. Mukherjee, S., Biswas, G.P.: Networking for iot and applications using existing communication technology. Egypt. Inf. J. **19**(2), 107–127 (2018)
10. Perrig, A., Stankovic, J., Wagner, D.: Security in wireless sensor networks. Commun. ACM **47**(6), 53–57 (2004)
11. Nordic Semiconductor.: nrf24l01+ single chip 2.4ghz transceiver preliminary product specification v1.0 (2008)

Text Detection Using Maximally Stable External Regions and Stroke Width Variation

Nishant Singh, Vivek Kumar(✉), and Charul Bhatnagar(✉)

Computer Engineering and Application, Institute of Engineering and Technology, GLA University, Mathura, India
{nishant.singh,vivek.kumar,charul}@gla.ac.in

Abstract. There is a proverb "an image is worth, than ten thousand words". So it is very difficult to explain a single image. But what happens if a single image itself contains some information in the form of text. Though it is easy to extract text from the structured image, it is difficult to retrieve it from unstructured image. Thus in this paper, we are providing an efficient and concrete algorithm to solve this problem. This algorithm consists of detecting candidate text region using maximally stable external regions (MSER). Then it removes false region based on basic geometric properties. Now, again removing false region based on stroke width variation (SWV) and finally merging all text regions for detection of the result. At last, recognition of detected text with the help of optical character recognition (OCR). All these methods are combined to give high performance of the proposed algorithm.

Keywords: Text retrieval · Maximally stable external regions (MSER) · Geometric properties · Stroke width variation (SWV) · Optical character recognitions (OCR)

1 Introduction

Detecting text from an image is one of the most typical tasks nowadays and it becomes much more difficult if the images are captured from camera. In this paper text is retrieved not only from the normal images that are available from the dataset, but also from the images that are captured by the camera [1]. Retrieving text from a good quality image is somehow easier than retrieving the text from the captured images. So it is very complex to find out the text from the captured image [2]. The complexity of finding text from an image also depends on the type of images. We can classify the text image as structured images, unstructured images, also as scene images and embedded images [3]. Structured images are those images in which the text is clearly visible, easily differentiable from the background. It does not contain any noise and the font style, font size, font family are linear in all the text. It has prior known location of text in it. The unstructured images are those images in which the text is not clearly visible, not easily differentiable from the background. It contains various kind of noise in an image and the font style, font size, font family are nonlinear in the given text of an image, the location of the text

N. Nain and S. K. Vipparthi (Eds.): ICIoTCT 2019, AISC 1122, pp. 358–365, 2020.
https://doi.org/10.1007/978-3-030-39875-0_38

is not known priorly. In Fig. 1, Scene, structured and unstructured images are shown. Scene images are those images in which the text is present implicitly in the image and embedded images are those images in which the text is present explicitly in the image [4].

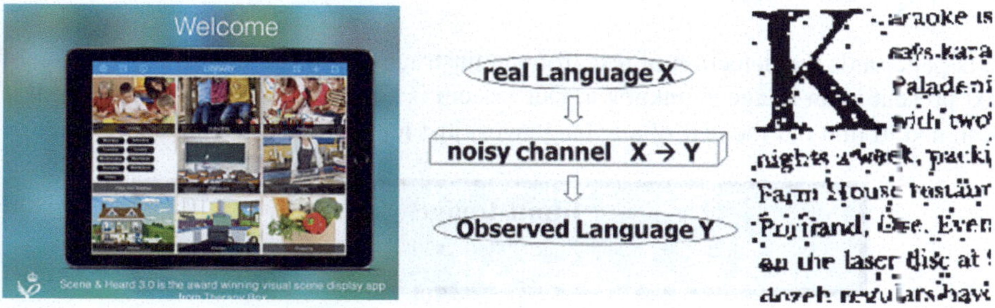

Fig. 1. Scene image, structured image and unstructured image

The latest technology is good enough to find out the text from clean background. So there is a need to design a system that is good enough to detect and extract the text from an image which includes rough and general background. Detection and extraction are two different terms used separately to perform its own task. Text Detection is basically used to retrieve the text from an image by using various methods like connected component, Fisher's discriminant rate (FDR) etc. Text Extraction is basically used for improvisation of the result which is retrieved from text detection methods. Techniques that are used for text extractions are wavelet transform, morphological component analysis algorithm, candy edge detector, etc. [5]. So here is our proposed algorithm that provides an efficient way to detect the text from an image and it also gives excellent results over captured images. In this paper, next sections consist of Related work, Proposed work, Results and Conclusion.

2 Related Work

Retrieving text from given image or video is one of the challenging tasks. In last few years, various methods have been proposed to detect and extract text from the given images. These methods have taken into consideration the various parameters such as color, intensity, edge, binarization, projection, reconstruction, connected component, etc. [6]. These parameters are used to distinguish the text region with the background region of the image.

Various methods have been used for detecting and extracting the text from an image. In Color clustering algorithm the pre step is to remove the noise from an image, if present. Now the input image is collected in terms of different color layers and gray components. This step is very important because the color of the text is generally different from the color present in the background. The text region is minimized with the help of connected components based on these layers. Another step aligning and then merging analysis (AMA) is used in which each row and column value is analyzed. This algorithm is very robust in locating mostly in English and Chinese languages. In texture based method, one feature based algorithm is used which construct the gray level co-occurrence matrix. It is basically used to determine the various features like homogeneity, dissimilarity, contrast, etc. [7].

3 Proposed Work

This paper aids the detection of text from an unstructured image means the location of text present in an image is unknown. Our system is generalized in it's orientation, it is designed to find out the text of any languages, just not restricted to English or Chinese

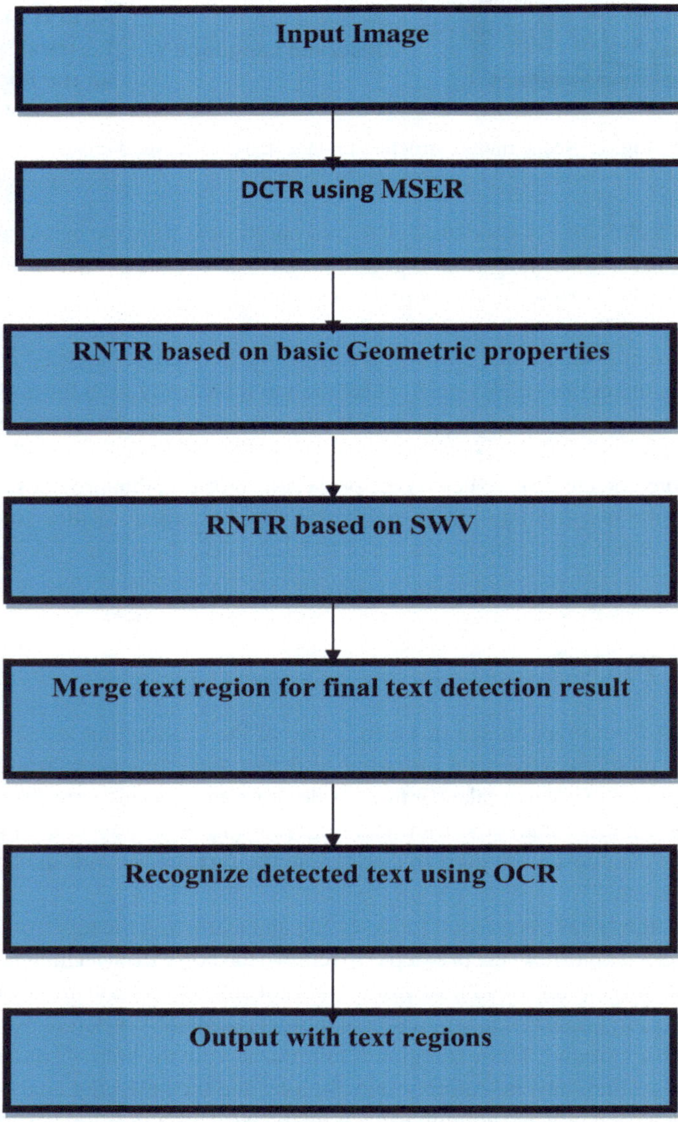

Fig. 2. Flow chart of algorithm

languages. The system is capable of detecting any text in horizontal, vertical, diagonal or in any other random direction and even if it is not uniform in their font size and family. The performance of the proposed work is booming as it performs both the task; first detecting the text based region from an input image and then from the resultant image removing the all non text based region.

The flow chart of the algorithm is shown in Fig. 2. In this algorithm the process is divided in five parts- Detect Candidate Text Regions (DCTR) using MSER Remove False region based on Stroke Width Variation (SWV), Remove Non Text Regions (RNTR) and Remove False region using basic geometric properties, Finally the text regions are to be merged for results and detected text is to be recognised using OCR [8].

3.1 Maximally Stable External Regions (MSER)

It is a feature detector, similar to the Scale Invariant Feature Transform (SIFT) detector, that extracts a number of co-variant regions [9]. It is a stable connected component of few level sets of an image. It works well for finding the text regions, because the same color and high contrast of texts give a stable intensity profile. Along with text regions, there are many false regions like non text region detected in the text. So as to remove all these non text regions we follow the next step. In a given input image the system finds the MSER regions along with non text regions as shown in Fig. 3.

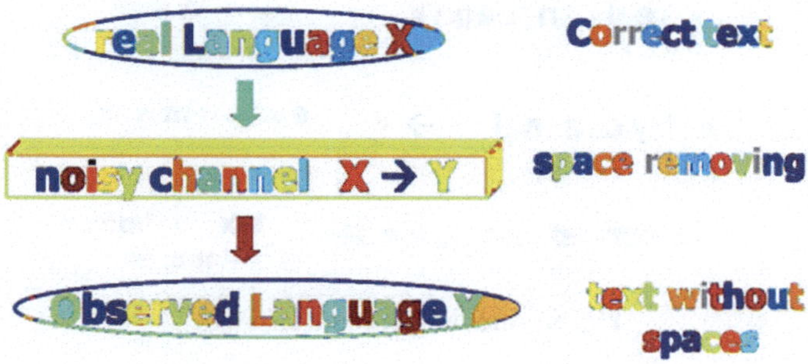

Fig. 3. MSER regions

3.2 Remove False/Non Text Region Based on Basic Geometric Properties

As seen in the above steps where in MSER method is used, it detects text and non text regions both. Thus, we are using a basic approach to remove non text based regions [10]. The geometric properties of texts that are good to differentiate between text region and non text region, are used to differentiate false regions using simple threshold. Some geometric properties are solidity, aspect ratio, extend, euler number, eccentricity etc. [11]. In this step regionprops method is applied to remove the non text region [12]. The output of applying RNTR based on basic geometric properties is shown in Fig. 4.

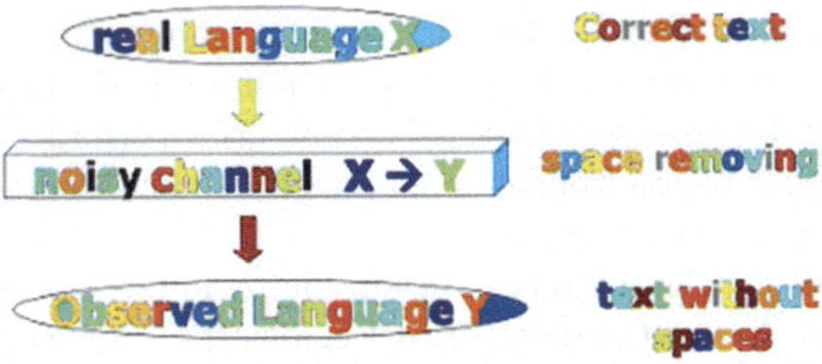

Fig. 4. After RNTR based on geometric properties

3.3 Remove False Region Based on Stroke Width Variation

To improvise result after applying geometric properties, we use SWV to remove non text region from the output image of step B. SWV is basically a measurement of the width of edges, curves and the lines that makes a character. Text regions have a small stroke width variation, whereas false and non text regions have a large variation [13]. For understanding how this stroke width variation works to remove the false regions, we need to estimate the width of one of the detected MSER regions. Here we are doing this by using a binary thinning and distance transform operation. The output of applying SWV to remove non text regions in an image is shown in Fig. 5.

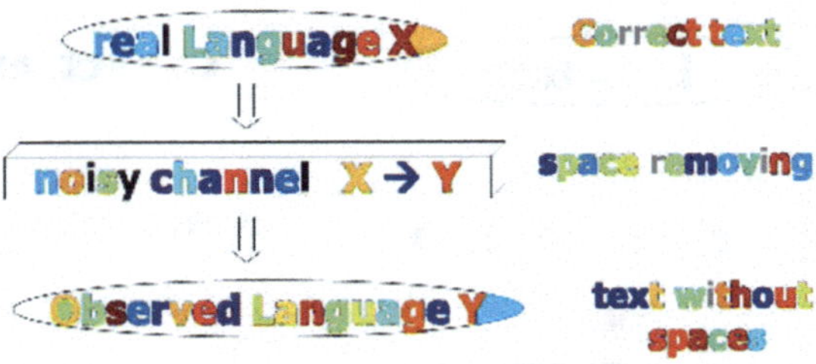

Fig. 5. After RNTR based on SWV

3.4 Finally Merging the Text Region for Result

In this step, all the detected text is composed of individual characters. To convert it into final text, all the detected characters are merged into words and then into text lines. It enables extraction of the final words in an image, that carry meaningful information instead of just single character. The algorithm which we are using to merge the individual text into words and lines is to find text region and then making a bounding box around these regions. To find the neighboring regions, expands the bounding box which is computed earlier with the help of regionprops. It makes the bounding box of neighboring text regions overlap the text regions that comes under the same word or line forming a chain of overlapping bounding box. The output is shown in Fig. 6.

Finally, all the overlapping bounded boxes are merged into a single bounded box. We do this by computing the overlap ratio between all bounding box pairs. It calculates

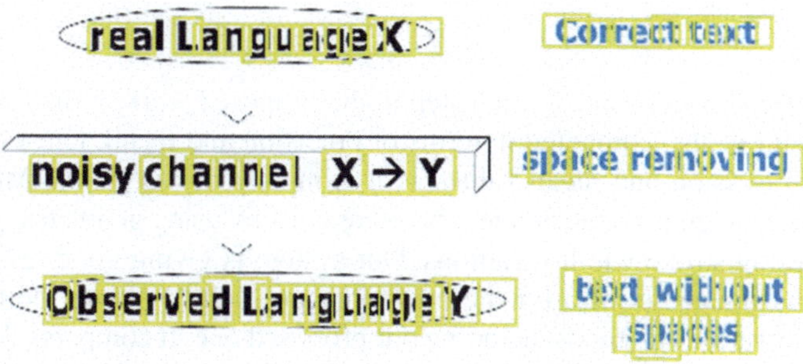

Fig. 6. Expanded bounding box text

the distance between every pairs of text regions. It is done in order to find groups of neighboring text regions by finding non zero overlap ratios. Once the Pairwise overlaps ratio is computed, a graph is used to detect text regions "connected" by a non zero overlap ratio. The final output after combining all text regions into one text regions, is shown in Fig. 7. Before showing the final result, the system removes non text detection by removing various bounding boxes made up of one text region, which removes the isolated regions. The output of this is shown in Fig. 7.

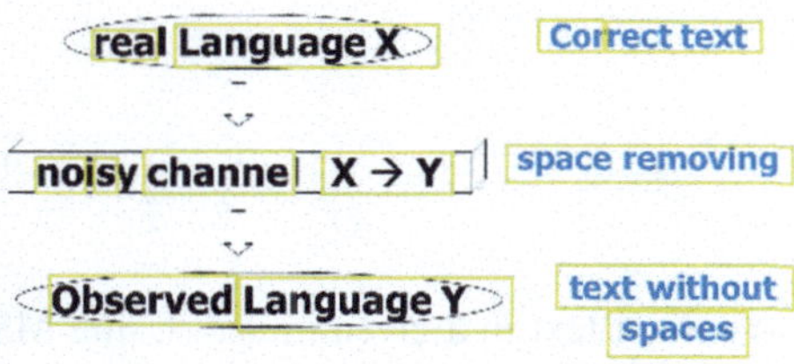

Fig. 7. Detected text

After detecting the final text region, the system uses the OCR to recognize the text within every bounding box. The output of OCR would be less noisy as shown in Fig. 8.

```
ans =

    'flgfl

     fig:-ved

     channel

     /Languafi

     Lea

     space removing

     Coz

     rect text

     text without

     spaces

     '
```

Fig. 8. Final text after OCR

4 Result

Though we have shown results at each step in the proposed work in the form of output image, the result is also computed in terms of precision and recall. Our system results contain good precision and recall values as it is mainly driven by the MSER feature detector and it also removes the non text based regions by using geometric properties or specially by using stroke width variations. Our system is giving good results not only in good quality image but also for the noisy images. Table 1 shows the text detector performance of various methods including our proposed one. It compares the method of Ofek [14], Thome [15] and Cord [4] in terms of precision and recall and shows that our approach has better results.

Table 1. Performance of text detection algorithm

Algorithm	Precision	Recall
Our proposed	.82	.68
Ofek [13]	.73	.60
Thome [14]	.63	.61
Cord [15]	.46	.39

5 Conclusion

This paper shows how to extract text in a given image. Using MSER method, it detect candidate text region and then it explains the use of geometric features followed by stroke width variation to remove all non text based regions. To get the final result of detecting text, it merges all the text regions and uses OCR to recognize the text individually. This paper not only detects the text region from an image, but also remove the non text based region which improves the quality of the system. Thus, our system is very well designed to detect the text even from the unstructured image.

References

1. Tsai, S., Chen, D., Chandrasekhar, V., Takacs, G., Cheung, N.M., Vedantham, R., Grzeszczuk, R., Girod, B.: Mobile product recognition. In: Proceedings of ACM Multimedia 2010 (2010)
2. Liang, J., Doermann, D., Li, H.P.: Camera-based analysis of text and documents: a survey. IJDAR **7**(2–3), 84–104 (2005)
3. Yao, C., Bai, X., Liu, W., Ma, Y., Tu, Z.: Detecting texts of arbitrary orientations in natural images. IEEE Part IV LNCS **8692**, 497–511 (2012)
4. Fabrizio, J., Cord, M., Marcotegui, B.: Text extraction from street level images. In: CMRT, pp. 199–204 (2009)
5. Yi, C., Tian, Y.: Assistive text reading from complex background for blind persons. In: Camera-Based Document Analysis and Recognition, pp. 15–28. Springer (2012)

6. Epshtein, B., Ofek, E., Wexler, Y.: Detecting text in natural scenes with stroke width transform. In: IEEE Conference on Computer Vision and Pattern Recognition, CVPR, pp. 2963–2970 (2010)
7. Saini, S., Marawaha, C.: Comparative study of text detection in natural scene images. In: IEEE International Conference on Recent Trends in Electronics Information & Communication Technology, RTEICT, pp. 1981–1985 (2016)
8. Mathworks (n.d.): Mathworks. in.mathworks.com. https://in.mathworks.com/products/matlab.html?s_tid=hp_products_matlab. Accessed 15 Sep 2018
9. Chen, H., et al.: Robust text detection in natural images with edge-enhanced maximally stable extremal regions. In: 18th IEEE International Conference on Image Processing, ICIP. IEEE (2011)
10. Neumann, L., Matas, J.: Real-time scene text localization and recognition. In: IEEE Conference on Computer Vision and Pattern Recognition, CVPR. IEEE (2012)
11. Gonzalez, A., et al.: Text location in complex images. In: 21st International Conference on Pattern Recognition, ICPR. IEEE (2012)
12. Li, Y., Lu, H.: Scene text detection via stroke width. In: 21st International Conference on Pattern Recognition ICPR. IEEE (2012)
13. Subramanian, K., Natarajan, P., Decerbo, M., Castañòn, D.: Character-stroke detection for text-localization and extraction. In: International Conference on Document Analysis and Recognition, ICDAR (2005)
14. Epshtein, B., Ofek, E., Wexler, Y.: Detecting text in natural scenes with stroke width transform. In: CVPR, pp. 2963–2970 (2010)
15. Minetto, R., Thome, N., Cord, M., Fabrizio, J., Marcotegui, B.: Snoopertext: a multiresolution system for text detection in complex visual scenes. In: ICIP, pp. 3861–3864 (2010)

Digital Image Steganography Using Modified LSB and AES Cryptography

Subhash Panwar[1](✉), Mukesh Kumar[2], and Sakshi Sharma[2]

[1] Department of IT, Government Engineering College, Bikaner 334004, India
subhash@gmail.com

[2] Department of CSE, Government Engineering College, Bikaner 334004, India

Abstract. Internet of Things (IoT) serves as part of our routine life activities nowadays and the secrecy of data has been compromised and cyber-crimes are increasing at an alarming rate therefore, the needs of a secure environment is vital in order to secure the transmitting data. This calls for a method to secure the digital messages being sent over the internet. In this paper we propose a modified concept on hiding the confidential data in an image. Here the Image Steganography is achieved using Modified LSB. Modified LSB uses a certain condition to replace the bits of the confidential data at the least significant bit position of the pixels in the image so Cryptography approach is achieved using AES. In this way, Steganography and Cryptography when used together ensures enhanced security of digital messages. Experimentally we found that the proposed approach is less prone to passive and active attacks. Also, as in nature it is hybrid approach which preserves the confidentiality of the information better than existing approaches.

Keywords: Internet of Things · Advanced Encryption Standard · AES · Cryptography · Image Steganography · Least significant bit · LSB · Steganography

1 Introduction

In today's modern scenario, the useful data is stored and shared via digital media. It requires the media as well as the communication channels to be safe enough to ensure security of the information. A technique known as Steganography is used for this purpose. Steganography is a word derived from Greek which means "covered writing". It hides the fact whether the data is alive or not, which is hidden inside a carrier medium which can either be text, image, audio or video. When a person sees the cover object, he cannot identify the existence of any information inside the cover object. The cover object, is thus, sent over a network to the desired destination and the intended recipient receives it [2]. When the pixels of an image hide the confidential data within themselves, it is referred to as Image Steganography (Fig. 1).

Digital images are stored in 8-bit or 24-bit formats. A 24-bit image is better at hiding information. The three primary colours: red, green, and blue make for all the variations

N. Nain and S. K. Vipparthi (Eds.): ICIoTCT 2019, AISC 1122, pp. 366–375, 2020.
https://doi.org/10.1007/978-3-030-39875-0_39

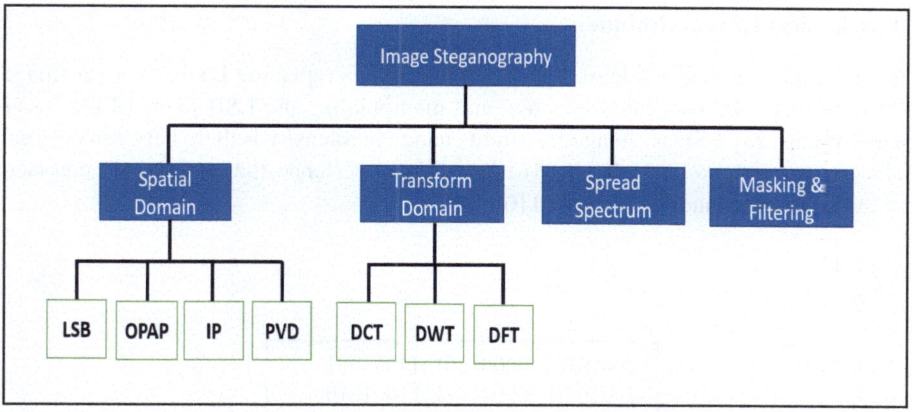

Fig. 1. Types of Image Steganography [3]

and shades in the colours in the pixel. Each primary colour is represented by 1 byte; 3 bytes per pixel are used in 24-bit images to represent a colour value. Hexadecimal, decimal, and binary forms are used as representation for 3 bytes. A white colour has hexadecimal value FFFFFF: representing FF for red, green and blue respectively. Its binary value is 11111111, 11111111, 11111111, and decimal value is 255, 255, 255 which are the three bytes making up white [4]. In this way, individual bits from the three bytes are used to hide the secret information. Various algorithms can be used to select which bits to be replaced with the secret data. The secret data to be hidden is subject to various transformations so that the message is unintelligible to intruders. This technique is known as Cryptography.

Cryptography is a word derived from Greek which means "secret writing". Cryptography is used as a security enhancement to Steganography. Cryptography is concerned with the alteration in the structure of the message. It converts the secret data into a non-readable format, which can only be deciphered by the intended receiver (Fig. 2).

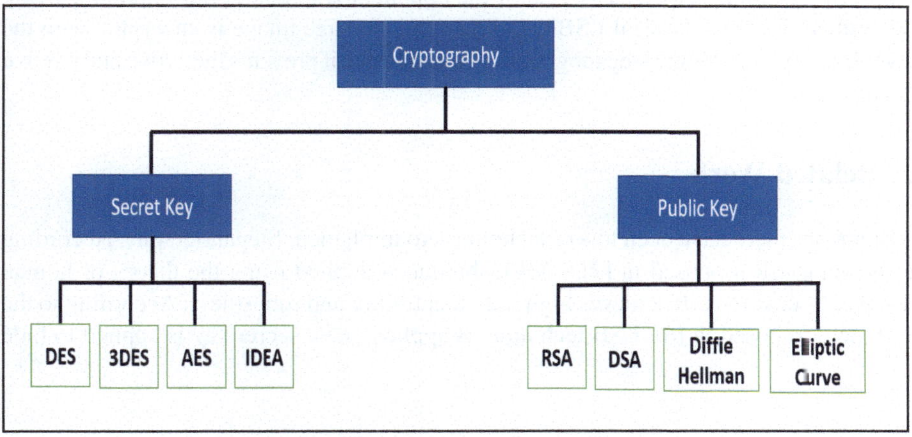

Fig. 2. Cryptography and its types [5].

1.1 Classical LSB Technique

The technique for classical least significant bit implies replacing LSBs of cover image with the bits of the concealed message and manipulating the LSB plane of the cover image. When only LSB is changed, a slight change in intensity is there between original and reconstructed pixel, which is hard to detect visually. Hence, the presence of a message inside the image cannot be identified [6].

Example:

Cover Image:

10010101	00001101	11001001
10010110	00001111	11001010
10011111	00010000	00101100

Message Image: 101101101

Steganographed Image:

10010101	0000110**0**	11001001
0001011**1**	0000111**0**	1100101**1**
10011111	00010000	11001011

A generic description of the steganographic process [6] provide us a formula as $C_I + M = S_I$. Here C_I = Cover Image, M = Message Image and S_I = Segno Image.

1.2 Active and Passive Attacks

An Active attack attempts to alter system resources or effect their operations. Active attack involves some modification of the data stream or creation of false statement. A Passive attack attempts to learn or make use of information from the system but does not affect system resources. Passive Attacks are in the nature of eavesdropping on or monitoring of transmission. The goal of the opponent is to obtain information is being transmitted. Here in classical LSB technique, the message image is encrypted with the cover image generates the steganographed image, hence it prevents the active and passive attacks.

2 Related Work

In literature, there have been lots of techniques to implement Steganography. According to the approach proposed in [7], LSB technique is applied using the theory of human visual cell sensitivity. It achieves high imperceptibility and robustness. According to the approach proposed in [8], LSB technique is applied and a secret key is applied to hide the information into a cover image. It provides better image quality and a better PSNR value. According to the approach proposed in [9], Least Significant Nibble is substituted with the secret data and AES encryption is also applied to the data. It achieves high PSNR values.

Data Encryption Standard [12] is based on a cipher known as the Feistel block cipher. This was a block cipher developed by the IBM cryptography researcher Horst Feistel in the early 70's. It consists of a number of rounds where each round contains bit-shuffling, non- linear substitutions (S-boxes) and exclusive OR operations. It encrypts the data in block size of 64 bits each. Same algorithm and key are used for encryption and decryption. Key is 56 bits long. The position of 8, 16, 24, 32, 40, 48, 56, 64 are discarded. DES is based on two fundamental attributes of cryptography Diffusion (Substitution) and Confusion (Permutation) consisting of 16 rounds. In each round key and data bits are shifted, permuted, XORed and sent through, 8 s-box. In the first round 64 bit plaintext is handed to initial permutation (IP). Then IP generates two halves left plaintext (LPT) and right plaintext (RPT). Each LPT and RPT goes through 16 rounds. At the last LPT and RPT are rejoined. Decryption is same process perform rounds in reverse order.

As an enhancement of DES, the 3DES (Triple DES) encryption standard was proposed. In this standard the encryption method is similar to the one in original DES but applied 3 times to increase the encryption level. It was used to remove the meet-in-the-middle attack occurred in 2 DES and the brute force attacks in DES. It also has the advantage of proven reliability and a longer key length that eliminates many of the shortcut attacks that can be used to reduce the amount of time it takes to break DES.

The most popular approach RSA [12] is public key encryption algorithm which is based on asymmetric key. Generally, it is used for the purpose of digital signature and security. The public and private key are generated using prime no. with mathematical calculus and large number multiplication. Plain text and chipper text are merged in block size data with combination 0 and n1. Number of n is considered 1024 bits. A separate keys are used for both encryption and decryption which are hidden from different end.

According to the approach proposed in [10], a comparative study is done between the various encrypting approaches like DES, AES and RSA taking into consideration, the encryption ratio, scalability and other factors. It is found that AES algorithm provides good results as compared to others.

Advanced Encryption Standard(AES) is an encryption standard recommended by NIST. 128 bits data block is encrypted using AES and it uses key length of 128, 192 or 256 bits. We have used key length of 128 bits.

The steps of AES algorithm are:

STEP 1. Derive the set of round keys from the cipher key.
STEP 2. Initialize the state array with the secret data.
STEP 3. Add the initial round key to the starting state array.
STEP 4. Perform nine rounds of state manipulation.
STEP 5. Perform the tenth and final round of state manipulation.
STEP 6. Copy the final state array out as the encrypted data.

AES Encryption process requires four types of operations.

STEP 1. Substitute bytes: Every byte is converted into a different value in this substitution operation.
STEP 2. ShiftRows: Rotation is performed on each row to the right by a certain number of bytes.
STEP 3. MixColumns: A new column is produced by separately processing each column of the state array. The old column is replaced by the new column.
STEP 4. XorRoundKey: The existing state array is taken and a XOR operation is applied with a portion of key.

AES Decryption requires inverse functions like InvSubBytes, InvShiftRows, InvMix-Columns[10, 11].

3 Proposed Technique

We have proposed a two-step implementation method for Digital image steganography. First, a secret message is transformed to cipher text by AES cryptography approach and further the cipher text is hidden inside the image using the modified LSB method.

3.1 Modified LSB Method

Figure 3 shows the major steps of proposed approach.

1. Until the cipher text length, convert each character of the cipher text to binary format according to the ASCII code and embed the cipher text being created from AES Encryption in the LSB's of cover image in the following order:

   B1, B2…Bn
   G1, G2…Gn
   R1, R2…Rn

   where B, G and R represent the blue, green and red components of a pixel.

2. The substitution is done on the LSB's of the cover image with the cipher text.
3. After the insertion is done in the LSB's, then find the 2nd and 3rd LSB's of the blue pixel.
4. Consider 2-bit combinations of (00, 01, 10, 11) in the 2nd and 3rd LSB's of blue pixel and count the number of blue pixels with the highest pattern.
5. Invert the LSB's of the blue pixels with the highest pattern.

The complete illustration of proposed approach with example string "BHARAT" is shown in Fig. 4.

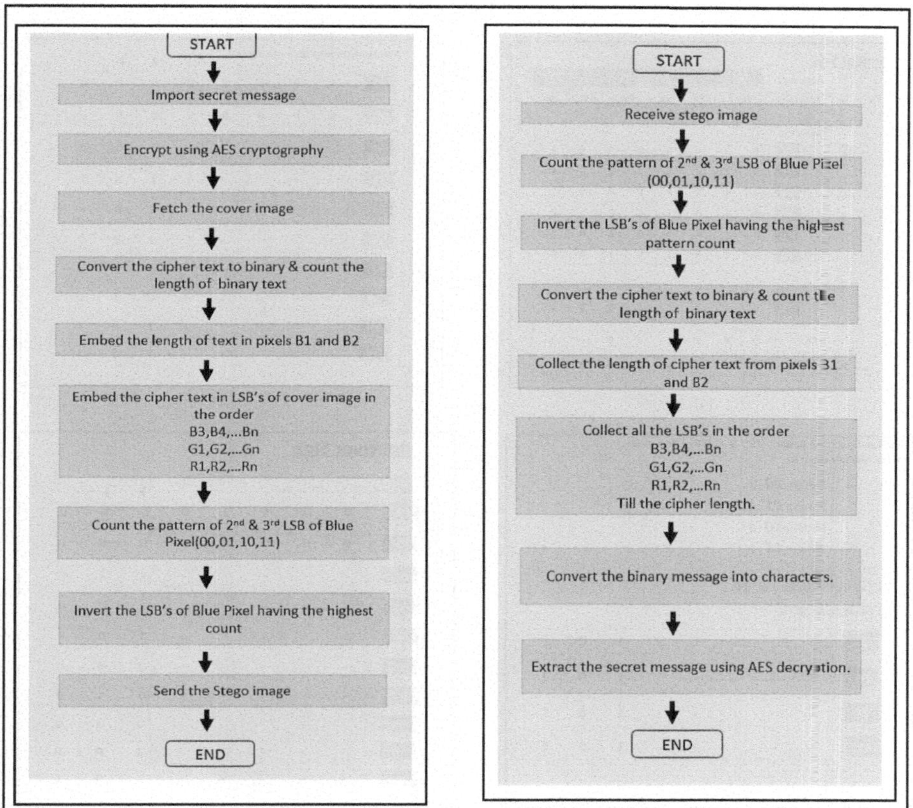

(A) Steps of hiding the secret message (B) Steps of Extracting the hidden message

Fig. 3. Steps of modified LSB

3.2 Experimental Setup

The proposed method ensures double layer security with AES Cryptography and modified LSB where embedding of secret message starts with the least significant bit of the blue component of the pixel and human eyes are least sensitive to the difference in blue colour. The approach is analysed using characteristics MSE and PSNR. The images shown in Table 1 are taken randomly from USC-SIPI image database which are used in research in theoretical foundation and image processing. This database is divided into volumes based on the basic character of the pictures. Images in each volume are of various sizes such as 256 × 256 pixels, 512 × 512 pixels, or 1024 × 1024 pixels. All images are 24 bits/pixel [13]. Table 1 shows the cover image and its respective stego image which is the resultant of proposed approach. Table 2 shows the MSE and PSNR of respective image which are selected randomly from USC SIPI database [13].

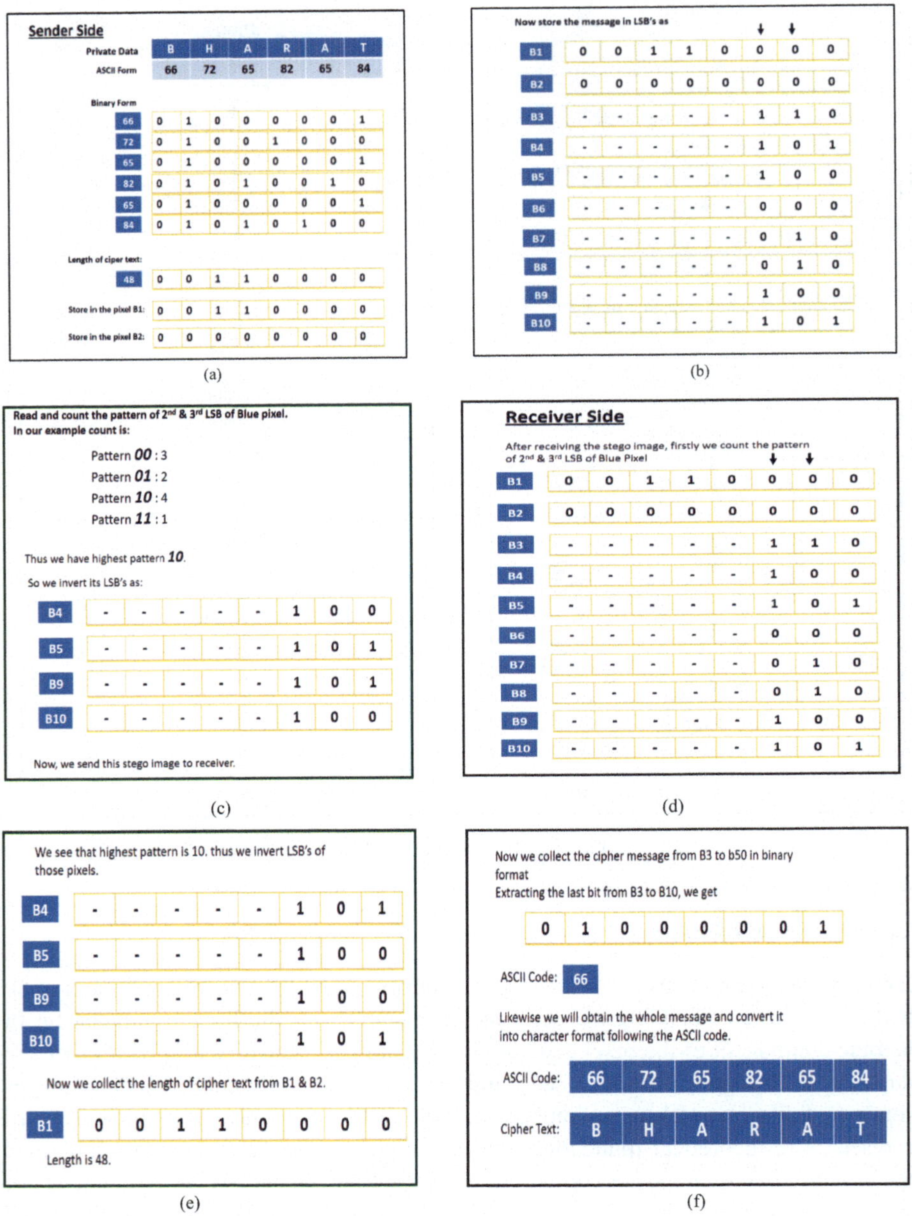

Fig. 4. Illustration of proposed approach using example string "BHARAT" sender side process Fig. (a) to (c), Receiver side process in Fig. (d) to (f)

Table 1. Result of modified LSB steganography

Name	Cover Image	Stego Image
Airplane		
Couple		
House		
San Diego		

4 Results

Image enhancement or improving the visual quality of a digital image varies from individual to individual and is very subjective. In order to compare the effects of image steganography algorithms on image quality, essential quantitative/empirical measures are adopted. Here we use two error metrics to compare the quality of cover image and the stego image. The MSE represents the cumulative squared error between the previous image and the new image thus formed, whereas PSNR represents a measure of the peak error. The lower the value of MSE, the less is the error, and best are the results.

MSE- Mean Square Error- It is the average of the squares of the errors or deviations that is, the difference between the estimator and what is estimated.

Table 2. Comparative analysis of MSE and PSNR metrics

Image (BMP)	Mean square error		Peak signal to noise ratio	
	Classical LSB	Modified LSB	Classical LSB	Modified LSB
Airplane	0.00042	0.00036	79.37	82.49
Couple	0.00148	0.00137	74.12	76.75
House	0.00154	0.00149	70.23	76.38
San Diego	0.00046	0.00037	76.56	82.35

PSNR- Peak Signal to Noise Ratio- PSNR is calculated in decibels (dB). 40 dB and above should be the value of a high quality stego image. PSNR value is defined by the mean square error (M.S.E) for two images, It is the ratio between the maximum possible pixel value MAX of the image and the mean square error where x as well as y are image coordinates, SGxy (stego image) and CVxy (cover image).

PSNR $=$ 10 log10 (*MAMSEX*), where MAX is the maximum fluctuation in the original image.

AES is used to encrypt the secret message and is more secure and less susceptible to cryptanalysis. AES is compatible to all international standards. Modified LSB is used to hide the secret message and it is very simple and easy to implement. It makes slight changes in the image, in the blue component of the pixel which are not detectable by bare eyes. Both the techniques, AES and LSB, when combined, their benefits add up and add an extra layer of security in our proposed approach. Modified LSB technique ensures that the presence of the message is not revealed and AES makes sure that the contents of the message are well hidden. The proposed approach is less prone to passive and active attacks. The approach is hybrid in nature and preserves the confidentiality of the information.

5 Conclusion

The proposed method is a good way to embed the hidden information without the risk of compromising its security. The secret information is concealed inside an image using a technique known as Image Steganography. For additional benefits, the content of the secret message is scrambled using Cryptography to avoid intrusion to unauthorised users and ensure safety during unfavourable situations. When the secret information is stored inside the original image, the new image thus formed is the stego image. AES

Cryptography provides good results, as it ensures that even though the existence of message is detected, it is not easy to determine its contents. Modified LSB technique also proves to be very helpful in preserving the security of the information inside the image. There are almost negligible chances for the unauthorized users to recognise significant and noticeable changes in the stego image. The amplitude of change in both the images is not human perceptible. The use of both the techniques gives a way to secure the information from illegal users.

References

1. Thangadurai, K., Devi, G.S.: An analysis of LSB based image steganography techniques. In: 2014 International Conference on Computer Communication and Informatics, Coimbatore, pp. 1–4 (2014)
2. Katzenbeisser, S., Petitcolas, F.A.: Information Hiding Techniques for Steganography and Digital Watermarking, pp. 17–18. Artech House, Boston (2000)
3. Cheddad, A., Condell, J., Curran, K., Mc Kevitt, P.: Digital image steganography: survey and analyses of current methods. Signal Process. **90**(3), 727–752 (2010)
4. Johnson, N.F., Jajodia, S.: Exploring steganography: seeing the unseen. Comput. Pract. **31**, 26–34 (1998)
5. Khicha, A., Kapoor, N.: Information System Security, pp. 2.33–3.34. Genius Publications, Jaipur (2013)
6. Akhtar, N., Khan, S., Johri, P.: An improved inverted LSB image steganography. In: International Conference on Issues and Challenges in Intelligent Computing Techniques (ICICT 2014), pp. 749–750 (2014)
7. Anand, J.V., Dharaneetharan, G.D.: New approach in steganography by integrating different LSB algorithms and applying randomization concept to enhance security. In: Proceedings of the International Conference on Communication, Computing & Security, pp. 474–476 (2011)
8. Karim, S.M.M., Rahman, M.S., Hossain, M.I.: A new approach for LSB based image steganography using secret key. In: Proceedings of 14th International Conference on Computer and Information Technology (ICCIT 2011), 22–24 December 2011, Dhaka, Bangladesh (2011)
9. Sheth, U., Saxena, S.: Image steganography using AES encryption and least significant nibble. In: International Conference on Communication and Signal Processing, 6–8 April 2016, India (2016)
10. Padmavathi, B., Kumari, S.R.: A survey on performance analysis of DES, AES and RSA algorithm along with LSB substitution technique. Int. J. Sci. Res. (IJSR) **2**(4), 170–174 (2013). India
11. Islam, M.R., Siddiqa, A., Uddin, M.P., Mandal, A.K., Hossain, M.D.: An efficient filtering based approach improving LSB image steganography using status bit along with AES cryptography. In: 3rd International Conference on Informatics, Electronics & Vision (2014)
12. Singh, A., Marwaha, M., Singh, B., Singh, S.: Comparative study of DES, 3DES, AES and RSA. Int. J. Comput. Technol. **9**(3), 1164–1170 (2013)
13. The USC-SIPI Image Database. http://sipi.usc.edu/database/

Design and Implementation of AES on FPGA for Security of IOT Data

Dinesh B. Bhoyar[1(✉)], Shelly R. Wankhede[1], and Swati K. Modod[2]

[1] Department of Electronics and Telecommunication Engineering, Yeshwantrao Chavan College of Engineering, Wanadongri, Nagpur, India
dinesh_bhoyar@rediffmail.com, shellywankhede72@gmail.com

[2] Department of Electronics and Telecommunication Engineering, Rajiv Gandhi College of Engineering and Technology, Wanadongri, Nagpur, India
swati06.mohod@rediffmail.com

Abstract. As we know that the world is been surrounded by the internet and interconnected to it and the different things or we can say the edge devices which helps us to connect to network and communicate to the different alternatives with the surroundings. In this paper we would be discussing about the different algorithms and their and disadvantages and how the advanced standard algorithm is been applied we has an great effect on cryptography and the different process carried out in the encryption and decryption process. In this paper the encryption and the decryption would be given in 128 bit format in hexadecimal, binary or any other format and the key expansion for getting the cipher text. in cryptography their are three classification for the expansion of the key which can be symmetric, asymmetric and the hash function in this paper we have discussed and the symmetric key is been given for both encryption and decryption process.as we know that we are using the 128 bit length which has 10 rounds and the initial round includes the state and the different transformation in the cryptograpy process which is the symmetric key expansion in which the same key is been applied to both encryption and decryption in binary, hexadecimal, etc. As it is lossless operation many applications can be derived using this algorithm as we are using symmetric key block cipher. We are limiting our paper with the 128 bit length of data in verilog language. For encryption and decryption the proposed paper describes the cryptograpy which would be having the fixed size. In this algorithm where we are using the symmetric key block cipher where the key would be given same with respect to encryption. Many applications it can be used as it is lossless operation. We limit our focus on 12b8 bit AES encryption and decryption where coded in VHDL coding. The proposed paper describes the private key cryptosystems which has a key with fixed size.

Keywords: AES · FPGA · Key · IOT · Cloud security · VHDL

1 Introduction

In this chapter introduction to Advanced Encryption Standard along with the drawbacks of certain encryption methods and the method used in proposed algorithm for to secure

N. Nain and S. K. Vipparthi (Eds.): ICIoTCT 2019, AISC 1122, pp. 376–383, 2020.
https://doi.org/10.1007/978-3-030-39875-0_40

the data and to classify the types is explained in 1.1, which is followed by Literature review of different algorithms in 1.2.

1.1 Overview

The vital and secret information is been hacked by wrong means in the communication and the communication is mostly through the internet so there are lot of chances of frauds and wrong means to hack the data and so the data should be protected and secured. Information encryption keeps up information secrecy, trustworthiness and validation. In the past years there was secrecy with the help of using the mystery keys now a days the different cryptography as been characterized which are public key and private key. As the data is crucial and private or we can say secret so it has to be protected. Top concern is that the data security is that how it can have the provision to provide an high level layered security. Mainly the surrounding depends on different software which can protect the secret data or mysterious data to be communicated to the other end the data which have to secured and be protected. In this world of digital technology there are different threats by means it can be hacked easily by the other user with help of wrong means. For security purposes such as privacy of multimedia and visual surveillance systems in which block cipher cryptography is used for protection. There are different algorithms for cryptography which are divided into two; Stream cipher and Block cipher. In AES there are 3 lengths in which data is been encrypted which are 128, 192 and 256 bit data as shown in the table given below (Table 1).

Table 1. Different key length

	Key length	Block size	Number of rounds
AES-128	4	4(128)	10
AES-192	6	4(128)	12
AES-256	8	4(128)	14

The basic diagram of the Encryption and Decryption processing which the original data i.e. the plain text is been converted into the cipher text and in the encryption process and in the decryption process the cipher text is been converted into the plain text.

1.2 Literature Review of Different Algorithms

The AES algorithm used cryptographic keys of 128, 192, and 256 bits to encrypt and decrypt data in blocks of 128 bits. The different algorithms are MARS, RC6, RIJNDAEL, SERPENT and TWOFISH. The conclusion was that the five Competitors showed similar characteristics. After having a study of the above algorithms Rijndael algorithm came to be best with respect to security, performance, efficiency, implementation and flexibility (Fig. 1).

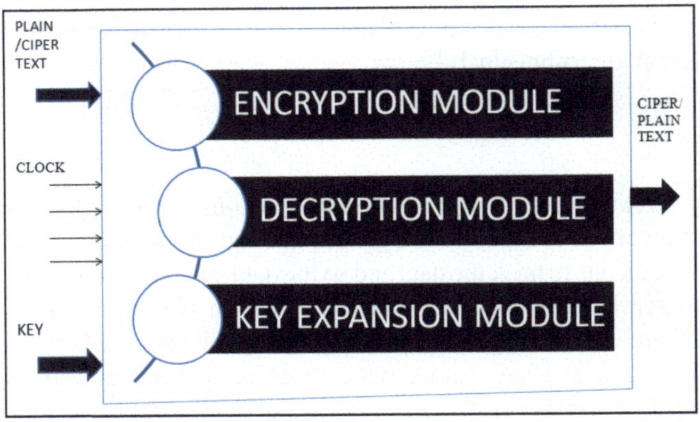

Fig. 1. Architecture of proposed algorithm

2 AES Encryption and Decryption

In this part we would give the encoding process for the AES algorithm. The algorithm divides different blocks of 128 bit which are the XOR block to provide the 1st round and so on which is been applied to input and output logical and no logical operations. The addition to the encryption and decryption process there is a key expansion process. The transformations that out are namely Sub Bytes, Shift Rows, Mix Columns, AddRoundKe The AES design has the original input and the secret key which would be the private key then by combining the original message and the cipher key we get the cipher text which could be an unintelligent form for the public. The input can be 192 bit and 256 bit also. The AES are only having three inputs which are 128 bit, 192 bit and 256 bit. We are using a 128 bit input then from cipher block; the output would be a cipher text. For 128 bit, 192 bit and 256 bit we have 10, 12, 14 rounds. There are regular rounds for each bit which are 9, 11, 13 and the final round which are different i.e. 10th, 12th and 14th (Fig. 2).

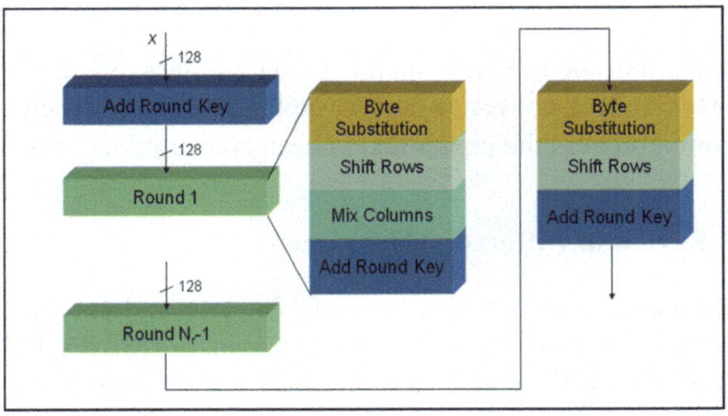

Fig. 2. AES encryption with different transformation

Decryption is an inverse procedure of encryption. It is used for obtaining the plain text from given cipher text. The key is same. In decryption the s-box is inverse of that which was used in encryption. With the help of VHDL the encrypted data which is the ciphered data, plain text can be obtained. Decryption is the inverse form of encryption.

The different tasks for decryption are as follows:

1. Add Round Key
2. Inv Mix Columns
3. Inv Shift Rows
4. Inv Sub Bytes

AES involves four steps which are follows:

Key Generation: Each round has its own round key that is derived from the original 128-bit encryption key in the manner described in this section it involves XORing of the round key with the state array.

Encryption: The plain text is given to encrypt i.e. in an encoded form which is the ciphered text.

Decryption: The ciphered text from the encryption process is been decoded and then the plain text is obtained.

Key Expansion: Most of the key expansions are performed based on

- ROT WORD (4 bytes: circular shift)
- SUB WORD (4 bytes: substitution)
- RCON
- XOR

There are some particular security prerequisites, including

- Authentication
- Privacy/confidentiality
- Integrity
- Non-repudiation.

3 Algorithm

The device that is the edge device generates the data. Firstly the data pushed to the cloud our data is the cloud offline data which is the hexadecimal data. There are different attacks such as spoofing, sniffing etc. It is very much safe to encrypt the data before transmission hence an hardware would be needed which is an AES performed on FPGA. For proper reception of encrypted data can cause attack so with the help of AES encryption and decryption (Figs. 3 and 4).

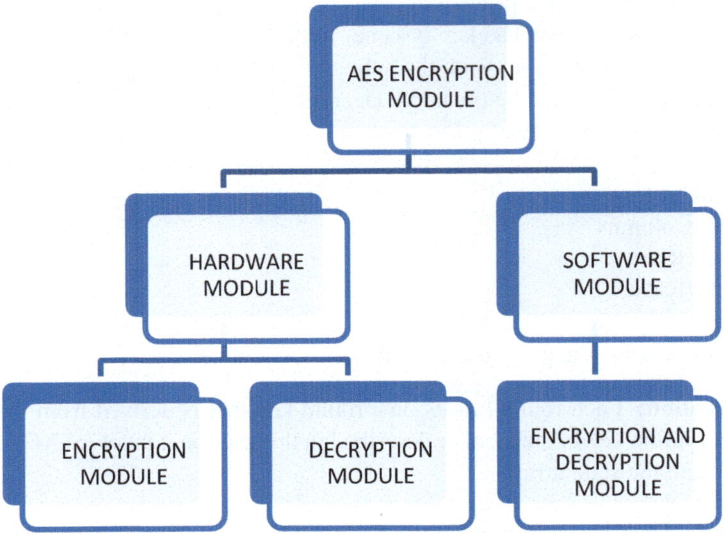

Fig. 3. Block diagram

PARAMETER	*AES 128(PREVIOUS WORK)*	*AES 128(PREVIOUS WORK)*	*AES128(PROPOSED WORK)*
KEY SIZE	128	128	128
TOTAL ROUNDS	10	10	10
SLICE REGISTERS	256	4800	951
SLICE LUTs	3559	24999	1459
LUT FF	459	40	957
BLOCK RAM	4	11	5
CLOCK CYCLE	41	10	41

Fig. 4. Design utilization of AES encryptor

4 Results of AES Algorithm

The security of the data transmitted to the cloud is double layer secured, it has better defense against attacks during transmission and during storage. The AES hardware has proven to be faster and efficient for the encryption of the data. The key length 128 bits sufficiently large for the IOT purpose at present. The AES encryption of the key helps in the maintaining security of the symmetric encryption. The proposed encryption algorithm is designed using Verilog Hardware Description Language and behavioral form of coding. The functionality, design and analysis is carried out on Xilinx ISim simulator (Figs. 5 and 6).

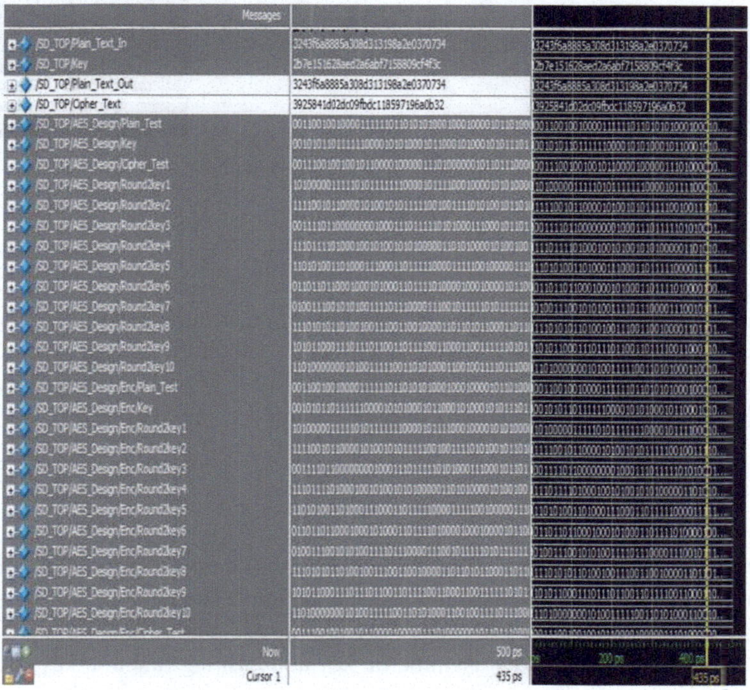

Fig. 5. Outputs with cipher text

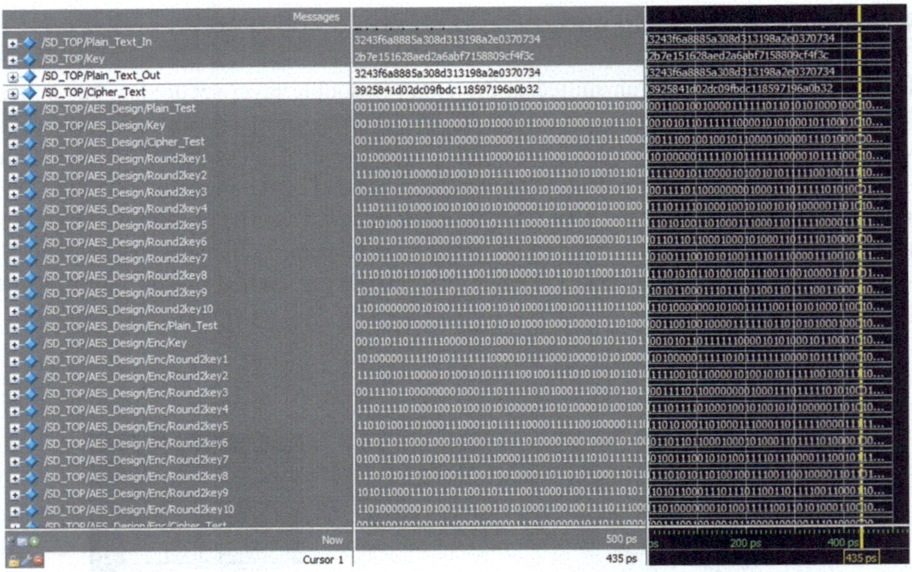

Fig. 6. Output after decryption i.e. the plain text

5 Conclusion

Verilog code is developed for the implementation of encryption and decryption. NIST provides some sample vectors with aminimal delay. Therefore AES is an efficient on FPGA with encryption and Decryption for every 128 bit The time varies from chip to chip and the calculated delay time can only be regarded as approximate.

6 Future Scope

The future scope for our proposed work would be to secure and protect the IOT data and securing the data from different threats, and the important data or vital information data which is been generated by the edge device. The algorithm which is been proposed future it can be implemented by asymmetric key and hashing function key encryption and decryption process. This would be made critical and complex for the users which would be hacking the information dat or secret data by using session key generation at the time for request. For reducing power and for faster processing an application specific integrated circuit should be designed. The optimization of the design can be carried out to improve the power efficiency and area efficiency [1].

References

1. El Bouchti, A., Bahsani, S., Nahhal, T.: Encryption as a service for data healthcare cloud security. In: Fifth International Conference on Future Generation Communication Technologies (FGCT 2016) (2016)
2. Jonwal, S.U., Shingare, P.P.: Advanced Encryption Standard (AES) implementation on FPGA with hardware in loop. In: International Conference on Trends in Electronics and Informatics, ICEI 2017 (2017)
3. Chandu, Y., Kumar, K.S.R., Prabhukhanolkar, N.V., Anish, A.N., Rawal, S.: Design and implementation of hybrid encryption for security of IOT data. In: 2017 International Conference on Smart Technologies for Smart Nation (2017)
4. Kumar, A., Kumar, M., Balramudu, P.: Implementation of AES algorithm using VHDL. In: 2017 International Conference on Computing Methodologies and Communication (ICCMC) (2017)
5. El-Deen, A.E.T., El-Badawy, E.A., Gobran, S.N.: Digital image encryption based on RSA algorithm. IOSR J. Electron. Commun. Eng. (IOSR-JECE) **9**(1), 69–73 (2014). e-ISSN: 2278-2834, p-ISSN: 2278-8735, Ver. IV
6. Stallings, W.: Cryptography and Network Security: Principles and Practices, 3rd edn. Prentice Hall, Upper Saddle River (2003)
7. Rivest, R.L., Shamir, A., Adleman, L.: A method for obtaining digital signatures and public-key cryptosystem. Commun. ACM **21**(2), 120–126 (1978)
8. Bhoyar, D.B., Dethe, C.G., Mushrif, M.M.: Two channel estimation methods for MIMO-OFDM System. Int. J. Comput. Eng. Intell. Syst. **5**(6), 46–53 (2014). ISSN: 2222-1719 (paper), ISSN: 2222-2863 (online)
9. Bhoyar, D.B., Raut, P.: Design FPGA based implementation of MIMO decoding in a 3G/4G wireless receiver. In: 2009 Second International Conference on Emerging Trends in Engineering and Technology (2009)
10. Bhoge, S.G., Chavan, M.D., Suryavanshri, C., Taksande, V.K.: Across layer approach for energy and communications efficient protocol of mobile and ad-hoc networks. In: International Conference of Communication and Signal Processing, pp. 1186–1189

Author Index

N. Nain and S. K. Vipparthi (Eds.): ICIoTCT 2019, AISC 1122, pp. 385–386, 2020.
https://doi.org/10.1007/978-3-030-39875-0

The manufacturer's authorised representative in the EU is Springer Nature Customer Service Centre GmbH, Europaplatz 3, 69115 Heidelberg, Germany. If you have any concerns regarding our products, please contact ProductSafety@springernature.com

Printed and bound by CPI Group (UK) Ltd, Croydon, CR0 4YY
25/09/2026
02222853-0004